Transforming Indian Agriculture

INDIA 2040

Productivity, Markets, and Institutions

Thank you for choosing a SAGE product! If you have any comment, observation or feedback, I would like to personally hear from you. Please write to me at <u>contactceo@sagepub.in</u>

—Vivek Mehra, Managing Director and CEO,
SAGE Publications India Pvt Ltd, New Delhi

Bulk Sales

SAGE India offers special discounts for purchase of books in bulk. We also make available special imprints and excerpts from our books on demand.

For orders and enquiries, write to us at

Marketing Department
SAGE Publications India Pvt Ltd
B1/I-1, Mohan Cooperative Industrial Area
Mathura Road, Post Bag 7
New Delhi 110044, India
E-mail us at <u>marketing@sagepub.in</u>

Get to know more about SAGE, be invited to SAGE events, get on our mailing list. Write today to <u>marketing@sagepub.in</u>

This book is also available as an e-book.

Transforming Indian Agriculture

INDIA 2040

Productivity, Markets, and Institutions

Edited by
Marco Ferroni

PRODUCTIVITY

MARKETS

INSTITUTIONS

SAGE
www.sagepublications.com
Los Angeles • London • New Delhi • Singapore • Washington DC

Jointly published in 2012 by

SAGE Publications India Pvt Ltd
B1/I-1 Mohan Cooperative Industrial Area
Mathura Road, New Delhi 110 044, India
www.sagepub.in

SAGE Publications Inc
2455 Teller Road
Thousand Oaks, California 91320, USA

SAGE Publications Ltd
1 Oliver's Yard, 55 City Road
London EC1Y 1SP, United Kindom

SAGE Publications Asia-Pacific Pte Ltd
33 Pekin Street
#02-01 Far East Square
Singapore 048763

2600 Virginia Avenue, NW, Suite 201
Washington, DC 20037 USA

Published by Vivek Mehra for SAGE Publications India Pvt Ltd, typeset in 9.5/14.5 Helvetica Neue LT Std and printed at Saurabh Printers Pvt Ltd.

Library of Congress Cataloging-in-Publication Data Available

ISBN: 978-81-321-1061-3 (HB)

The SAGE Team: Rudra Narayan, Shreya Lall

Contents

List of Figures, Tables, and Boxes

Figures

List of Abbreviations

ACABC	Agriclinics and Agribusiness Centers
Ag	Agricultural
AIBP	Accelerated Irrigation Benefits Programme
AICC	All India Congress Agrarian Reforms Committee
AKIS	Agricultural Knowledge and Information Systems
AP	Andhra Pradesh
APMA	Agricultural Produce Marketing Act
APMC	Agricultural Produce Marketing Committee
AS	Assam
ATEC	agro-technical extension and service center
ATMA	Agricultural Technology Management Agency
BAIF	Bharatiya Agro Industries Foundation
BCM	Billion cubic meters
BH	Bihar
BPL	Below Poverty Line
BRGF	Backward Regions Grant Fund
Bt	Bacillus thuringiensis
CA	Commission Agent
CAG	Comptroller and Auditor General
CCA	Cultivable Command Area
C-DAP	Comprehensive District Agriculture Plan
CESS	Tax/fee
CGWB	Central Ground Water Board
CIGs	Commodity Interest Groups
CIMMYT	Centro Internacional de Mejoramiento de Maíz y Trigo
CMSA	Community Managed Sustainable Agriculture
CPI(AL)	Consumer Price Index (Agricultural Labor)
CPI(IW)	Consumer Price Index (Industrial Workers)
CRP	Community Resource Person

CSO	Central Statistical Organization
CWC	Central Water Commission
DAC	Department of Agriculture and Cooperation
DAC	Department of Agriculture and Cooperation
DAESI	Diploma in Agricultural Extension Services for Input Dealers
DAHD	Department of Animal Husbandry and Dairying
DARE	Department of Agricultural Research and Education
DDO	District Development Officer
DDP	Desert Development Plan
DICONSA	Financial Services for the Rural Poor Program
DM	District Magistrate
DoA	Department of Agriculture
DPAP	Drought Prone Areas Programme
DPR	Detailed Project Report
DRDA	District Rural Development Agencies
DSC	Development Support Center
DSCL	DCM Shriram Consolidated Ltd.
DWDU	District Watershed Development Units
EPZ	Export Processing Zones
FAO	Food and Agriculture Organization of the United Nations
FAOSTAT	Food and Agriculture Organization of the United Nations
FBO	Farmer-based Organization
FCI	Farm Chemicals International
FDI	Foreign Direct Investment
FFS	Farmer Field School
FICCI	Federation of Indian Chambers of Commerce and Industry
FIGs	Farmer Interest Groups
FMCG	Fast-moving Consumer Goods
FPS	Fair Price Shops
FRDC	Fisheries Research and Development Corporation
FSE	Center on Food Security and the Environment
FSI	Stanford's Freeman Spogli Institute for International Studies
GB	Governing Board
GDP	Gross Domestic Product
GDSP	Gross State Domestic Product
GDSPA	Gross State Domestic Product from Agriculture
GJ	Gujarat

GM	Genetically Modified
GMOs	Genetically Modified Organisms
GOI	Government of India
GOP	Grand Old Party
GU	Gujarat
ha	hectare
HLL	Madhya Pradesh by Hindustan Lever Ltd
HP	Himachal Pradesh
HY	Haryana
ibid	in the same place
ICAR	Indian Council for Agricultural Research
ICRIER	Indian Council for Research on International Economic Relations
ICRISAT	International Crops Research Institute for the Semi-Arid-Tropics
ICT	Information and Communication Technology
IFFCO	Indian Farmers Fertilizer Cooperative Ltd
IFPRI	International Food Policy Research Institute
IIIT	International Institute of Information Technology
IIM	Indian Institute of Management
IKSL	IFFCO Kisan Sanchar Limited
IMAGE	Integrated Model to Assess the Global Environment
IMT	irrigation management transfer
IRADE	Integrated Research and Development
IRRI	International Rice Research Institute
ISOPOM	Oil Palm Development Programme
ITC	Indian Tobacco Company
IWDP	Integrated Watershed Development Programme
IWMI	International Water Management Institute
KA	Karnataka
KCC	Kisan Call Centers
KDP	Kecamatan Development Program
KN	Kamataka
KR	Kerala
KVKs	Krishi Vigyan Kendras
kWh	Kilo Watt Hours
Labr	Labor
LCU	Local Currency Unit
LFPR	Labor Force Participation Rates

M&E	Monitoring and Evaluation
MA	Maharashtra
MANAGE	National Institute of Agricultural Extension Management in Hyderabad
MC	Management Committee
MGNREGA	Mahatma Gandhi National Rural Employment Guarantee Act
MGNREGS	Mahatma Gandhi National Rural Employment Guarantee Scheme
MH	Maharashtra
MKV	Mahindra Krishi Vihar
MMA	Macro Management of Agriculture
MMI	Major and Medium Irrigation
MOA	Ministry of Agriculture
MoFP	Ministry of Food Processing
MOSPI	Ministry of Statistics and Programme implementation
MOU	Memorandum of Understanding
MP	Madhya Pradesh
MS	Mandal Samakhyas
MSAMB	Maharashtra State Agriculture Marketing Board
MSP	Minimium Support Price
MSS	Maharogi Sewa Samiti
MSU	Michigan State University
MW	Mega Watt
MWR	Ministry of Water Resources
MWRRA	Maharashtra Water Resources Regulatory Authority
NADEP	Narayan Deotao Pandharipand's composting method
NAIS	National Animal Identification System
NAS	National Accounts Statistics
NATP	National Agricultural Technology Project
NCAER	National Council of Applied Economic Research
NCEUS	National Commission for Employment in the Unorganized Sector
NDDB	The National Dairy Development Board
NDP	Net Domestic Product
NFHS	National Family Health Survey
NFSM	National Food Security Mission
NGOs	Non Government Organization
NHB	National Horticultural Board
NHM	National Hortriculture Mission
NIRD	National Institute of Rural Development

NLT	Nokia Life Tools
NPM	Non pesticidal farming
NRAA	National Rainfed Area Authority
NREGS	National Rural Employment Generation Scheme
NRLM	National Rural Livelihood Program
NSS	National Service Scheme
NSSO	National Sample Survey Organization
NSSs	National Sample Surveys
NWDPRA	National Watershed Development Project for Rainfed Areas
O&M	Operations and Maintenance
OR	Orissa
PB	Punjab
PDA	Personal Digital Assistant
PIA	Project Implementing Agency
PIM	Participatory Irrigation Management
PMGSY	Pradhan Mantri Gram Sadak Yojana
PPP	Public-Private Partnership
PRADAN	Professional Assistance for Development Action
PRIs	Panchayati Raj institutions
PUWR	Potentially Utilizable Water Resources
R&D	Research and Development
REDS	Rural Economic and Demographic Surveys
RGSY	Rashtriya Gram Swaraj Yojana
RJ	Rajasthan
RKVY	Rashtriya Krishi Vikas Yojana
RML	Reuters Market Light
Rs	Rupees
Rs/ha	Rupees/hectare
SAMETI	State Agricultural Management and Extension Training Institute
SAP	State Agriculture Plan
SAU	State Agricultural University
SC/ST	Scheduled Castes/Scheduled Tribes
SEBs	State Electricity Boards
SERP	Society for the Elimination of Rural Poverty
SEWP	state extension work plans
SFSA	Syngenta Foundation for Sustainable Agriculture
SGSY	Swarn Jayanti Gram Swarojgar Yojana

SHG	Self-help Group
SIC	Site Implementation Committee
SIL	Syngenta India Limited
SLNA	State-level Nodal Agencies
SLSC	State Level Sanctioning Committee
SMS	Short Message Service
SPV	Special Purpose Vehicle
SRI	System of Rice Intensification
SSP	Sardar-Sarovar Project
STD/BCP	Subcriber trunk dialing
STD/PCO	Subcriber trunk dialing/public call office
T&V	Training and Visit
TAC	Technical Audit Cell
TE	Triennium Ending Average
TFP	Total Factor Productivity
TFPG	Total Factor Productivity Growth
TISS	Tata Institute of Social Sciences
TKK/TKS	Tata Kisan Kendra/Tata Kisan Sansar
TMO	Technology Mission on Oilseeds
TN	Tamil Nadu
UN	United Nations
UP	Uttar Pradesh
USD	United States Dollar
USDA	United States Department of Agriculture
USGS	United States Geological Service
VO	Village Organizations
VOP	Value of Output
WALMI/DSC	Water And Land Management Institute/Development Support Center
WAPCOS	Water and Power Consultancy Services Ltd
WB	West Bengal
WIE	Stanford Woods Institute for the Environment
WSS	Warsi Sewa Sadan
WUAs	Water Users Associations
ZRS	Zonal Research Station
ZS	Zilla Samakhya

Foreword

Montek Singh Ahluwalia

Agriculture has been a central concern of Indian policy makers for two very good reasons. First, a large proportion of the population derives the major part of their employment from agriculture and they are among the poorest. The percentage has been falling over time but it remains well above fifty percent. Faster growth in agriculture is seen as a critical factor in raising incomes of this category and thereby reducing the incidence of poverty. The second reason why agriculture has special importance relates to food security. A large country like India inevitably must plan for near complete self-sufficiency in food. This is not to say that there should be no reliance on imports, and indeed India does import substantial amounts of edible oils and pulses, but imports of food must at best be viewed as a supplementary source of supply and the country must produce enough food to be able to meet the essential needs of its population.

Viewed from these perspectives, India's recent experience with agriculture presents a mixed picture. The Tenth Plan (2002–03 to 2006–07) had set a target growth rate of 4 percent for agricultural GDP, but the actual achievement was only 2.4 percent. The Eleventh Plan (2007–08 to 2011–12) again set a target growth rate of 4 percent, and there was an improvement in performance with agriculture growing at 3.4 percent, which is lower than the target but much better than the growth rate achieved in the previous plan. As far as food security is concerned, India's production of foodgrains has increased substantially and has led to an accumulation of substantial foodgrain stocks. India has also emerged as an exporter of rice. Production of vegetables, fruits, milk, and eggs has increased much faster than foodgrains, but demand is rising even faster with rapid and inclusive growth, which is reaching a larger and larger proportion of the population.

The experience of the past five years shows many other positive developments which augur well for the future. Public investment in agriculture, which had declined as a percent of agricultural GDP, has increased and private investment has been especially buoyant. As a result, total investment in agriculture has increased from 15 percent of agricultural GDP to over 20 percent. There is clear evidence that Indian agriculture is diversifying successfully beyond foodgrains into higher-value production such as horticulture, dairying, poultry, and fisheries. India is the world's leading producer of milk and pulses and in the ranks of the top five producers of wheat, rice, eggs, vegetables, and melons. However, as noted above, the growth in consumer demand calls for even higher growth rates in these sectors.

There is no doubt that Indian agriculture can and must do much better. While the immediate task is to raise the rate of growth of agricultural GDP to 4 percent, a target that is reiterated in the Twelfth Plan and now has a much better likelihood of being achieved, we need to be more ambitious looking further ahead. There is no reason why the target should not be set higher for the next twenty years. India's agricultural land productivity remains low, especially in the dryland rainfed parts of the country. Studies show that productivity per hectare can be raised by between 50 and 100 percent in many crops, even with available technology, provided that access to key inputs such as water can be improved and there is adequate dissemination of knowledge about productivity enhancing cultivation practices and access to reliable seeds. Actually, given the limited potential of increasing total arable land, growing water constraints, and rising rural wages, the key to increasing total agriculture output and sustaining a rise in rural incomes will be increased productivity. More efficient use of agriculture inputs will also be good for the environment.

How to achieve this end result is the subject of this book in which the authors explore issues that need to be addressed over a longer time horizon of thirty years. The basic postulate that India's transition to rapid growth must be accompanied by dramatic improvements in agricultural performance and associated policies is indisputable. What is interesting is that the transformation will involve many structural changes. As the book points out, Indian agriculture in 2040 will be characterized by mostly small farms, but these will have to be modern farms which use state of the art technology to achieve efficient agricultural production. While farms must have some minimum size to be able to use modern technology, they do not have to be large. Even small and modest sized farms can use advanced biological and mechanical technologies for crops, horticulture, livestock, and acquaculture. The average size of farms in China is even smaller than in India yet China currently boasts much higher yields in most crops.

The market linkage between the farms and the consumer will also have to be modernized. This is especially important as the faster growing subsectors—vegetables, fruits, milk and dairy products, poultry, and fish production—are all vulnerable to wastage unless supported by a very efficient logistic chain, starting with post-harvest management and continuing all the way to final consumption. This is perhaps the area where there has been the least action thus far, although it is now beginning.

The transformation needed will require a substantial effort from the government in many areas. Perhaps the most important relates to better management of scarce water resources. The scope for large irrigation projects is limited, and in any case, their economic impact is less impressive than was at one stage supposed. However, water conservation and groundwater recharge are activities which have demonstrably high payoffs. The government also has a big role to play in improving connectivity in rural areas through better roads and telecom connectivity, which can favourably impact farmer market connectivity. Finally, it is imperative that we raise both the quality and effectiveness of all public services directed at agriculture, including extension, used by the farmers.

The government also has a crucial role to play in supporting agricultural research aimed at the development of better varieties. This is particularly important in view of the looming threat of climate change, which is likely to affect temperature levels and also water availability, both of which could have significant adverse effects on crop productivity. Given the long time lag between the initiation of research and the generation of results that can be used in the field, there is an overwhelming need to devote more resources to this area. Equally important as providing more resources is the need to restructure the research system to incentivize good quality research. The scope for public private partnerships in research also needs to be explored innovatively.

An important difference between the transformation that is now necessary and that which occurred at the time of the Green Revolution is that success will not come from any single breakthrough, as was the case with high-yielding wheat hybrids in the late 1960s. Higher productivity will be achieved through a multiplicity of interventions in different crops, often tailored to the special circumstances of each agro climatic zone. The role of the private sector will be much larger in marketing output and linking farmers to different markets and also in seed production.

I hope this timely book will provoke a discussion of longer term policy options in agriculture over the next three decades. I congratulate the authors for the work they have done and the editor for collecting the distinguished authors and producing a volume that is both relevant and thought provoking.

Acknowledgments

This book originates from a study and series of thought-provoking background papers commissioned by Centennial Group, with financial and technical support from the Syngenta Foundation for Sustainable Agriculture, which initially proposed the study. The study was managed by Harinder S. Kohli, Praful Patel, and Anil Sood of the Centennial Group, Washington, DC, and is a follow-up to the Centennial Group's report "India 2039: An affluent society in one generation," published in 2010.

Part I is based on the summary report of a project led by Hans P. Binswanger-Mkhize and Kirit Parikh, and background papers written by the following authors (in alphabetical order): Richard Ackermann, G.V. Anupama, Cynthia Bantilan, Pratap S. Birthal, Alice Chiu, Partha R. Das Gupta, Uttam Deb, Klaus Deininger, Marco Ferroni, Probal Ghosh, Derek Headey, P.K. Joshi, Suneetha Kadiyala, Bart Minten, Hari K. Nagarajan, A.V. Narayanan, Kailash C. Pradhan, Rahul Raturi, Thomas Reardon, N.C. Saxena, J.P. Singh, Sudhir K. Singh, Alwin d'Souza, and Yuan Zhou. The summary report was written by Harinder S. Kohli.

Part II consists of six chapters based on background papers prepared by a group of experts on Indian agriculture, which were chosen because they each represent the "building blocks" referred to in Part I of the book. The core authors of the book are: Hans P. Binswanger-Mkhize (India 1960–2010: Structural Change, the Rural Non-farm Sector, and the Prospects for Agriculture); Pratap S. Birthal, P.K. Joshi, and A.V. Narayanan (Agricultural Diversification in India: Trends, Contribution to Growth, and Small Farmers' Participation); Richard Ackermann (Improving Water Use Efficiency: New Directions for Water Management in India); Marco Ferroni and Yuan Zhou (Review of Agricultural Extension in India); Partha R. Das Gupta and Marco Ferroni (Agricultural Research for Sustainable Productivity Growth in India); and Thomas Reardon and Bart Minten (The Quiet Revolution in India's Food Supply Chains).

The study was closely coordinated with the Planning Commission of India, and with the Ministry of Agriculture. Shenggen Fan of the International Food Policy Research Institute; C.H. Hanumantha Rao of the Centre for Economic and Social Studies, Hyderabad; and M.S. Swaminathan of the Economic Advisory Council of the Prime Minister all served as members of the Advisory Committee of this study, and their guidance and insights are gratefully acknowledged.

The study team is grateful to the following institutions and their invaluable staff for their significant contributions to the research done for this study: the International Crops Research Institute for the Semi-Arid Tropics (ICRISAT), Integrated Research and Action for Development (IRADe), the National Center for Applied Economic Research (NCAER), the International Food Policy Research Institute (IFPRI), and the Syngenta Foundation for Sustainable Agriculture.

Finally, the study team is deeply grateful to Honorable Union Minister Sharad Pawar; Deputy Chairman of the Planning Commission, Montek Singh Ahluwalia; and the senior officials at the Ministry of Agriculture and the Planning Commission of India for their support and encouragement, as well as helpful comments, suggestions, and critiques without which this study would not have been possible. Any remaining weaknesses and errors remain the responsibility of the Centennial Group, Syngenta Foundation and the authors.

Centennial Group and the study team deeply appreciate the financial support from Syngenta Foundation for Sustainable Agriculture that made this project possible.

The final manuscript was compiled, copy-edited, and formatted by Katy Grober of Centennial Group, under the general guidance of Marco Ferroni (editor), and with much support from Natasha Mukherjee (formerly of Centennial Group). Harinder S. Kohli and Anil Sood worked closely with Marco Ferroni in finalizing this book. Charlotte Hess, Drew Arnold, and Dana Sleeper, all of Centennial Group, provided valuable assistance in formatting the manuscript.

Introduction

Marco Ferroni, Harinder S. Kohli, and Anil Sood

India's recent performance in agriculture has been favorable. With agricultural production growing over the past 30 years, India has stepped into the ranks of the top five countries in key agriculture products: wheat, rice, cattle, eggs, vegetables, and melons. It has become the world's leading producer of milk and pulses. And it has been a net exporter of agriculture products every year since 1990–91. Smallholder farmers are shifting toward high-value outputs. Agriculture investment as a share of agriculture GDP rose from 13 percent in 2004–05 to over 18 percent in 2008–09; private investment increased significantly even while public investment was stagnant; and the private sector has moved into agricultural research and extension services. Retail in food products is modernizing rapidly.

Yet there is widespread consensus that, relative to the rest of the economy, agriculture is lagging and that it can and must do much better to support India's overall high economic growth and dynamism.

An earlier Centennial Group study published in 2010[1] postulated that continuing rapid economic growth could transform India into an affluent country by 2039, with average living standards comparable to those in many European countries (such as Portugal). Indeed, with average per capita incomes of USD20,000 a year, India could transition from poverty to affluence in one generation. But such rapid economic growth will require bold and sustained policy and institutional reforms, and success at the economy-wide level will lead to major challenges for India's agricultural sector.

India will face a rapid expansion of food demand and major shifts in its composition, thereby requiring an acceleration of agricultural growth rates beyond the current 4 percent target rate, or a rise in imports. With limited land and water resources, the acceleration of agricultural growth requires a significant acceleration of productivity growth, and much higher water use efficiency in the context of continued irrigation growth. The public and private institutions responsible for research and extension and for irrigation will have to adapt to these requirements. Rapid agricultural and rural non-farm growth is important and will be driven by a combination of agriculture and spillovers from the urban economy. Subsidies will need to be reformed to be more efficient and will ultimately need to be accommodated within India's fiscal headroom. India's numerous

---

1 India 2039: An Affluent Society in One Generation.

agricultural and rural development programs will need to be streamlined and reformed to deliver higher impact. Such major reforms will challenge both the central government and the states, which have the major responsibility for implementing agricultural and rural development programs. Effective implementation will overshadow new policies as the decisive factor.

The vision of Indian agriculture in 2040 spelled out in this book addresses these challenges. The vision is that of a more efficient sector supplying the food needs of an affluent, highly urban India. Under this scenario, India would remain a top world producer in most agricultural products and could strengthen its export performance in some of them. Such a scenario can be achieved only with bold institutional, policy, and program changes encapsulated by four necessary, inter-linked, and simultaneous sub-transformations: (i) from traditional grains to high-value crops and livestock products; (ii) from production based on low labor costs, widespread subsidies, and price support to efficiency and productivity-driven growth; (iii) from wasteful to efficient water use; and (iv) from public support and protection to ever greater involvement of the private sector throughout the value chain.

This vision is based not on mechanical projections but on what we believe Indian agriculture must transform into in order to match the economy's progress as a whole. This vision is not a fantastic construct; it is based on the experiences of other successful developing and developed economies, as well as India's own experience and planning documents.

Our review indicates that with a few notable exceptions—such as the lack of progress in making national markets for agricultural inputs and outputs more vibrant and efficient, and the persistence of high and often inefficient, inequitable, and environmentally damaging subsidies—India currently has most of the policies and strategies necessary to achieve the vision of an affluent and modern agricultural sector.

The worry is that even though good policies exist, they are often accompanied by others—notably hefty subsidies for water, electricity, and fertilizer—that prevent the proper development of a national market for agriculture inputs and outputs and slow the modernization of the value chain between farmers and consumers. More fundamentally, India's well-designed and well-intentioned policies have not delivered the expected results because of shortfalls in their on-the-ground implementation. The encouraging news is that, in numerous instances, a number of states have carried out successful initiatives that showcase desired reforms. The challenge now is to replicate these experiences and scale them up across the country, and thus reap the full benefits of these policies nationally.

With that in mind, this book proposes a set of recommendations that should be implemented on a priority basis. These recommendations are as follows: (i) make public programs much more focused and effective; (ii) recognize water as a critical, long-term constraint to India's agricultural growth and give top priority to significantly improving the efficiency of water use; (iii) promote new high-yield seeds and related technologies, including mechanization, to improve yields and productivity; (iv) improve the effectiveness of agricultural research and extension; (v) support further improvements of the farm-to-market value chain and reduce spoilage; and (vi) improve markets and incentives related to agriculture through reforms of prices, trade, and subsidies.

The purpose of this book

This book explores the future and presents the audacious question: what could the agricultural sector in India look like 30 years from now and how should it look if it is to successfully meet the needs of the country's affluent society? To answer this question, Centennial Group, with financial and technical support from the Syngenta Foundation for Sustainable Agriculture, commissioned a group of experts on Indian agriculture to produce a series of thought-provoking background papers. These papers ranged from an examination of specific, resource-focused concerns (such as water for irrigation), to institutional issues affecting the performance of agriculture.

The papers were presented and thoroughly discussed at a three-day conference "The Long-term Future of Indian Agriculture and Rural Poverty Reduction" in New Delhi, 27–29 April, 2011. The conference was attended by eminent policymakers from the Planning Commission and the Ministry of Agriculture; researchers and academics from Indian and international universities and research institutes; and representatives from the private sector and non-governmental organizations. While based on these papers, the book offers a bold vision for the longer term development of the sector. It presents the ideas developed in these various papers in a cohesive whole in an attempt to link Indian agriculture with the overall economy.

Looking back to gauge the future

In order to contemplate the book's underlying question about the future of Indian agriculture, it is helpful to look back to the past 30 years, partly because the backward look underlines how dramatically India's agricultural sector has already changed. Further, it helps focus the analysis because more than three decades ago, India's agricultural sector similarly stood at an important juncture at which point it was about to experience a significant transformation.

Today, India is no longer synonymous with crushing poverty, recurring famines, and starving people. It has become a country that has attained food self-sufficiency—and actually has done one better than that: it has become a net exporter of agricultural production. On the global front, it has become a leading agricultural producer. In terms of cereals production, it holds third place after the US and China; second after China in both wheat and rice production (all 2007 data), and first place in milk production.

Beginning in the mid-1960s, the Green Revolution led to the transformation of parts of India's agricultural sector. The Green Revolution utilized more readily available inputs (e.g., seeds, fertilizers, mechanization, including irrigation) to make technology more widespread, and also promoted agricultural research and extension messages. Agricultural growth took off: the sector grew at an annual rate of roughly 3 percent.

Where is the agricultural sector today?

The Green Revolution helped jumpstart a process of structural transformation of the agricultural sector.

Trends that bespeak of such transformation are very much in evidence today: agricultural production is diversifying—away from traditional crops and staples such as wheat and rice, towards horticultural and animal food products. Their share in the value of output of the agricultural sector, including animal husbandry and fisheries, is now close to 50 percent, which is 17 percentage points higher than in the early 1980s. The private sector is becoming an increasingly important participant in both agricultural production and agricultural marketing.

As India's economic growth continues and its prosperity increases, the country will experience a rapid, historically unprecedented rise in the demand for food. Even if the agricultural sector has performed sufficiently over the past decades, the looming question is whether it can supply at the levels necessary to keep abreast of the anticipated demand. Simply stated, agricultural performance that is doing "well enough" might not be good enough.

According to Dev (2011), increasingly persistent food inflation figures from the past several years indicate that the supply of these higher-value agricultural commodities is lagging compared to demand. It may be argued that these inflationary trends are a transitory phenomenon, but the Reserve Bank of India (2011) has posited that food inflation is on its way to becoming a structural problem in the country. It goes even further by hypothesizing that protein inflation—the price inflation of pulses, milk, eggs, meat, and fish—is, and will continue to be, especially acute (Reserve Bank of India, 2011).

The recent inflation in non-cereal food is a serious concern. India's poor still get more than 55 percent of their daily calories from cereals. Despite positive developments in the agricultural sector, the nutritional standing of many Indians remains weak. There is still a need for diet diversification to improve nutrition among the poor, which, in turn, requires deeper agricultural diversification, particularly among small and marginal farmers and those in resource-poor areas.

Another serious concern is the realization that India is reaching its physical resource constraints—thereby severely limiting the scope to expand agricultural production without significant improvements in productivity. India's water availability is especially worrisome. Water availability is already constrained, and is likely to worsen, as groundwater stores are depleted and the effects of climate change negatively impact India's hydro-geological situation. Land degradation has reportedly increased, in the form of depleted soil fertility, erosion, and waterlogging.

As the structural transformation of India's economy proceeds, the share of agriculture in GDP will likely continue to decline. Already it has declined significantly from above 40 percent in the 1960s to around 9 percent in 2010. And yet the agricultural and non-agricultural rural economy continues to provide a significant share of total employment, estimated at around 52 percent in 2011.

Hence, the livelihood of a significant share of Indian households depends on the performance of the agricultural and non-agriculture rural economy, both now and in the foreseeable future.

The premise of this book is that Indian agriculture must undergo an important transformation on both the demand and the supply side, along three primary directions by: (i) emphasizing higher-value outputs; (ii) increasing productivity; and (iii) redefining the roles of the public and private sectors.

The organization of this book

The book is divided into two parts. Part I "Transforming India's agriculture: productivity, markets, and institutions—an overview" was discussed at a high-level policy dialogue in New Delhi in January 2012. Part II, on the current state of Indian agriculture and areas of reform for its transformation, consists of six chapters, which were chosen because they each represent the "building blocks" that we refer to in Part I of the book.

These six chapters are the necessary foundation of the agenda that will transform India's agriculture to meet the needs and challenges of the future. On one end of the spectrum, they concern issues related to inputs, specifically the crucial input of water/irrigation for agriculture. On the other end of the spectrum, they represent institutional concerns: specifically addressing the institutional weakness of agriculture services provided by public institutions and enhancing the role of private sector players.

Chapter 1 "India 1960–2010: Structural Change, the Rural Non-farm Sector, and the Prospects for Agriculture" is written by Hans P. Binswanger-Mkhize, Adjunct Professor at China Agricultural University's College of Economics and Management in Beijing. This chapter examines past and likely future agricultural growth and rural poverty reduction in the context of the overall Indian economy, and notes that while the growth of India's economy has accelerated sharply since the late 1980s, agriculture has not followed suit. Additionally, the chapter draws attention to the rise of the rural population and especially the rural labor force in absolute terms. Despite a sharply rising labor productivity differential between non-agriculture and agriculture, limited rural-urban migration, and slow agricultural growth, urban-rural consumption, income, and poverty differentials have not been rising. This is because urban-rural spillovers have become important drivers of the rapidly growing rural non-farm sector—the sector now generates the largest number of jobs in India. Rural non-farm self-employment has become especially dynamic with farm households rapidly diversifying into the sector to increase income. Binswanger-Mkhize identifies this growth of the rural non-farm sector as a structural transformation of the Indian economy, but concludes that it is a stunted one. Nevertheless, non-farm sector growth has allowed for accelerated rural income growth, contributed to rural wage growth, and prevented the rural economy from falling dramatically behind the urban economy. The chapter further discusses how the bottling up of labor in rural areas means that farm sizes will continue to decline, agriculture will continue its trend to feminization, and part-time farming will become the dominant farm model. It also presents factors on which continued rapid rural income growth is contingent, such as continued urban spillovers from accelerated economic growth, and a significant acceleration of agricultural growth based on more rapid productivity and irrigation growth.

Chapter 2 "Agricultural Diversification in India: Trends, Contribution to Growth and Small Farmers' Participation" presents a positive story on the current state of India's agriculture. It is written by P.K. Joshi, the Director for South Asia at the International Food Policy Research Institute (IFPRI), and Pratap S. Birthal and A.V. Narayanan, researchers at the International Crops Research Institute for the Semi-Arid Tropics, and describes the current trend of diversification—out of traditional grains and into higher-value crops. The authors conclude that it is the transformation of the production portfolio (along with technological change) that has sustained India's agricultural growth in recent decades. The share of horticultural crops in overall growth of the crop sub-sector increased from 26 percent in the 1980s to 47 percent in the 1990s, but declined to 39 percent in the 2000s. Most of the increase came from the diversification of land away from less profitable crops, mainly coarse cereals and pulses. Diversification contributed one-third to the agriculture growth in the 1990s, marginally more than in the 1980s and 2000s. Most remarkably, the authors find that India's success with agricultural diversification has come from the small farm sector. Increased demand for high-value food commodities has become a very important opportunity for small farmers. The challenge for the future will be to assure that the opportunities related to production are not undermined by smallholder disadvantages where marketing and the value chain between farmer and consumers is concerned.

Chapter 3 addresses one of the most important areas of reform to increase agricultural growth: how to improve water use efficiency. Written by Richard Ackermann—a noted water economist with a distinguished career in academia and international development organizations—the chapter "Improving Water Use Efficiency: New Directions for Water Management in India" takes a purely evidence-based and pragmatic approach to identifying practical and demand-driven ways to resolve some of the deep-seated problems in the water sector. There is now a large enough body of irrefutable evidence of what works and what does not work in the Indian institutional environment related to water, and how farmers have responded—whether by design or by default—that allows one to make robust recommendations going forward. The chapter argues that past emphasis on the development of surface water resources must be changed to greater reliance on groundwater sources and on development of water markets.

Chapter 4, written by Marco Ferroni and Yuan Zhou, "Review of Agricultural Extension in India," is concerned with the manner in which agricultural extension in India can be improved. During the Green Revolution period, extension—along with improved seeds, fertilizers, and irrigation—increased productivity and enhanced agricultural development. In the period since then, the public provision of extension has fallen short of expectations. Research-extension-farmer linkages are absent or weak in many instances, while there are also duplications of efforts among a multiplicity of extension agents without adequate coordination. In order to boost total factor productivity growth in the agricultural sector in India, tailored extension solutions are needed. Ferroni and Zhou offer a list of recommendations for all providers of extension—NGOs, the for-profit private sector, and the public sector—that could revamp this important contribution to the agricultural sector.

Chapter 5, an essay by Partha R. Das Gupta, a practitioner who has devoted his life to agricultural development in India, and Marco Ferroni, is entitled "Agricultural Research for Sustainable Productivity Growth in India" and argues for a renewed commitment to agricultural research in the country. Beginning in the 1960s, the national agricultural research system made historic early contributions by way of improved varieties of cereals, pulses, oilseeds, fiber crops, sugarcane, potato, horticultural, and plantation crops. Besides evolving high-yielding crop varieties, improved soil and water management technology, and pest and disease control techniques to name but some of the solutions that were developed, the national system played a significant role in taking new technologies to farmers' fields. Despite these successes, by the mid-1990s, a sort of 'technology fatigue' set in, even in the agriculturally progressive regions that practiced intensive agriculture. According to policymakers, this was partly due to a 'business as usual' attitude that had crept into the public research system. Today, the aim is to rediscover ways to put public-sector research back on its path of former progress, and supplement it with private sector research and product development.

Finally, Chapter 6 discusses ways in which the efficiency of the farm-to-market value chain can be improved. In "The Quiet Revolution in India's Food Supply Chains," Thomas Reardon and Bart Minten, from Michigan State University and the International Food Policy Research Institute (IFPRI), respectively, show how modern and traditional supply chains are rapidly transforming rural areas. The authors assert that it is the private sector (with participants from both the modern and traditional sectors) that is the most important actor in this 'quiet revolution' that is shaping food security in India. The common perception is that India's agricultural markets are largely publicly-driven—the government's direct role, as a buyer and a seller, is only 7 percent of the food economy of India. Despite important policy changes that have spurred a quiet revolution, a number of constraints still prevent supply chains from operating at their optima.

Concluding thoughts

Given its continental size and diversity, India experiences great inter-regional differences from the national average; there is considerable variation in the trends and contributions of different parts of the country to agricultural growth.

The Green Revolution helped parts of Indian agriculture to prosper, but, in retrospect, it appears that the boost from the Revolution mainly benefited the high potential areas, such as Punjab and Haryana. The dry lands and rain-fed areas, and those areas that might be called the 'periphery', such as the Eastern states of Bengal, and Orissa, did not benefit to the same extent—they continue to lag behind the other regions. There is, however, the example of Gujarat, the positive outlier. Its agricultural sector has prospered—and continues to outshine any of the other Indian states—despite its natural resource limitations.

India's Green Revolution represents a process by which public resources were used to create the foundation upon which private actors were able to flourish. Fifty years later, people are calling for a Second Green Revolution—to jumpstart a similar process, but with the added objective to benefit the parts of India's agricultural sector that were bypassed during the first round. This would serve the needs of a society that is becoming increasingly middle class and more urban.

The challenges for the sector's future are both complex and daunting. Most observers of Indian agriculture believe that its problems have less to do with policies than with actual policy and program implementation. Fortunately, at all links of the agricultural value chain—from extension services to storage facilities and marketing arrangements—private actors have started to complement public institutions, and are often providing better services. While the private sector is moving quickly to transform research, extension, and value chains according to its comparative advantage, most public sector institutions and programs for agriculture require urgent reforms, often along lines long recognized but not yet implemented. The public sector institutions must also concentrate more on areas that are not of interest to the private sector, and collaborate more with the latter.

Equally important, the enormous challenges of agricultural growth, natural resource management, and social services for rural areas must be resolved with greater citizen empowerment and decentralization. Such reforms have long been discussed in India, but initiatives have so far failed to bring them about. Reforms will have to be driven primarily by the states, with support from strong incentives and perhaps further legislative interventions from the center as well.

The reforms will not come about without pressures from below as well as from the very top. Once again, like two generations ago, India's agriculture stands at an important crossroad.

The leadership must make the critical decision about which direction to take, then follow up that decision with effective implementation of the necessary policy and institutional reforms. Hopefully, this book will make a contribution, however modest, to this process.

Part I

Productivity, Markets, and Institutions

Overview
Section 1

Harinder S. Kohli and Anil Sood

India's recent performance in agriculture has been fairly favorable. With agricultural production growing over the past 30 years, India has stepped into the ranks of the top five countries in key crop production categories, including wheat, rice, cattle, eggs, vegetables, and melons. It has become the world's leading producer of milk and pulses. And it has been a net exporter of agriculture products every year since 1990–91. Smallholder farmers are shifting toward high-value output. Agriculture investment as a share of agricultural GDP rose from 13 percent in 2004–05 to 18 percent in 2008–09; private investment increased significantly even when public investment was stagnant, and the private sector has moved into agricultural research and extension services. Retail in food products is modernizing rapidly.

Yet there is a widespread and justifiable consensus that, relative to the rest of the economy, agriculture is lagging and that it can and must do much better to support India's overall high economic growth and dynamism.

The Centennial Group reports find that continuing rapid economic growth would transform India into an affluent country by 2039, with average living standards comparable to those in many European countries today (such as Portugal). Indeed, with average per capita incomes of USD20,000 a year, India could go from poverty to affluence in one generation. But such rapid economic growth will lead to challenges for India's agricultural sector.

India will face a rapid expansion of food demand and major shifts in its composition. Agricultural growth rates would have to accelerate, perhaps even beyond the current 4 percent target rate, or imports would need to rise. With limited land and water, the acceleration of agricultural growth requires a significant acceleration of productivity growth, and much higher water use efficiency to enable continuing irrigation growth. The public and private institutions responsible for research and extension, and for irrigation, will have to adapt to these requirements. Rapid agricultural and non-farm growth are important and will be driven partly by agriculture and partly by spillovers from the urban economy. Subsidies will need to be reformed to be more efficient and ultimately will need to be accommodated within India's fiscal headroom. India's numerous agricultural and rural development programs will need to be streamlined and reformed to deliver higher impact. Such major reforms will challenge both the central government and the states, which have the major responsibility for implementing agricultural and rural development programs. Effective implementation, more than new policies, will be the decisive factor.

The vision of Indian agriculture in 2040 spelled out in this book addresses these challenges. The vision is that of a more efficient sector supplying the food needs of an affluent, highly urban India. Under this scenario, India would remain a top world producer in most agricultural products and could strengthen its export performance in some of them. Such a scenario can be achieved only with bold institutional, policy, and program changes encapsulated by four necessary, interlinked, and simultaneous transformations:

- From traditional grains to high-value crops and livestock products. The agricultural sector continues to move away from the production of traditional cereals to the production of higher value horticultural crops and livestock products. As India becomes more affluent, domestic demand will continue to shift toward higher value crops.

- From production based on low labor costs to efficiency and productivity-driven growth. Agriculture moves away from traditional production patterns based on low labor costs and subsidized inputs to commercial production that uses improved and appropriate technologies, including modern information techniques.

- From wasteful to efficient water use. The use of water—Indian agriculture's critical constraint over the long term—is highly inefficient. Despite dwindling surface water and groundwater resources, all economic sectors—especially agriculture—use water as if it were an abundant resource. This is clearly unsustainable.

- From public support and protection to an even greater involvement of the private sector in the value chain. The private sector plays a much stronger role throughout the value chain, including food grain distribution. Public support will continue to be critical for marginal groups (such as small farmers and poor households).

This vision is based not on mechanical projections but on what we believe Indian agriculture must transform into in order to match the economy's progress as a whole. The vision is not a fantastic construct; it is based on the experiences of other successful developing and developed economies, as well as India's own experience and planning documents.

Our review indicates that with a few notable exceptions—such as the lack of progress in making national markets for agricultural inputs and outputs more vibrant and efficient, and the persistence of high and often inefficient, inequitable, and environmentally damaging subsidies—India today has on the books most of the policies and strategies necessary to achieve the vision of an affluent and modern agricultural sector.

The worry is that even though the good policies exist, they often are accompanied by others—notably hefty subsidies for water, electricity, and fertilizer—that prevent the proper development of a national market for agriculture inputs and outputs and slow the modernization of the value chain between the farmers and consumers. More fundamentally, India's well-designed and well-intentioned policies have not delivered the expected results because of shortfalls in their on-the-ground implementation. The encouraging news is that in numerous instances a number of states have carried out successful pilots that showcase desired reforms. The challenge now is to replicate these experiences and scale them up across the country, and thus reap the full benefits of these policies nationally.

With that in mind, this book puts forth a set of building blocks and recommendations that should be implemented on a priority basis. These building blocks and recommendations are summarized below:

Make public programs much more focused and effective
- Sharply reduce the number of centrally sponsored schemes and public programs.
- Convert the current fragmented silo approach into block grants.
- Decentralize, as soon as is practical, down to Panchayati Raj institutions (accompanied by capacity building), the implementation of the public programs, with increased accountability for results at all levels.
- Use the independent evaluation bodies, such as the newly established Evaluation Office of the Planning Commission, to monitor and evaluate impacts. The feedback from such evaluations must be fed into the design, budgeting, and monitoring of future schemes and programs.

Recognize water as a critical, long-term constraint to Indian agricultural growth and give top priority to significantly improving the efficiency of water use
- Shift the emphasis from open surface irrigation (such as canals) to more efficient, sustainable groundwater-based irrigation (including underground pipelines).
- Promote the emergence of water markets and greater involvement of private wholesale providers.
- Actively promote measures to artificially recharge groundwater aquifers.
- Expand the public financing of on-farm or local water storage facilities.
- Phase out electricity subsidies, as part of broader subsidy reforms; provide electricity for fixed hours for electricity feeders for groundwater extraction (separate from domestic electric supply).
- Support expansion of micro-irrigation and drip irrigation.

Promote new high-yield seeds and related technologies, including mechanization, to improve yields and productivity
- Promote more and better seeds, including *Bt* and other GM seeds.

- Tackle regulatory bottlenecks that restrict the availability of improved seeds, both at the national and farm level, while keeping strong safeguards to protect public health and safety.
- Move to market-based output and input prices.
- Adopt banking practices that promote financial inclusion and the availability of needed credit, particularly to poor, marginalized farmers and others active in the rural economy.
- Encourage the private sector to step up investments in developing and marketing new technologies.

Improve the effectiveness of agricultural research and extension
- Step up the investment in agricultural research from 0.7 percent of agricultural GDP to 2 percent (as envisioned in 12th Five Year Plan).
- Review and implement the recommendations of the Swaminathan and Mashelkar Committees to restructure the Indian Council of Agricultural Research (ICAR).
- Build and better exploit synergies between public and private sectors in agricultural research, including through public-private partnerships.
- Accelerate the scaling up of the Agricultural Technology Management Agency (ATMA) and the Krishi Vigyan Kendras (KVKs), replicating successes in selected states across the country.
- Encourage participation of the private sector in extension, including by leveraging information technology, and e- and mobile- (m-) applications.
- Support professional NGOs in providing extension services—with a strong focus on the small farmer and remote areas.

Support further improvements of the farm-to-market value chain and reduce spoilage
- Remove remaining regulatory barriers (such as mandi monopolies) to private sector expansion in transport, storage, and marketing—by accelerating the adoption and implementation of the Agricultural Produce Marketing Committee (APMC) Model Act in all states.
- Encourage higher investments by the private sector to improve productivity and competition throughout the value chain, and permit farmers to retain a larger share of their output prices while simultaneously lowering consumer prices.
- Open retailing to foreign direct investment to bring in additional capital, technology and management know-how to enhance competition and improve supply chains.
- Reform food grain policy and institutions.

- Replace subsidized food distribution with conditional cash transfers that can be used for a broad set of foods rather than just food grains, using newly introduced Aadhar cards (along the lines of food stamps), and adopt a system that excludes the identifiable well-off.
- Shift the procurement, imports, and storage of most agricultural products to the private sector.
- Refocus the Food Corporation of India (FCI) on managing buffer stocks.

Improve markets and incentives related to agriculture through reforms of prices, trade, and subsidies

- Remove regulatory and administrative barriers to achieve truly nationwide markets for agricultural inputs and outputs.
- Phase out subsidies that encourage environmentally damaging patterns of input use.
- Eliminate remaining restrictions on land rental and lease markets, and further improve rural land administration to improve the land sales markets, and the ability of farmers to use land as collateral for loans.
- Integrate Indian agriculture into the global economy well before 2040.

A Vision of Indian Agriculture in 2040

Overview
Section 2

Harinder S. Kohli and Anil Sood

Our vision of what India's economy in 2040 should and can look like, with a well-performing, modern agricultural sector, will require patterns of agricultural growth that are primarily driven by diversification, the market, rapid technical change and productivity growth, and improvements in water use efficiency and irrigation combined with much more effective implementation of officially adopted policies and programs.

By 2040, Indian farms will comprise modern, market-oriented small farms that use state-of-the-art technology to achieve efficient patterns of production. Farms will use advanced biological and mechanical technology for crops, horticulture, livestock, and aquaculture. Machinery rentals, water markets, and cooperation in marketing and input supply, as well as contract farming, will realize economies of scale. The living standards of the entire rural population in 2040 will have improved dramatically: everyone will have access to protected water supplies, modern sanitation, and 24/7 electricity supplies. Almost all citizens—including women—will be literate, having attended at least primary school. Easy access to modern telecommunications will allow even rural Indians to access the latest information. Key aspects of agriculture and the food economy will include the following:

Many farmers will conduct their agricultural operations part-time

By 2040, Indian agriculture will be dominated by part-time farm households, with a certain number of full-time farmers at the top—in keeping with ongoing trends and the pattern of other countries that have already experienced the commercialization and modernization of their agricultural sectors. Women will play an increasing role in these farms as managers and workers, partly because their opportunities in the rural non-farm and urban sectors are more limited than those of men. The part-time farmers will derive more of their incomes from non-farm activities than from agriculture; small farmers will focus much more on horticulture, milk, eggs, and fish.

Rural incomes and wages will be a multiple of what they are today—agricultural wages could increase manyfold from current levels over the next 30 years. The combination of rising agricultural and rural non-farm incomes will support rapid income growth in rural areas, including rapid rural wage growth: rural-urban incomes and consumption ratios will be improving, or at least not deteriorating, and rural poverty will decline rapidly.

The composition of food demand will change

Agricultural production will be driven by consumer demands for a highly diversified basket of high-value, often information-intensive, commodities (that is, horticultural and dairy products), while the demand for traditional grains and other subsistence crops will decrease in relative (though not absolute) terms. To cater to this high-value demand, agricultural production will become highly diversified across regions, states, and villages.

Consumer demand will drive the already observable trends toward traceable agricultural output and quality control—providing additional income opportunities for the agricultural sector.

The private sector will be the most important actor in value chains and in agricultural services, while the public sector will attend to parts of the agricultural sector where the private sector is less interested

By 2040, the relative roles of the private and public sectors will have changed significantly. Given the highly diversified agricultural production and the vast variations in India's soil, water, and weather, the private sector will provide—according to its comparative advantage—most location-specific inputs and services to farmers. The public sector will become more focused on remote regions with poor agricultural endowments and less favorable prospects for rural development and local job creation. It will also play a sophisticated role as a regulator and provider of necessary public goods.

Based on continuing trends, private service providers and agribusiness companies will have substantially expanded their offer of mobile applications and information services on technologies and practices, as well as agricultural markets. Mobile business transactions and both professional and social networks will have become a normal part of life in all segments of the value chain.

Public agricultural and rural programs will be consolidated and significantly decentralized to best address the location-specific needs

Agricultural and rural development programs of the center will be consolidated from the hundreds of central and Centrally Sponsored Schemes to a sharply reduced set of block grants that will provide much more flexibility for implementers at state, district, block, and village levels. Many of them will also empower the final beneficiaries, who will do more in planning and implementing the schemes. Roles and accountabilities will be clarified and strengthened, along with monitoring and impact evaluation. As a consequence, the implementation of agricultural and rural development programs will be significantly improved, more transparent, and less a source of corruption.

Irrigation—with water a critical constraint to India's agriculture—will have become an efficient modern sector

Water management will have become more efficient: canal irrigation systems will have shifted to more demand-driven modes of providing water in a timely and controlled manner, primarily through using pipes rather than open canals. Private or cooperative water providers will develop and run these networks, rather than the irrigation departments and civil servants. The ground-water situation will be rendered less critical through pervasive strategies—at both the public and private levels—that improve the efficiency of water use at all levels, ranging from water harvesting to groundwater recharge and improved micro-irrigation technologies. Electricity supplies for irrigation will have become reliable, and subsidies reduced, or replaced, by direct payments. Command-and-control interventions to control groundwater extraction may be required only in some of the most critical watersheds.

Remaining constraints in the marketing of agricultural commodities will have been eliminated and prices will be market-driven with few price controls—if any

In 2040, India will have one single national market that allows the unhindered flow of all agricultural products and inputs. The existing constraints on agricultural marketing through regulated markets will have been eliminated by 2040, and marketing and value chains will have modernized at an accelerated pace from the farm to the retail outlets. Flourishing competition in marketing will reduce the current excessive markups in the value chain—and thus assist in combating food inflation while offering the farmers a greater share of the market value of their outputs. All classes of farmers will have a wide range of choices along the value chain, including the choice of retailers and processors that demand high-quality outputs. Information to enable such choices will be available through the many m- and e-applications. Most prices will be market-driven, with few, if any, price controls.

Indian agriculture markets will be more integrated with global markets, with domestic prices aligned with broader global trends

By 2040, the country's economy will be even more integrated with the global economy—as its international profile and its policies, including those in agriculture, will draw intense interest worldwide. With India as the world's leading producer and consumer of most food products, domestic prices will be aligned with developments in international markets.

The effects of climate change will pose significant new challenges: agricultural yields will decline and real agricultural prices will rise, unless the trend is neutralized by science-based technical change

The consensus is that global temperatures are rising; 2040 may be hotter than today. In such a world, billions will experience significantly altered water supplies. The flow of rivers from the Himalayas will likely be disrupted. The Indus River could lose as much as 8 percent of its flow, and the Brahmaputra 20 percent, by 2065—a frightening prospect given that more than 250 million people now rely on these rivers (Hepburn and Ward, 2010). With current irrigation techniques, these losses in river flow could mean that 25 million fewer people (roughly 10 percent of the current population) could rely on the Indus for food production; the Brahmaputra would feed between 28 million and 41 million fewer people (Hepburn and Ward, 2010), underlining the need for sharp improvements in water use efficiency.

As a result of climate change, significant reductions in crop yields are expected in most countries, unless counteracted by adaptive practices and more technical change. In India, the potential decline in yields is estimated at roughly 14 percent, under a scenario where the international community is unable to take meaningful action to curtail climate change (Kohli et al., 2011). With action by developed countries alone or concerted action by developed and emerging countries, it is estimated that agricultural yields in India may still suffer dramatically, falling around 11 percent, unless agricultural research investment is stepped up in the context of reforms of the national agricultural research system, and breakthroughs to compensate for the effects of climate change are achieved.

The impacts from climate change will be significant between now and 2040, but they will be even greater in the years beyond. This projection means that it will be crucial for agriculture to adopt adaptation measures and science-based technical change.

Summing up, therefore, the challenges to be overcome are daunting. But the action implications accompanying our vision of Indian agriculture in 2040 seem feasible and are in any case unavoidable if India is to achieve the agricultural transformation and inclusive growth required to become an affluent society within a generation. The rest of this Overview discusses the main action implications by first looking at the legacies of the past and then investigating in some detail the building blocks of the strategy we propose.

Legacies of the Past and Key Challenges

Overview
Section 3

Harinder S. Kohli and Anil Sood

A look back at the past 30 years provides vital insights. Agricultural production has been growing—but not at its currently targeted level of 4 percent. It has fallen short of the target, at around 3 percent during the latest 11th Plan period, even if it has recovered from around 2 percent during the 10th Plan period.

Over the past 30 years, India's agricultural production has been growing, with livestock and nonstaples up more sharply than agriculture as a whole (Figure 1a). The shares of fruits and vegetables in agricultural production have been growing, while cereals have been shrinking (Figure 1b). Despite increases in absolute volume of production, agriculture's share of GDP has been falling steadily, from 35 percent in 1980 to 15 percent today.

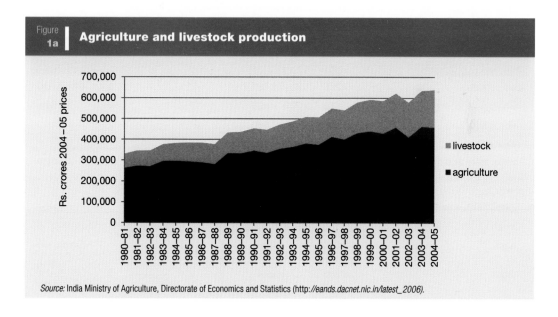

Figure 1a | Agriculture and livestock production

Source: India Ministry of Agriculture, Directorate of Economics and Statistics (http://eands.dacnet.nic.in/latest_2006).

On the plus side, India enjoys a significant footprint in global agricultural production, ranking in the top five in most relevant categories of crop production: rice and wheat, fruits and vegetables, commercial crops, livestock, and animal products (Table 1 and Figures 2a and 2b).

I/3

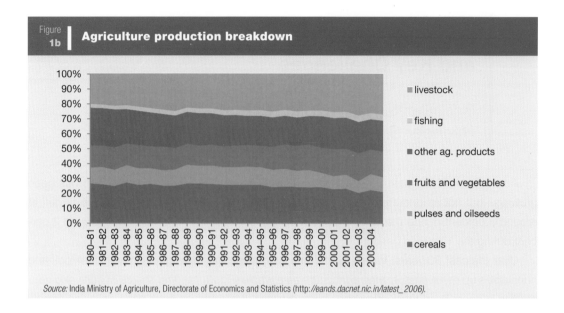

Figure 1b | Agriculture production breakdown

Legend:
- livestock
- fishing
- other ag. products
- fruits and vegetables
- pulses and oilseeds
- cereals

Source: India Ministry of Agriculture, Directorate of Economics and Statistics (http://eands.dacnet.nic.in/latest_2006).

Table 1 | India's position in global agriculture, 2007

food production (million tons, unless otherwise indicated)	India	World	% share	India's rank	behind
total cereals	260	2,351	11.1	3	China, US
wheat	76	606	12.2	2	China
rice (paddy)	145	660	21.9	2	China
total pulses	14	56	25.4	1	
vegetables and melons	77	909	8.5	2	China
livestock and animal products					
cattle (million heads)	177	1,357	13.0	2	Brazil
milk	106,100	679,207	15.6	1	
eggs	2,670	63,411	4.2	3	China, US
meat	6,508	269,149	2.4	5	China, US, Brazil, Germany

Source: MOA (2010).

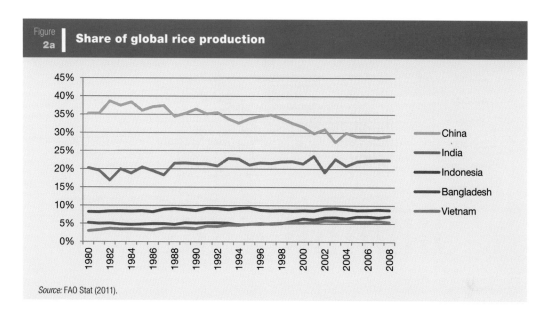

Figure 2a | Share of global rice production

Source: FAO Stat (2011).

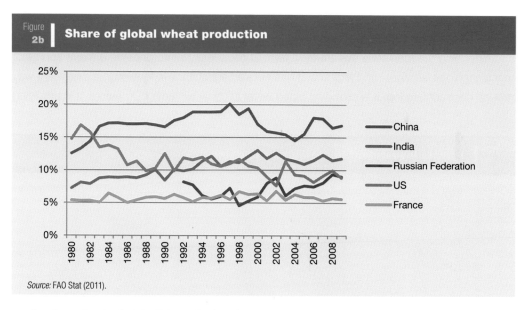

Figure 2b | Share of global wheat production

Source: FAO Stat (2011).

Another plus is that India's agricultural exports have increased steadily, from 5 percent in 1990 to 16 percent today, with agricultural imports valued at around 11 percent of production—making India a net exporter of agricultural products (Figure 3).

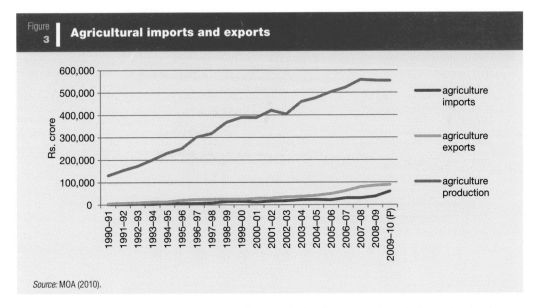

Figure 3 | **Agricultural imports and exports**

Source: MOA (2010).

Yet another plus is that since the 10th Plan period, private investments have risen substantially as a percentage of agricultural GDP (Figure 4). In addition, the private sector has emerged as a key driver of many components of agricultural and rural development. The private sector is transforming the marketing system from farm to consumer. It has become a major source of new technology, services, and logistics. It has entered agricultural extension in a significant way through contract farming. It is providing piped water in canal systems to irrigators. It is providing

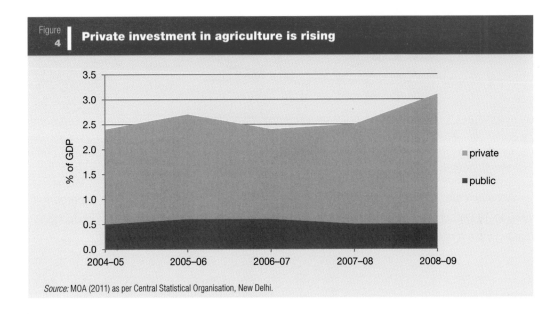

Figure 4 | **Private investment in agriculture is rising**

Source: MOA (2011) as per Central Statistical Organisation, New Delhi.

agricultural credit through contract farming and microfinance. And in places, it is assisting in the administration of land record systems.

So why is India not reaching its agricultural production growth target of 4 percent? In the late 1990s and early 2000s, Indian agriculture went through a crisis that had its origin, to some extent, in poor rainfall, but more because of declining agricultural prices, low public and private investment, and a slump in productivity growth. These negative factors have now been reversed (except that rainfall performance cannot be controlled), and agricultural growth has accelerated to around 3 percent, still short of the 4 percent Plan target. However, irrigation growth has also declined. It will take continued high investment levels and other stepped-up sources of productivity growth to further accelerate agricultural growth.

A critical element will be further accelerating growth in total factor productivity (TFP)—a measure of the efficiency of all inputs to the production process. Studies show that TFP in Indian agriculture did not increase during the 1960s and the 1970s, but it rose 2 percent a year in the 1980s, as the Green Revolution spread to all regions and most crops. It then declined sharply to almost zero around 2000, but started to recover in 2003 to reach an unprecedented high level of close to 4 percent in 2006–07 (Figure 5).

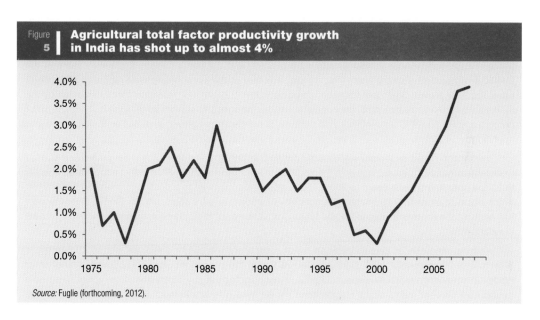

Figure 5 | Agricultural total factor productivity growth in India has shot up to almost 4%

Source: Fuglie (forthcoming, 2012).

This recent rise may indicate that India has overcome its slump in agricultural productivity growth. China also had no TFP growth in the 1960s and 1970s, but its TFP growth was close to 3 percent in the 1980s—and then shot up to an astonishing 4.2 percent in the 1990s, slipping back to 3 percent in the first seven years of the 2000s. This means that China had TFP growth close to or far above 3 percent for nearly 30 years, by far the world's most prolonged period of such rapid growth (Box 1).

Box 1 | **The structural and agricultural transformations in China and India**

Chinese economic growth started accelerating in 1984, in the wake of the economic reforms that started around 1980. In each decade since, agricultural growth was faster than in India, reaching 4.9 percent a year between 1980 and 2009. It slowed in the first decade of this century, but less than in India, to 3 percent a year.

Between the 1980s and 2009, the share of agriculture in GDP dropped from around 30 percent to about 10.6 percent (about 15 percent in India). Because of the large migration from rural to urban areas, the agricultural labor force declined from about 64 percent to 40 percent in the same period, which is in sharp contrast to the much slower decline in India.

After the mid-1980s, nonagricultural labor productivity started to move ahead of agricultural labor productivity at an accelerating rate. The ratio of nonagricultural to agricultural labor productivity has now risen to about 6.0:1, compared with India's 4.2:1. Clearly, both in the gap between the agricultural output and labor share, and in the intersectoral productivity differential, China has also has not yet reached the turning point in its structural transformation where agriculture and non-agriculture start converging.

Crop productivity started to rise rapidly through the 1990s and the first decade of the 2000s, growing at more than 5 percent during the three years to 2009. Labor productivity grew at nearly 4.5 percent during the 1990s, slowing a bit to 2.9 percent in the 2000s. Total factor productivity (TFP) rose at nearly 2.5 percent in the 1980s, compared with a little below 2 percent in India. It rose to 4.2 percent in the 1990s (while TFP growth in India slowed to below 1.5 percent) and 2.7 percent in the first seven years of the 2000s, a full percentage point faster than in India. No other country or subregion has recorded TFP growth rates above or around 3 percent for nearly three decades. The fast TFP rates are partly explained by the marked slowdown in the growth of the population and in the agricultural labor force in the last decade to 0.7 percent and 0.1 percent, respectively, and partly by the exceptional growth in public agricultural research spending.

Cropland in China expanded significantly during the 1980s. Since then, it first stagnated and started to decline at nearly 1 percent a year in the first decade of this century, a decline that accelerated sharply to 2.2 percent in the three years to 2008. These trends are mirrored in a decline in cropland per worker, which has now fallen to 0.25 hectare per agricultural worker, only 39 percent that of India. Over the past half century, the share of cropland irrigated rose from 30 percent to 52 percent, compared with 37 percent in India today.

In China, the share of cereals declined much more sharply than in India—from around 40 percent in 1980 to only 20 percent in 2009. During the entire period the shares of pulses and oilseeds and of other crops did not change much. The share of horticulture (including starchy tubers) rose from about 16 percent to about 26 percent today, slightly higher than the 22 percent share in India. Since the early 1980s, the share of livestock rose from 25 percent in 1983 to about 42 percent by 1997 and then stayed much the same. This compares with a share of livestock of about 28 percent in India today. In India, the bulk of livestock consumption is eggs, poultry, and especially milk, while in China, the share of milk is negligible and the bulk of livestock consumption is meat, poultry, and eggs. As a consequence, China now imports significant quantities of feed grains and oilseed cakes.

China's structural transformation is following more closely the past patterns of the developed world and of East Asia, with rapid rural-urban migration of labor, labor scarcity for agriculture, and some land

| Box 1 | **The structural and agricultural transformations in China and India** |

consolidation toward larger farms. Four factors account for the difference with India:

- First, the population growth rate has declined much more sharply than in India, slowing the growth of the labor force.

- Second, the share of manufacturing in the economy is more than twice as high, and the sector has grown very rapidly.

- Third, irrigation has grown faster and covers a greater proportion of agricultural land than in India.

- Fourth, for the last three decades, agricultural TFP has grown much faster than in India.

These factors together mean that China's agricultural growth has been significantly faster than in India for the past three decades, although these two growth rates have converged to 3 percent a year. As a consequence of all these factors, we do not see the bottling up of labor in rural areas that has been characteristic of India, but instead a rising labor scarcity. Agricultural mechanization is progressing rapidly. The farm population and labor force is getting older, more dominated by females.

How different are the water sectors in China and in India?

The differences between China and India with respect to water merit some reflection. First, a simple yet stark difference: electricity—including that for agricultural uses—is metered (both at the point of use and at the transformer) and charged throughout China. If nothing else, this sets a natural limit to the amount most farmers are willing to spend in their search for very deep groundwater. A part-time farmer-electrician functions as a commission agent of the township electricity bureau to collect the fees.

Somewhat harder to describe are the structure and workings of government, and in particular the water sector institutions, at the six different government levels in China. The most prominent feature is the substantial authority of the local government at the township and village level. This level must support itself entirely out of local taxes and is therefore directly accountable to the local population for the quality of services it provides. It forms a kind of anchor for the enormous, extremely fragmented national bureaucracy. Even by Indian standards, the water sector is large—with some 40,000 staff deployed in water bureaus throughout each province, not counting locally funded village staff.

The vertical and horizontal fragmentation would be worse if it were not for a bottom-up initiative coming from townships and villages to consolidate all water-related agencies into water resource bureaus, then broaden their roles by renaming them water affairs bureaus. This first reform step is expected to yield much greater benefits once the functions of the Ministry of Water Resources at the top have been consolidated, and the vertical fragmentation is reduced if not removed.

Purists comparing the Chinese system with that in some industrialized countries gripe at the lack of organization into river basin agencies. But moving in that direction at this stage of Chinese development would take the only effective part of the government—the local government—out of the picture. The "nine dragons" (meaning many masters) will no doubt continue to manage China's water for quite some time. But the process set in motion—particularly from the township and village on upward—implies a genuine and overdue reorientation of the large water bureaucracy from water resource development to water resource management. In pursuing this gradual reform, the authorities decided to maintain the size of the bureaucracy but imbue it with a service-oriented spirit.

Source: Binswanger-Mkhize and others (2011); Ackermann (2011).

The bottom line: TFP growth in India increased in the last decade even as low water use efficiency and a slowdown in irrigation continued to be major sources of concern. The challenge is to make sure TFP growth continues to increase and a fundamental turn-around in water use efficiency is engineered. What rate of TFP growth is needed to sustain specified desired levels of agricultural and economic growth?

A number of considerations can be assembled to begin to address this question. Continuing population growth coupled with rapid economic growth will drive up food demand and change its composition. Net sown area has been around 140 million hectares for many years in India and is not expected to increase significantly. Agricultural output can increase only through investment, expansion of irrigation, more appropriate input use, and technical progress. Since intensification will soon run into diminishing marginal returns, and water availability is limited, technical progress will become the ultimate source of agricultural growth and is needed to accelerate TFP growth.

Of course, in an open economy, the rising food demand could be met by imports instead of domestic production growth. But natural and political economy constraints limit the proportion of food that can be imported without putting the food security of the Indian population at risk. Using a dynamic programming model of the Indian economy (described in some detail in Annex I.1), Parikh and Binswanger-Mkhize find that with specified restrictions on imports, higher agricultural and TFP growth are necessary in land- and water-constrained India to supply the increasing demand for agricultural commodities driven by population growth and higher per capita incomes. The authors find that under the model's assumptions, future growth rates of 8–10 percent of the Indian economy would be commensurate with, and require, agricultural growth rates in excess of 4–5 percent. This (and even higher rates of agricultural growth) can be achieved by a combination of sharply accelerated rates of agricultural productivity growth and yields, higher water use efficiencies, continued rapid growth in irrigation, and moderately higher agricultural prices (needed to spur investment).

Framework to Achieve India's Agricultural Transformation

Overview Section 4

Harinder S. Kohli and Anil Sood

As argued above, our vision of Indian agriculture by 2040 requires the sector's transformation toward high-value products, efficient market-, private sector- and productivity-driven growth, and judicious water use. To create a modern agricultural sector that matches India's economywide transformation, we propose a new agricultural revolution in which the efficiency of input use and technical change are the sector's dominant objective—achieved through reforms on both the physical and the administrative fronts. How can this hugely complex task be achieved? Our analytical and prescriptive framework rests on six building blocks:

Building block 1: Improving productivity in crops and livestock through new technologies and a more efficient use of inputs—including land, fertilizer, irrigation, state-of-the-art technology, seeds, and investment.

Building block 2: Improving water use efficiency.

Building block 3: Improving the efficiency of farming services and practices—including agricultural research and extension services.

Building block 4: Improving the efficiency of the farm-to-market value chain—including storage, transport, and infrastructure—by removing the remaining constraints on the private sector.

Building block 5: Improving agricultural pricing and subsidy reforms.

Building block 6: Improving public program implementation by consolidating existing fragmented programs and by decentralizing away from centrally sponsored programs to states and villages.

Building block 1: Improving productivity in crops and livestock through new technologies and more efficient use of inputs—including land, fertilizer, irrigation, state-of-the-art crop and livestock technology, seeds, and investment

Productivity improvements in Indian agriculture are vital to accelerate the country's agricultural growth in ways that sustain overall economic growth ambitions, meet the growing demand for agricultural commodities and food, offset the impacts of climate change, and compensate for rising farm wages.

New and better technology for crops and livestock

For India to improve agricultural productivity and increase yields, it needs widespread use of quality seeds of well-performing crop varieties. While traditional approaches to plant breeding will continue to gradually produce new and better seeds, biotechnology in a broad sense of the term is an indispensable resource in this context. Molecular breeding, in particular marker assisted selection, backcrossing and recurrent selection are mainstay activities by private sector seed companies and agribusinesses. *Bt* and other GM seeds are part of the solution. Improved biotech seeds can provide a win-win way out of productivity, resilience, and sustainability bottlenecks and lead to higher yields. Biotech seeds can be developed for important traits that include herbicide tolerance; viral, fungal, and insect resistance; abiotic stress tolerance; induction of male sterility; and delayed fruit ripening, among other characteristics.

For example, herbicide-tolerant rapeseed, soybean, and sugar beet allow post-emergent herbicides to substitute for tilling of land, boosting carbon sequestration, equivalent to 3.9 million tons of carbon in 2008, or 6.4 million cars on the road. They also reduce fuel use for tractors. And they mitigate the need for pesticides, reducing pesticide safety hazards and adverse environmental impacts, where they exist (Binswanger-Mkhize and Parikh, 2011).

Given the lack of evidence on negative consequences from *Bt* and other GM crops, and the significant potential productivity, food security, and sustainability benefits, the corresponding regulatory frameworks and their implementation deserve rethinking. A number of biotech products are passing through various stages of regulatory review in India at this time, but it is hard to predict when some of them might receive approval.

For livestock and fisheries, past productivity improvements have been as impressive as for crops. Further improved breeds will come from investments in breeding and product development. New veterinary drugs and vaccines will come primarily from the private sector, as will machinery and improved feed mixes. Fodder constraints for milk production will require improved water use efficiency and improved fodder crops.

Building block 2: Improving water use efficiency

The current water debate in India is almost exclusively focused on whether surface water schemes are preferable to groundwater use and on how to build new infrastructure to increase the water supply. But this focus misses the real issue, which is how the country can deliver water reliably and use it more efficiently. The focus should not be on water resource development (the product) but on water resource management (the service). The reality is that further surface water development will not reduce the demand for groundwater unless it can meet the requirement of timely water delivery without discriminating against tail-enders and small farmers.

Expanding irrigation and efficiency of water use

With limited cultivable land, increasing water availability is critical to increase yields, improve the productivity of other inputs (such as fertilizer), and enlarge cropped area through multiple cropping. Yet despite several decades of large and expensive government programs to expand traditional surface irrigation systems, the area under surface irrigation in 2007–08 was the same as in 1980–81—and since 1990–91 surface water irrigation area has gone down (Figure 6). Also striking is the fact that the area irrigated by groundwater wells is now more than twice that from canals even though most of these wells are private ones built by farmers at their own cost. Alarming trends have been the slowdown in the expansion of irrigated areas from that source and the decline in water tables in large areas.

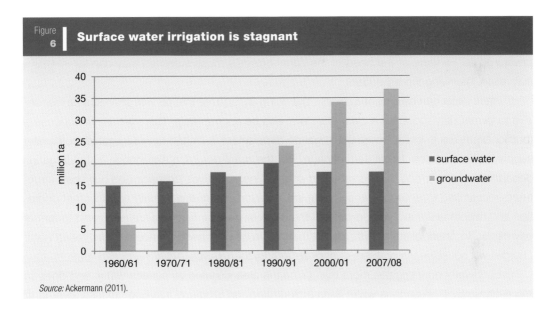

Figure 6 | **Surface water irrigation is stagnant**

Source: Ackermann (2011).

How much additional land is India likely to irrigate in the coming years? Estimates show that if India, which currently has about 63 million hectares of irrigated land, continues on its current path it could reach 90 million hectares by 2039 (Directorate of Economics and Statistics, 2011).

Prospects to expand irrigated land are constrained largely by the difficulties of constructing new dams, canals, and distribution systems, and the accompanying political and societal frictions—such as interstate water disputes and displaced populations.

Future expansion will become even tougher, given that climate change is expected to drastically alter India's surface water. Moreover, with more than 5,000 large dams in India, new and large surface water storage options are economically questionable. Peninsular rivers are already heavily regulated, and additional dams on Himalayan rivers would not add significantly more irrigation. Two cases underscore the difficulties:

- The Sardar-Sarovar project on the Narmada River is more than 10 years behind schedule and has already cost more than 10 times its estimated cost (Ministry of Water Resources, 2006).

- With the Ganges system, only about a fifth of its usable water resources could be captured even if it were feasible to build all possible dams under consideration—a virtual impossibility given the current unfavorable climate for new dam construction.

There also are enormous operational problems with surface water irrigation. Water is not equitably distributed. Distributor canals are not properly maintained. And the tail-enders do not receive their fair share of water. The fundamental problem: most of India's canal systems are designed for extensive rather than intensive cultivation. While this orientation might be desirable for maximizing the social productivity of water, it is certainly not consistent with maximizing of private goals. Farmers at the head of the canal draw as much water as they can for intensive cultivation, depriving the tail-enders.

Surface water can be provided reliably, but doing so requires that service providers be responsive to farmer demand, have a clear mandate, and be accountable. There are some successful (public) examples in India, where the irrigation department is limited to accountable, bulk water supply, within a system that it can manage. But the more common experience is that irrigation departments have strong bureaucratic and engineering incentives to design and build new structures—but limited incentives to ensure that the irrigation potential is actually used, or that operation and maintenance are carried out—all the more since their budgets are not linked to farmer payments. So, financial constraints are not the reason for poor performance, and increased water charges will not solve the problem of poor service by themselves (Ackermann, 2011).

Groundwater now irrigates more than twice the area irrigated by surface water, and does so more efficiently than surface water (less evaporation, no siltation). It is used extensively—in all countries, not just India—because it is available when and where needed, and in easily controlled amounts. This is especially true for India, where declining farm sizes (75 percent of farms are smaller than one hectare) increase the urgency to maximize returns to scarce land.

Indeed, the use of groundwater may have been the single most important factor contributing to poverty reduction in India.

Extensive groundwater use has, however, come at a tremendous cost—one that will have severe social and environmental consequences if not checked. Groundwater tables in northwestern India and in many hard rock areas of peninsular India have dropped dramatically. Limiting groundwater extraction by restricting the number of subsidized pumps is what the Model Groundwater Act seeks to do. But the act is extremely difficult to implement in a setting where there are tens of millions of small-scale independent irrigators. International experience is not encouraging: even in Spain it has been only partly successful (Shah, 2009).

Improving efficiency

Innovative, indirect instruments are required to improve water use efficiency in agriculture and address the country's groundwater challenge. There are three ways to do this: making groundwater use sustainable, reorienting surface irrigation infrastructure to meet the needs for reliable, on-demand water, and reducing water consumption through improved agricultural productivity.

Using groundwater sustainably: One of the most promising levers to manage groundwater abstractions involves separating agricultural from domestic electricity feeder lines. By providing reliable electricity for a limited number of hours, some states have significantly reduced electricity consumption while economizing on groundwater draft. Take Gujarat for example—when the supply of electricity where farm wells are supplied power was restricted to scheduled fixed hours and stable voltage through separate feeders, power supply to agriculture fell from 16 billion units in 2001 to 10 billion units in 2006. Groundwater drafts fell 20–30 percent. And the Gujarat government's electricity subsidies came down from USD786 million in 2001–02 to USD388 million in 2006–07 (Shah and Varma, 2008).

Artificial groundwater recharge has many advantages. It is local. It does not disadvantage downstream users. It does not displace people. And it involves little, if any, evaporation loss. Of course, it requires energy to pump it up, but it should nevertheless be the first line of defense. The scope is substantial: 36.4 billion cubic meters of water can be stored, which translates to irrigating 3.65 million additional hectares of land. If efficiency of water use were higher, it could irrigate 5 million hectares. Programs are now emerging in western and peninsular India to replenish groundwater.

Other methods for sustainable groundwater use range from popular movements to build water harvesting structures to large-scale water recharge programs using the millions of old dug wells in the 100 districts most affected by groundwater depletion. These efforts will need to be supplemented with high-quality technical support to ensure that they are effective and consistent with improving water balances from a river basin perspective. In some areas of peninsular India, information and education campaigns show promise in persuading users to limit groundwater abstractions to sustainable levels (Ackermann, 2011).

Water use efficiency can also be increased sharply by accelerating the introduction of drip and sprinkler irrigation and the use of less water in paddy rice production. Gujarat has the most effective program to support these investments that could be emulated elsewhere.

Reorienting surface irrigation infrastructure: By default, seepage from many of the large surface irrigation systems (reservoirs, tanks, and canals) is already replenishing groundwater stocks. This can be planned in such a manner as to optimize surface water and groundwater use. Especially in the eastern Ganga basin, conjunctive management—combined with extending a dual electric grid to regions now using diesel and kerosene—could significantly and equitably improve agricultural production. In addition, irrigation departments can be responsible for bulk water supplies from rivers and main canals from which private irrigation service providers draw water and deliver it up to tens of kilometers away through surface or buried pipes. This provides reliable and even pressurized water for micro-irrigation and drinking while reducing the stress on groundwater. Using pipes instead of open canals (using the canals as trenches for new piped water systems) avoids wasting valuable farmland, reduces transaction costs, and saves significant amounts of water by preventing evaporation.

A greater role for the private sector: Faced with inefficient public irrigation programs, the private sector is getting more involved—a move that should be allowed and further encouraged. Private actors (including water providers and those overseeing common-property resources, such as water user associations) can make more rational decisions because they operate at decentralized, location-specific levels.

Private operators have already become distributors of water from canal irrigation in pipes over long distances, especially in Maharashtra. This model should be generalized across India, with the irrigation departments concentrating on bulk water supply while the private sector distributes water on demand via pipes to irrigators who will pay the private provider for the service.

Self-regulation by farmers sharing a small aquifer can be effective, particularly when empowered with information on the state of the aquifer. Recent reports speak highly of efforts by more than 500 farming communities in seven drought-prone districts of Andhra Pradesh to self-regulate groundwater abstraction at low cost, while increasing farm incomes (Binswanger-Mkhize and de Regt, 2009). This approach involves intensive farmer education through "barefoot hydrologists" who raise awareness about the groundwater situation and the effects of pumping groundwater at different times. As many as one million farmers participate in crop water budgeting. The key to the program's success is the unique hydro-geological setting for the communities participating: the groundwater in this hard rock area of Andhra Pradesh is confined to relatively small aquifers, so that actions of individual farmers or communities directly affect water availability.

But water user associations will not be able to function if they remain dependent on the irrigation departments. They must become fully autonomous, with rights to set and collect water charges and retain revenues. They will also need sustained technical and management support. Moreover, even with the widespread establishment of (private) water markets, public sector interventions will remain important. Policymakers should rethink the subsidies that have encouraged inefficient water use—such as those for electricity that promote excessive groundwater pumping.

Building block 3: Improving the efficiency of farming services and practices— including agricultural research and extension services

Reforms need to focus on stepping up public and private agricultural research, improving public and private extension services, and strengthening public-private partnerships and the role of professional NGOs.

Agricultural research

From the 1960s to the 1980s, public R&D investments in agriculture paid off handsomely in the remarkable successes of India's Green Revolution. But in the 1990s, this type of R&D funding slowed, the result of questions about its relevance and focus, among other factors. At about the same time, agricultural productivity growth began to decline. "Technology fatigue" made itself felt in key agricultural settings where yields had previously grown rapidly. Public agricultural research institutions began to become complacent, producing useful solutions at a declining rate and losing some of their relevance in light of the challenges at hand.

Private agricultural research began to pick up some of the slack and, indeed, made important and growing contributions over the years. But the private sector goes where it sees a business case and thus may fail to focus on the needs of poorer and more marginal farmers. The private sector thrives on high-value, high-margin seeds such as vegetables, cotton, maize, and sunflower and focuses on hybrids in particular where (to maintain vigor) farmers periodically need to buy new seeds. Rain-fed agriculture is neglected by both private and public agricultural research to this day—an important omission, since a sizeable share of Indian agriculture (currently about 55 percent of total net cropped area) will remain rain-fed for a long time to come. Strengthening rain-fed agriculture should therefore be a high national priority involving seeds development, the right kinds of fertilizer solutions, agricultural extension and other services, including farm credit and, above all, crop insurance to provide the kind of risk mitigation options farmers need before they can invest in productivity-enhancing purchased inputs such as improved seed.

Many studies confirm and document the received wisdom that investment in agricultural research is rewarded by high economic and social returns. We tend to agree with this but would point out that an open economy can also import some of its technological needs, as India has essentially done with agricultural biotechnology during the last two decades. Stepping up agricultural research investment is still a good idea, given agriculture's challenges today and going forward and the numerous opportunities to reduce yield gaps that exist. Looking to 2040 and considering India's low average yields in crop and livestock activities today, it is hard to escape the conclusion that agricultural research capability and preparedness needs to be enhanced. Increased public investment in agricultural research should, however, be accompanied by institutional reforms of the national agricultural research system with a view to improving the relevance, efficiency and effectiveness of its functions. The recommendations of the Swaminathan and Mashelkar Committees on the reorganization of ICAR should be reviewed and reforms along these lines implemented. Last but not least, proactive forms of partnership and cooperation between ICAR and private firms (both Indian and foreign) should be explored and implemented as an indispensable means to accelerate the development of solutions to productivity challenges today and in the future. The public sector should rethink its attitude and reach out to the private sector, explore partnership possibilities with specific companies (or groups of them) for specific purposes, and communicate the social gains possible and achievable with win-win public-private partnerships where costs and rewards are shared equitably among the participants and account-ability is transparent.

Agricultural extension

Agricultural extension deals with disseminating new technologies and improving practices to reduce yield gaps between potential and actual yields in farmers' fields. In India, as elsewhere, extension has a mixed record. However, we can see many dynamic changes and innovation in the public, private, and NGO sectors, and through mobile applications—especially interactive next-generation platforms. As in marketing and agricultural research, many exciting innovations are spreading. A dynamic and pluralistic system of extension that addresses the needs for differ-ent types of farmers is emerging.

Agricultural extension services from the public sector continue to be weak, with little account-ability for verifiable results. By many accounts, the private actors—especially the input dealers and suppliers—are providing more effective extension than public providers, especially in high-potential areas and for medium to larger farmers. NGOs have also entered the field in promising ways but on a relatively small scale. The key unresolved issues for agricultural extension are how to reach the increasing number of small and part-time farmers, many of them women, and to reach poorer farmers where the private sector has little interest. Coverage will only diminish as farm sizes become even smaller and part-time farming, especially led by women, becomes more common. The productive potential of small and part-time farmers could be multiplied with the right technologies, services, mentoring, and market access.

The challenge is to expand coverage to all farmers operating under conditions where there is potential for growth through agriculture. This requires stepped-up contributions from all exten-sion providers—NGOs, the for-profit private sector, and the public sector. Public and private extension systems should complement each other and operate in partnership rather than at cross-purposes.

The centrality of markets should be recognized and with this the need to incorporate sourcing of inputs and prospects for product sales into the extension agenda. This can be done in coop-eration with the private sector, such as rural business hubs where they exist. To reach into the lower strata of farm capability, the public sector should expand its cooperation and partnership with NGOs that can act as effective "retailers" of extension and agricultural support services at the block and village level. The public sector should also expand partnerships with the private sector to bring the latter into work with smaller farmers. Public-private partnerships can take advantage of technology that is developed in the private sector more than in the public sector (for example, mechanization, some hybrids, agrochemicals, and some transgenic seeds). Finally, high priority should be attached to training input dealers, given their importance as sources of extension advice for farmers. The work of MANAGE (National Institute of Agricultural Extension Management) is a model that could be followed.

India has undertaken major institutional reforms to address the weaknesses of the public sector in agricultural extension. A prominent recent one involves the Agricultural Technology Management Agency (ATMA), which has been scaled up in parts of the country after a successful pilot. But implementation bottlenecks have emerged because of constraints in qualified manpower, insufficient technical and financial support, lack of a framework for implementing public-private partnerships, and weak links to extension units such as Krishi Vigyan Kendras and others. It is encouraging that under the 2010 modified guidelines of ATMA, important institutional and organizational issues are being addressed. The responsibility to bring these systems to full effectiveness belongs to the states, and unless they step up their leadership and finance, as Gujarat has done, little will happen.

The comparative advantage of the public sector is to focus on the lagging states and the lagging areas in states where the private sector has no incentive to be present.

Building block 4: Improving the efficiency of the farm-to-market value chain—including storage, transport, and infrastructure—by removing the remaining constraints on the private sector

The private sector is leading improvements in markets and in the value chain that would lead to higher farm-gate prices and lower consumer prices. But the remaining constraints on the private sector and farmers' own organizations must be addressed.

A "quiet revolution" has been transforming rural traditional markets, with the private sector—whether modern or traditional—as the main driving force (Minten and Reardon, 2011). In fact, the rise of modern private retail food chains in India over the past six years has been among the fastest in the world, growing at a 49 percent nominal rate per year on average, and bouncing back to growth after a dip in the recent global slowdown. The great majority of modern private retail (around 75 percent) has arisen in 2006–10. These changes are especially vibrant downstream (in the retail segment) and midstream (in the food processing and wholesale/logistics segments). The government's direct role (as buyer and seller) is only 7 percent of India's food economy.

Despite such impressive changes, significant marketing barriers remain—showing that India's progress could be even higher and reach its intended agricultural output growth target of 4 percent if implementation barriers are effectively removed. One key barrier: poor roads and a lack of electricity in the poorer regions like Eastern Uttar Pradesh. Another barrier: foreign direct investment constraints in food retail (unlike most of Asia, and especially China and Southeast Asia), which means forgoing investment capital and expertise, and the benefits of increasing competition. But this constraint may be moderate because domestic retail investment is far more vigorous In India than in many other countries.

Also highly problematic are the remaining policy-based limitations to direct procurement from farmers by retailers, processors, and modern wholesalers. These include either partial or slow liberalization of wholesale markets (APMC reform); limits on private sector procurement, storage, and sales to traders (for example, Storage Control Orders under the Essential Commodity Act); and the regulatory and fiscal uncertainty and transaction costs (like double taxation for interstate movements). Compounding matters are the well-intentioned but distorting series of marketing acts, such as the Essential Commodity Act, which restricts the movement and storage of agricultural products, the Agricultural Produce Marketing Act, under which agricultural marketing takes place through a licensed trader system, and the Small-Scale Industry Reservation, under which most food processing was reserved for small firms until 1997 (Minten and Reardon, 2011).

Another priority area concerns waste and spoilage rates, which are reportedly large and can be reduced substantially in the spirit of achieving greater outputs with fewer or better-used inputs.

The private sector has a comparative advantage in agricultural marketing. Even though the private sector has a logical comparative advantage in marketing infrastructure and agricultural transport, the public sector has a vital role.

Government interventions were intended to integrate markets, reduce speculation, and address perceived exploitation of farmers and consumers by the private traders. The APMC objective was to regulate agricultural markets and establish a large number of market yards. While some have argued that the regulated marketing system has served farmers well over time (Acharya, 2004), this view is questioned owing to several problems with the regulated system. These include farmers being prohibited from selling outside the marketing yard, the large area served per market yard, the presence of bureaucrats in managing APMCs, the creation of barriers to entry for newcomers, and the use of market fees as a source of government income rather than for reinvestment in market infrastructure (Acharya, 2004).

Many states are already amending the Agricultural Produce Marketing Committee acts

The APMC acts have been revised in several states to allow for more competition and greater participation of the private sector, thus raising the prices that farmers receive. This positive development needs to be followed through: the improved profitability of higher value commodities should be regarded as the catalyst of a virtuous cycle that enhances private sector involvement. But the process needs to be completed. The distortions in food grain markets put a large wedge between the consumer price and the producer price. So, producers get less than they should and consumers pay more than they need. As one study notes, "well-functioning private markets require completing the reform of the APMC act, permitting direct purchase from farmers, abolishing restrictions on storage and movement, permitting warehouse receipts and opening import and export by private traders. Farmers and consumers can be protected from the excesses of markets through appropriate but minimal interventions by the government" (Shahidur et al., 2008).

Building block 5: Improving agricultural pricing and subsidy reforms

In the recent past, international agricultural prices have been rising sharply—a move that appears to be structural. Over time, India's domestic consumer prices will (and should) rise to international levels. As economic growth unfolds, the negative effects of higher prices on consumers will have to be weighed against the advantages of higher prices for producers. As real agricultural wages increase, perhaps by several hundred percent, the incomes of the poorest and most disadvantaged groups in India, the agricultural workers, will improve; the countervailing element is that higher agricultural wages will reduce farm profits and agricultural growth.

Agricultural price levels will have to be managed the same way as the issues associated with agricultural growth and food security: by increasing the rate of TFP growth in agriculture and the rate of growth of irrigation and water use efficiency; by the diversification of agriculture toward higher valued commodities; and in the case of rising agricultural wages, through agricultural mechanization.

Higher farm-gate prices will enable greater investments

The higher international prices have already penetrated the Indian economy, and have been accompanied by significant increases in support prices over the past few years. While these hurt poor consumers, they also spur agricultural investments and improve farmer incomes. It will be important to continue to protect poor consumers (both rural and urban) through the public distribution system.

Higher agricultural prices will necessitate a fresh look at food subsidies and the public distribution system

The reform of food subsidies is under discussion as part of the preparation of the Rights to Food Act. The fertilizer subsidy has already been made more efficient by linking it to the nutrient content of fertilizer. Its transformation into a cash transfer is under preparation. Reforms of electricity subsidies have been attempted by several states at great political costs and with little lasting impact. In OECD countries, reforms of agricultural input and output subsidies have not succeeded unless they were replaced by more direct income transfers.

The general principle of subsidy reforms for agriculture and food should be to transform the subsidy into some form of cash entitlement, possibly linked to the purchase of inputs or food. When this is done, subsidized input or food prices that differ from market prices disappear. So do all associated opportunities for diversion and rent-seeking, as well as the cost of acquisition, storage, and physical distribution of the inputs or the food, which can instead be left to the private sector. The income transfer corresponding to agricultural subsidies can be targeted to farmers, perhaps on the basis of their agricultural area, and be capped to a maximum to avoid excessive subsidies for the better-off.

In the case of the food subsidies, in addition to these principles, we offer the following recommendations. Instead of targeting the poor, which inevitably leads to great difficulties in defining poverty and to large errors of exclusion and inclusion, the food subsidy should be a universal entitlement that excludes the clearly identifiable well-off and rich, such all those who pay income tax, those who own motorized vehicles, and those working in an organized sector, including government, with monthly emoluments of more than a specified amount. Cash transfers (or food stamps) should cover a broader set of foods than food-grains in order to improve the perceived low impact of the program on nutrition. The cash transfers should be to the women of the household through Aadhar cards. The problem of traders charging high market prices in remote areas can be resolved by encouraging cooperative societies and even fair price shops that stick to prices announced every week by the government. Mexico, for example, has 22,000 cooperative DICONSA stores in remote areas that sell food and other necessities, and compete with private traders.

The minimum price support system could be redesigned by giving a much greater role to private traders who would purchase and hold stocks with government support, as is common in most OECD countries. Combined with the reforms of the food subsidies, this would then allow a redirection of the mandate of the Food Corporation of India (FCI) to the management of a much reduced buffer stock.

Input subsidies are distorting prices in an environmentally damaging way

Input subsidy reforms are needed because the fiscal costs have become heavy (Table 2), and because they are inefficient, lead to deleterious effects for the environment, and are not well targeted to the poor and small farmers. The fertilizer subsidies may encourage overuse of fertilizer and damage soils. The electricity subsidy encourages inefficient water use and ground water depletion. The subsidies are large and pervasive, covering fertilizer, food, irrigation, and electricity. In 2008–09, they reached 2.2 percent of GDP. Arguably, it would be better to shift these funds to more productive investments and public goods.

Table 2	**Subsidies are large, pervasive, and not shrinking (% of GDP)**	

	2004–05	2009–10
fertilizer (total)	0.4	0.7
indigenous fertilizer	0.3	0.2
imported fertilizer	0.0	0.1
sale of decontrolled fertilizer with concession to farmers	0.1	0.5
Irrigation	0.3	NA
other subsidies given to marginal farmers and farmers' cooperative societies in the form of seeds, development of oilseeds, pulses, cotton, rice, maize, and crop insurance schemes, price support schemes, and so on	0.1	NA
total inputs	0.9	> 1.5
electricity[a]	0.5	NA
food subsidy	0.7	0.7

Notes: NA is not available.

[a] includes all subsidies to Electricity Boards and Corporations. Separate estimates of electricity subsidies accountable exclusively to the agricultural sector are not available.

Source: India Ministry of Agriculture, Directorate of Economics and Statistics (http://eands.dacnet.nic.in/latest_2006.htm).

Agriculture would benefit from better functioning land markets in rural areas

Current constraints on land sales and rentals reduce farmers' ability to get access to land. Reforms to improve land markets will encourage needed investments and ensure land right security, allowing land to be used as collateral. Land rentals have been an important avenue for land access for poor, land-scarce, and landless households, helping reduce poverty. It has been shown that those who rent land obtain higher returns to their labor than are available in the casual labor market.

In theory, land markets transfer land to more efficient producers, who increase their incomes. In practice, this has occurred; but weather shocks have encouraged distress sales by poor—not necessarily inefficient—households. In areas where employment guarantee schemes were operating, the distress sales were reduced, and it is expected that the Mahatma Gandhi National Rural Employment Guarantee Act (MGNREGA) will further prevent such sales. With such safeguards in place, constraints on land sales among land reform beneficiaries and in tribal areas can be safely eliminated.

Further actions to improve the functioning of land markets in rural areas include providing land to landless and land-poor people; eliminating remaining constraints on land rentals; strengthening land inheritance rights for women; clarifying and recording rights in marginal areas traditionally outside the system and in tribal areas; and improving land administration in rural areas through computerization and spatial records.

Building block 6: Improving public program implementation by consolidating existing fragmented programs and by decentralizing away from centrally sponsored programs to states and villages

Institutional reforms of public sector agricultural programs will need to consolidate fragmented public programs and decentralize centrally sponsored schemes to the state or even the village. First-round reforms should consolidate fragmented, overlapping public programs. Second-round reforms should change public programs from centralized silo structures to block grants that go down to the state or possibly even local levels. Such decentralization would align incentives to ensure accountability and the flow of funds for the effective implementation of agricultural programs.

Consolidating public programs

Funds from the center and the states are fragmented into a large number of overlapping centrally sponsored schemes, a development that the government identifies as a major problem to tackle. "This [proliferation] has led to poor implementation, duplication, lack of convergence, and suboptimal results. There is an urgent need to transform the system and sharply reduce the number of schemes. This will enable more focused and effective implementation. A Committee under Shri B.K. Chaturvedi, member, Planning Commission, has been appointed to review the entire gamut of Centrally Sponsored Schemes" (Government of India, 2010).

Reforms to consolidate have been outlined, but they now need to be implemented. The draft proposal of the review commission recommends that the centrally sponsored schemes in the six departments that deal with agriculture and rural development be reduced from 41 to 18. Other recommendations include:

- Review physical norms, including variations for northeast, tribal, or coastal areas, or other special areas. Allow for variations of norms at the state level.
- Revise financial norms every two years, with revisions linked to the wholesale price index.
- Reform procedures for fund transfers to the state budgets, bringing an end to all direct transfers to districts or other independent bodies or societies by the end of the 12th Plan.
- Share experiences among states and with the center.
- Strengthen monitoring and evaluation of all centrally sponsored schemes on a regular basis, by ministries and independent evaluators. Use the services of the Independent Evaluation Office of the Planning Commission.

The transfer of funds to communities requires a number of accounting and accountability innovations, which amount to a form of local and community governance reform. Funds would be transferred in tranches to community bank accounts and become the property of the communities, eliminating the need to use government procurement and accounting methods. Instead, the community would be required to obtain three bids from three providers, following simple guidelines to evaluate the bids. Accountability within the community would be ensured by a variety of mechanisms, including community finance and audit committees, checks signed by at least two members, project financial accounts open to the general assembly, and random audits.

Decentralizing public programs

Globally, agricultural and rural development strategies have shifted—from top-down prescriptions from planners and policy implementers to communities as full participants with increasing responsibilities. In the process, community participation has evolved from mere consultation to participation in project planning, co-financing, and running—and then to the community empowerment model. Under this model, responsibility for planning and implementation is entirely devolved to communities, along with the government financial resources.

With greater devolution of powers, communities and local governments are more willing to co-finance projects and raise revenues, leveraging central and state resources. Local priorities and preferences, and local knowledge, including that of women, receive greater weight, leading to better adaptation of and satisfaction with government programs. The greater accountability of local government and service providers to citizens in turn enhances government responsiveness, reduces absenteeism, improves program quality and timeliness, reduces costs, and improves accountability (women are particularly good at ensuring accountability). In addition, local governance is needed to deal with the complex problems of significant anticipated growth in competition for land, water, and other natural resources. Stronger local governance and people's empowerment should help produce the accelerated agricultural growth India needs in an environmentally and socially sustainable way.

Institutions at the lower levels (such as the Panchayati Raj institutions) are not yet ready to ensure the expected results from effective local governance

While many functions have been partly or fully transferred to the Panchayats through the 1993 constitutional amendment, the transfer of funds remains limited. Few functionaries have been transferred to the Panchayat system, especially at the village level, or made accountable to it. Panchayats are reluctant to use their fiscal powers, which reduces their fiscal fitness and accountability. Most system funds come from plan expenditures of the center and the states, and are fragmented into a large number of overlapping centrally sponsored and state schemes—centralizing power in the ministries at the national and state level and effectively clawing back the powers that were to be devolved.

Funds from centrally sponsored schemes can flow from central departments to state departments, to the Panchatyati Raj system, or to parallel state and district bodies, and from there down to the corresponding entities at the block or the village level or to the communities. Adding to the confusion, the programs are managed through a large number of government entities. Funds arrive in narrow silos that are not fungible across local development objectives, making convergence of programs at the local level almost impossible. In addition, funds arrive in bits and pieces at separate times, with unpredictable delays. Many government programs are implemented either departmentally or through parallel bodies. The sole exception has been the MGNREGA, which is implemented through the village Panchayat. Planning requirements are often excessive, and planning at the district level is neither participatory nor effective.

A common reason given for the inadequate decentralization to the Panchayats, despite the passage of the 1993 constitutional amendment, is that most Panchayati Raj Institutions do not have the capacity to adequately carry out their implied responsibilities. To overcome this legitimate concern, the government bodies in charge at both the national and state levels must intensify their efforts to build up the administrative and implementation capacities of Panchayati Raj Institutions. This task must be assigned the highest priority.

Overcoming bureaucratic and political opposition to greater decentralization, community empowerment, and consolidation of central programs will require political leadership at the center and in the states

Decentralization and participation have been advocated in the 11th Plan as the only way to resolve India's difficulties in implementing its agricultural and rural development programs. There is broad consensus on the Panchayati Raj reforms required among scholars and the Administrative Reforms Commission, Finance Commission, and Planning Commission. Reforms include strengthening the legal framework, streamlining fiscal relations between the central government, states, and local governments, and developing a competent professional staff of the Panchayati Raj Institutions. Yet there has been little follow-up to these insights and recommendations.

The reality is that constitutional and legislative reforms are often not enough. They have to be complemented by administrative and fiscal reforms, and most important, strong direct support from the head of government. Over time, the opponents to decentralization find that a reformed system provides them with more opportunities for developmental and political success.

*   *   *   *   *   *

To conclude, most observers of India's agriculture believe that its problems have less to do with policies than with actual policy and program implementation. Fortunately, at all links of the agricultural value chain—from extension services to storage facilities and marketing arrangements—private actors have started to complement public institutions, and are often providing better services. While the private sector is moving quickly to transform research, extension, and value chains according to its comparative advantage, most public sector institutions and programs for agriculture require urgent reforms, often along lines long recognized but not yet implemented. The public sector institutions must also concentrate more on areas that are not of interest to the private sector, and collaborate more with the latter.

Equally important, the enormous challenges of agricultural growth, natural resource management, and social services for rural areas must be resolved with greater citizen empowerment and decentralization. Such reforms have long been under discussion in India, but initiatives have so far failed to bring them about. Reforms will have to be driven primarily by the states, with support in the form of strong incentives and perhaps further legislative interventions from the center as well. The reforms will not come about without pressures from below as well as from the very top.

National Food Security, Productivity, Irrigation Growth, and Trade— The Model, Assumptions, and Results

Hans P. Binswanger-Mkhize and Kirit Parikh

Trends that motivate the model structure

Continuing population growth and rapid economic growth will rapidly drive up food demand and change its composition. Net sown area has been around 140 million hectares for many years. Agricultural output can increase only through investment, expansion of irrigation, more appropriate input use, and technical progress. Since intensification will soon run into diminishing returns, and since water availability is limited, technical progress will become the ultimate source of agricultural growth. What rate of total factor productivity in agriculture will be needed to sustain agricultural and economic growth?

In an open economy the rising food demand could be met by imports, but natural and political economy constraints limit the proportion of food that can be imported without putting the food security of the huge Indian population at intolerable risk. The threat of climate change is generally considered to increase that vulnerability of Indian agriculture, just as in the world as a whole. Climate change could lead to global price increases and make reliance on imports less acceptable. Will accelerating productivity growth and sustained expansion of irrigation support the higher agricultural growth needed? Will domestic agriculture be able to provide the required food in the long term—say over the next three decades? Or will limits to agricultural growth impose limits to economywide consumption and income growth? What would be the role of imports?

Model features

Parikh and Binswanger-Mkhize developed a multisectoral, dynamic programming model that has the needed structure and features to address the issues of food security, productivity and irrigation growth, and trade over the next 30 years. It has 28 sectors, of which 15 are agricultural. Crop production from irrigated and nonirrigated lands is distinguished so that there are 40 production activities. The model covers the whole economy, captures macro feedback, and ensures macro balances. It has 20 consumption classes, 10 rural and 10 urban. Of these classes, five are at much higher consumption than observed today, classes into which people will be moving as their incomes rise. Each class has its own expenditure system. Income distribution is determined every period endogenously, depending on the level of aggregate consumption and prescribed parameters of the log normal income distributions for rural and urban consumption. Rural people migrate to urban areas depending on the relative GDP from agriculture and non-agriculture.

The production activities are based on the social accounting matrix for the year 2003–04. Rural consumption depends on income from agriculture as well as from rural non-farm activities. India has recently seen rapid growth in rural non-farm employment. But data are not available that would permit us to endogenously determine the level of non-farm activity in rural areas. Parikh and Binswanger-Mkhize's model therefore used an indirect approach where total rural consumption expenditure is related to agricultural GDP, on the ground that rural non-farm activities are still linked significantly to agricultural growth. Thus overall rural consumption is related to agricultural GDP, while urban consumption is related to non-agricultural GDP, using statistically estimated relationships. The estimated multipliers for agriculture and non-agriculture are consistent with the rising share of non-farm income in rural consumption.

Once the aggregate consumption is determined for rural and urban areas, total population is distributed between urban and rural populations based on an exogenously stipulated urban-rural consumption parity ratio. The model assumes a per capita consumption parity ratio of 2.35 as reflected in the social accounting matrix of 2003–04. Because the National Sample Survey data already show an urban-rural consumption parity ratio of 1.75, Parikh and Binswanger-Mkhize let the urban-rural parity ratio decline from 2.34 in 2003 to 1.75 by 2039.

A particularly important feature of the model is a demand system that can predict the consumption behavior of income classes at much higher income levels at which income elasticities of demand for food will be much lower than today. Parikh and Binswanger-Mkhize were able to estimate a nonlinear demand system based on National Sample Survey (NSS) and Central Statistical Organization (CSO) data without having to make ad hoc assumptions about consumer behavior at very high income levels. The model then uses these estimates to generate linear approximations of the demand system for each separate consumption class, which together approximate the nonlinear demand system in a piecewise manner.

Consumption estimates derived from the NSS data are lower than the CSO national accounts data. Parikh and Binswanger-Mkhize ensure that the demand systems are made consistent with the national accounts data by making it consistent with the social accounting matrix for 2003–04 in which private consumptions reflect the higher CSO consumption estimates. This has been done by adjusting upward the committed consumption levels for each expenditure class.

The model in most scenarios maximizes the present discounted sum of private consumption over 10 time points four years apart. In one scenario it maximizes the present discounted value of GDP. The base year is 2007, and the last year is 2039. The constraints ensure commodity balances, capacity constraint of each production activity, balance of payments, land constraints, and upper bounds on trade for different commodities. Investment is constrained by the availability of domestic savings with a marginal savings rate of 35 percent and on foreign inflows, and by the availability of different types of investment goods. In addition, when the model maximizes GDP, it imposes a discount rate of 3 percent and minimum growth of private consumption of 3 percent. When the model maximizes consumption, it instead stipulates an upper bound on the growth rate of private consumption in each year. Government consumption growth is set at 9 percent.

Net sown area is kept constant at 140 million hectares, and the ratio of net irrigated area to net sown area increases from 0.45 in 2003 by a prescribed rate that varies from scenario to scenario. In the reference scenarios it grows at 1 percent per year. Thus the net irrigated area increases to 90 million hectares in 2039. From 1980 to 2007 India has added only 20 million hectares of net irrigated area, and if it continues to grow it can reach 90 million hectares by 2039. This would be too optimistic if the intersectoral competition for water and environmental concerns becomes even more important as a consequence of the lowered estimates of available water in India, if the stagnation in development of surface irrigation schemes continues, and if efficiency of use cannot be greatly enhanced. Nevertheless, Parikh and Binswanger-Mkhize have used this less pessimistic growth rate of irrigation to reflect the emphasis being placed in five year plans on improvements in water use efficiency and rapid development of irrigation. In two alternative scenarios the model explores the impact of lower and higher levels of irrigation development.

The model incorporates import and export constraints on all sectors. For example, in the reference run, the imports of wheat and rice are limited to 3.0 percent of domestic availability (or 3.1 percent of domestic production), close to self-sufficiency. This reflects the strong policy preferences of India's policymakers for national food security that are consistent with the large size of India's population. Coarse cereal imports are limited to 10 percent, milk and milk product imports are limited to 6 percent, and animal products and forestry products to 30 percent while imports of all other agricultural commodities are limited at 15 percent of availability (17.6 percent of production). The import constraints imposed on other sectors of the economy are much wider than seen historically. The import constraints on food commodities are relaxed significantly in some scenarios compared to what India has had in the past three decades, but such higher import bound may be considered to reflect India's increasing openness.

Model scenarios and findings

Scenarios are not predictions, but tools to explore the economic consequences of alternative assumptions. In the reference run, total factor productivity growth in agricultural sectors is set at 2 percent, a rate that was achieved in the 1980s and again in 2003–07, but that, like the irrigation assumption in the base run of 90 million hectares, may be a bit on the optimistic side. In the non-agricultural sector TFP growth is set at 3.0 percent. It may be noted that over and above the prescribed TFP growth rates, Parikh and Binswanger-Mkhize stipulate fuel use efficiency growth of 1.5 percent and electricity use efficiency growth of 1.0 percent, while the efficiency of use of wheat, rice, and other agricultural commodities as intermediate inputs grows at 1.5 percent. Thus the overall productivity growth would be significantly higher than the weighted average of the TFP growth of 2 percent for agriculture and 3 percent for non-agriculture assumed in many of the scenarios.

To achieve a widely shared growth scenario, Parikh and Binswanger-Mkhize maximize consumption in most of the scenarios. In the "Reference Scenario" GDP agriculture grows at 4.25 percent, total consumption at 7.7 percent, and GDP at 8.4 percent. Between 2007 and 2039 total consumption per capita would rise nearly 11-fold while GDP would rise 13-fold. The number of people below the poverty line would decline from 359 million in 2007 to 4 million in 2039, all of which would be in rural areas.

With these growth rates, the share of agriculture in GDP declines to 5 percent by 2039, and the share of GDP from food grains sector to only 1 percent. The industry share increases from 30 percent to 45 percent at the end of the period, while the share of services decreases from 54 percent to 40 percent. The per capita food consumption increases very moderately from 139 kg/person in 2007 to 142 kg/person in 2039. The per capita consumption of rice (52kg/person in 2003 to 38 kg/ person in 2039) and coarse cereals (32 kg/person in 2007 to 20 kg/person) decreases, while that of wheat (45 kg/person in 2007 to 62 kg/person in 2039), other lentils (3 kg/ person to 7 kg/person) and pulses (7 kg/ person to 15 kg/person) increases. The share of food grain consumption in total consumption of agricultural commodities declines, while the shares of fruits and vegetables, vegetable oils and oilseeds, plantains, milk and milk products, eggs, meat and fish, and other crops increase. The most significant increase is seen in horticulture and especially in milk and milk products, which rise from 17 percent of total consumption expenditures to 31 percent. These trends have been captured in the Parikh-Binswanger-Mkhize vision for the farm sector.

With the same assumptions, much higher economic growth rates are possible if instead of maximizing consumption growth, GDP growth is maximized. The model then sharply increases investment growth at the expense of consumption growth. If a lower bound is imposed on the annual consumption growth rate of 3 percent (compared with the 7.7 percent achieved when maximizing consumption), the economic growth rate rises to an enormous 15.25 percent, while the agricultural growth rate also rises slightly to 4.42 percent and poverty declines to 111 million people. Because the non-agricultural sectors grow much faster than the agricultural sectors, rural incomes lag behind, migration shoots up to 592 million people, and the rural population declines to 40 percent of the total. Clearly, the high rate of remaining poverty in the face of extremely rapid growth would be unacceptable to India's policymakers, who would much prefer the widely shared growth achieved by maximizing consumption.[1]

The much higher agricultural growth rate needed in the reference scenario can be achieved by a combination of sharply accelerating the growth rate of agricultural productivity, achieving higher water use efficiencies in agriculture, returning to historically higher growth rates of irrigation, and relaxing the tight bounds on agricultural imports. Increasing the agricultural TFP growth rate to 3 percent, which China has achieved over the last three decades, leads to an agricultural growth rate of 5.6 percent, which can support an economywide growth rate of 1.4 percent. Conversely, increasing irrigation growth to 1.5 percent, as achieved in previous decades, will lead to an irrigated area of 108 million hectares that can support an agricultural growth rate of 4.9 percent and an economic growth rate of 9.4 percent. Finally, increasing food grain imports to 69 million tons by 2039 and increasing other agricultural imports would support an economywide growth rate of 9 percent. Higher imports of food would allow India to focus more on agricultural products where it has comparative advantage and therefore would accelerate agricultural growth, rather than slow it down. If the high agricultural productivity growth is combined with the high growth of irrigation and the high food import constraints, agriculture could grow at 6.3 percent per year, consumption per capita at 11.6 percent, and GDP at 11.7 percent.

The relationship between the economywide growth rate and the agricultural growth rate in the Parikh-Binswanger-Mkhize model that arises when consumption is maximized the consequence of the joint imposition of constraints on agricultural imports and irrigation, and the high emphasis placed on consumption growth. It implies not only that the volume of agricultural commodities is fixed, but also that consumers compete sharply with the non-agricultural sectors for agricultural output as intermediate inputs. When the economy grows faster, with given import bounds, higher agricultural growth rates are needed both to supply the more rapidly growing demand for food by consumers and the demand for agricultural products as intermediate inputs in the non-agricultural sector. Conversely, the higher agricultural growth rate then permits a higher economywide growth as the demand for intermediate inputs is no longer a constraint. Parikh and Binswanger-Mkhize especially emphasize the first interdependence, as the second one from agricultural growth to economywide growth may partly, but not fully, be a consequence of their linear model structure. The results from the different scenarios suggest that the current growth target

---

1 A growth rate greater than 15 percent over more than 30 years is of course highly unrealistic, but any intermediate growth rate of consumption between 7.7 percent in the reference scenario and the Growth First scenario could be achieved by varying the minimum consumption growth rate in the model between 3 percent and 7.7 percent. The model illustrates several further points. The total factor productivity growth rate of 3 percent, combined with the other productivity improvements, is able to support very high growth rates if growth is maximized. The constraints on imports of non-agricultural commodities and on foreign investments are also more than ample to achieve high growth. And finally the fact that in spite of slow consumption growth the agricultural sector grows faster than in the reference run suggests that the use for agricultural commodities as intermediate inputs in the rapidly growing non-agricultural sector is a major demand force for agricultural output.

of 4 percent for agricultural growth may be insufficient to support the economic growth target of 9–10 percent to which India aspires.

On the downside, under the low import scenario, if the growth rate of total agricultural factor productivity were to slow to 1.5 percent, or if irrigation growth were to slow to 0.5 percent to achieve only 70 million ha, the agricultural growth rate would slow down to 3.6 percent in both scenarios, with consumption per capita growing between 6.2 percent and 6.4 percent and GDP between 7.3 percent and 7.4 percent. Of course if both irrigation and technical change were growing at the slow rate of growth, the impacts would be more severe, and the growth of agriculture would most likely go down to around 3 percent.

Under widely shared growth, the model scenarios suggest that the Indian population may continue to be predominantly rural. Across the scenarios, of the total population of 1,511 million people, the urban population ranges from 529 to 635 million, and is the higher the higher the rate of economic growth. In 2039 the share of the rural population in the total population ranges from 65 percent when the rate of productivity growth or irrigation growth is at the pessimistic levels to 58 percent when we combine fast technical change and irrigation growth with higher import constraints. Only in the growth first scenario is this conclusion altered, with the rural population share falling dramatically to 40 percent.

The Model illustrates the very high gains from higher productivity growth in agriculture. Similarly, the significance of increasing irrigation through water harvesting, efficiency increase using micro-irrigation and better management of expanded conventional storage systems cannot be overstressed. Small gains in agricultural growth and somewhat larger gains in economic growth are feasible with expanded constraints on food imports that would lead to around 70 million tons of food grain imports by 2039. World markets would most likely be able to provide the increased import requirements, and perhaps even higher ones. The question is whether the country would or should accept such import dependence, a significant issue for agricultural policy in the future.

Part II

India 1960–2010: Structural Change, the Rural Non-farm Sector, and the Prospects for Agriculture

Chapter 1 | *Hans P. Binswanger-Mkhize*

Overview[1]

This chapter[2] looks at past and likely future agricultural growth and rural poverty reduction in the context of the overall Indian economy. The growth of India's economy has accelerated sharply since the late 1980s, but agriculture has not followed suit. Rural population and especially the labor force are continuing to rise rapidly. Meanwhile, rural-urban migration remains slow, primarily because the urban sector is not generating large numbers of jobs in labor-intensive manufacturing. Despite a sharply rising labor productivity differential between non-agriculture and agriculture, limited rural-urban migration, and slow agricultural growth, urban-rural consumption, income, and poverty differentials have not been rising. Urban-rural spillovers have become important drivers of the rapidly growing rural non-farm sector—the sector now generates the largest number of jobs in India. Rural non-farm self-employment has become especially dynamic with farm households rapidly diversifying into the sector to increase income.

The growth of the rural non-farm sector is a structural transformation of the Indian economy, but it is a stunted one. It generates few jobs at high wages with job security and benefits. It is the failure of the urban economy to create enough jobs, especially in labor-intensive manufacturing, that prevents a more favorable structural transformation of the classic kind. Nevertheless, non-farm sector growth has allowed for accelerated rural income growth, contributed to rural wage growth, and prevented the rural economy from falling dramatically behind the urban economy. The bottling up of labor in rural areas, however, means that farm sizes will continue to decline, agriculture will continue its trend to feminization, and part-time farming will become the dominant farm model. Continued rapid rural income growth depends on continued urban spillovers from accelerated economic growth, and a significant acceleration of agricultural growth based on more rapid productivity and irrigation growth. Such an acceleration is also needed to satisfy the

1 Acknowledgements: The research in this chapter was generously supported by the Centennial Group, Washington DC, and the Syngenta Foundation for Sustainable Agriculture, Basel, Switzerland. The analytical work was also supported by Integrated Research and Development (IRADe), New Delhi, India. The sections in this chapter on the structural transformation of the Indian economy are based on Binswanger-Mkhize and d'Souza, 2011. The Center on Food Security and the Environment (FSE) is a joint center between Stanford's Freeman Spogli Institute for International Studies (FSI) and Stanford Woods Institute for the Environment.

2 Hans P. Binswanger-Mkhize was commissioned by Centennial Group to complete the study *The Future of Indian Agriculture and Rural Poverty Reduction.* The materials presented in the chapter are based on a paper and presentation given at Stanford University on May 10, 2012, which can be accessed at http://foodsecurity.stanford.edu/events/structural_change_and_the_future_of_indian_agriculture/'.

increasing growth in food demand that follows rapid economic growth and fast growth of per capita incomes.

Introduction

All across the industrialized world, prior to rapid economic growth and structural transformation, agriculture accounted for the bulk of the economic output and labor force. Because productivity in the non-agricultural sector was higher than in the agricultural sector, the share of agriculture in total GDP fell short of its share in the labor force. As industrial growth took off, industry became even more productive, and the productivity differential with agriculture increased. As a result of rapid economic growth the share of agriculture in GDP fell much faster than the share of agricultural labor, and the inter-sectoral differential in labor productivity widened. Farm incomes visibly fell behind incomes earned in the rest of the economy. *"This lag in real earnings from agriculture is the fundamental cause of the deep political tensions generated by the structural transformation"* (Timmer, 2009, p. 6, emphasis in original).

During structural transformation employment grows rapidly in the non-agricultural sector and labor is pulled out of agriculture at a speed that depends on the labor intensity of industry and services. Convergence is driven by rapid agricultural productivity growth that allows for a reduction of labor input per unit of output. A turning point is reached when the labor productivity differential between the sectors starts to diminish and the share of labor in agriculture starts to decline faster than its share in output. Korea from the late 1960s is a typical example, as illustrated in Figure 1.1, that shows the share of agriculture in employment and in GDP, and the difference between them, plotted against GDP per capita.

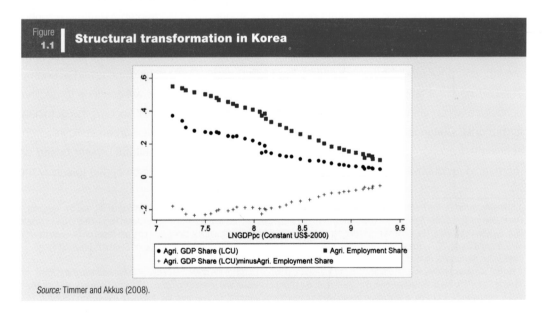

Figure 1.1 | Structural transformation in Korea

Source: Timmer and Akkus (2008).

This chapter deals with the following topics: It first characterizes the structural transformation in India and China. It then looks at the Indian case in greater detail, first its agricultural growth and productivity growth, and then at employment, unemployment, and wage trends. The next section asks the question why, in the presence of rapid growth of the differential in labor productivity between the non-agricultural sector and the agricultural sector, and in the presence of limited rural-urban migration, has there not been a rising divergence in rates of poverty, and in per capita incomes and consumption? The next section on the rural non-farm sector shows that this is explained by the rapid growth of the rural non-farm sector, especially rural non-farm enterprises of farmers, and associated employment growth. After summarizing the findings on employment and poverty trends across sectors of the economy, the chapter shows that the structural transformation in India is a stunted one, in which workers move primarily from the agricultural sector to the rural non-farm sector, rather than to more secure jobs with pension and health benefits in the urban economy. The final section develops a vision for agriculture and rural poverty reduction, and discusses policy implications on how, under the constraints of limited urban labor absorption, rural incomes can nevertheless be increased.

Structural transformation in India and China

Compared to international experience India's structural transformation has been slow and atypical, mainly on account of a low share of manufacturing in the economy and of its disappointing growth and employment performance. At the same time, the share of the agricultural sector in GDP has declined and the remaining industrial sectors and services have shown growing GDP shares. Absorption of labor in the urban economy has been slow, and rural-urban migration has been far less than could have been expected in a rapidly growing economy. Therefore, the difference between the share of agriculture in the economy and its share in the labor force has widened significantly (Figures 1.2a and 1.2b). At the same time, the accelerating growth of the economy since the 1980s did not lead to an acceleration of the agricultural growth rate. As a consequence of high non-agricultural growth, low agricultural growth, and continued growth of the agricultural labor force, labor productivity in the non-agricultural sector and the agricultural sector has widened at an accelerating rate, and their ratio now stands at over 4.2. These data show that India is still far away from a turning point in its structural transformation, where the shares of agriculture in GDP and in the labor force are starting to converge, and the productivity differential between the non-agricultural and the agricultural sector starts to narrow.

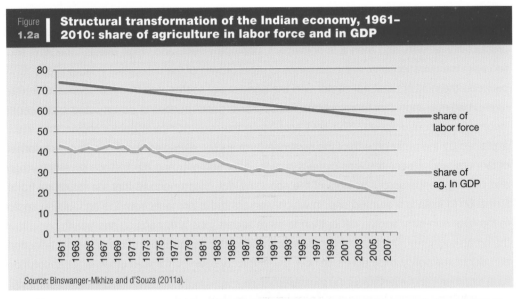

Figure 1.2a | **Structural transformation of the Indian economy, 1961–2010: share of agriculture in labor force and in GDP**

Source: Binswanger-Mkhize and d'Souza (2011a).

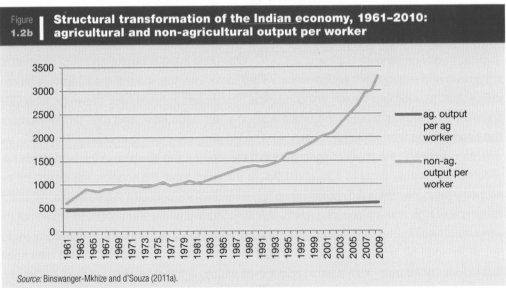

Figure 1.2b | **Structural transformation of the Indian economy, 1961–2010: agricultural and non-agricultural output per worker**

Source: Binswanger-Mkhize and d'Souza (2011a).

Figure 1.3a shows the same relationships for China. In Figure 1.3b the intersectoral productivity differential in China has been rising even faster than in India, and reached a ratio of nearly 6 to 1. The agricultural share in GDP has been declining even faster, but so has its share in employment. As such the absolute difference between the shares has started to decline.

The slow decline in the agricultural labor force is a consequence of the still relatively high rate of population growth in India, 1.6 percent in the past decade, and the relatively slow rate of urban-rural migration.[3] In China, on the other hand, the population growth rate has now declined to almost zero, and rural-urban migration has involved around 220 million workers in the past two decades.

II/1

| Figure 1.3a | **Structural transformation of the Chinese economy, 1978–2010: agricultural shares of labor and output** |

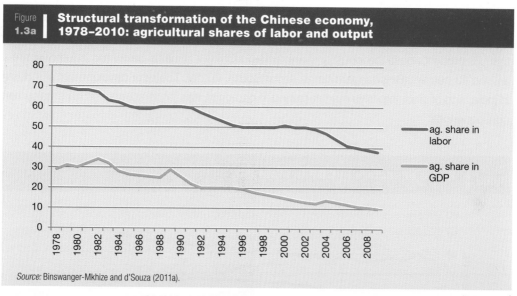

Source: Binswanger-Mkhize and d'Souza (2011a).

| Figure 1.3b | **Structural transformation of the Chinese economy, 1978–2010: labor productivity in non-agriculture and agriculture** |

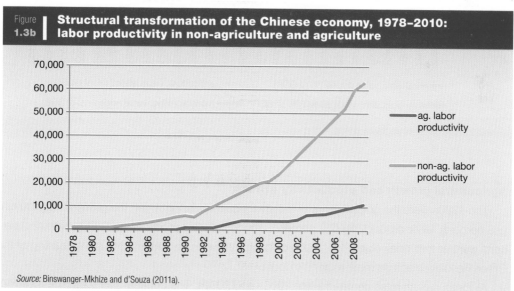

Source: Binswanger-Mkhize and d'Souza (2011a).

3 Urban populations grew at the rate of 2.76 percent in the intercensus period from 2001–2011. Of this, natural population growth accounted for 44 percent, while the remaining 56 percent were accounted for by reclassification of rural areas to urban areas and by rural-urban migration. A large share (not yet quantified) of this second component is accounted for by the reclassification of rural to urban areas which proceeded at a rapid rate. Migration clearly is a fairly low contributor to urban population growth (Bhagat, 2011).

The relatively slow rural-urban migration rate is a consequence of the low share of manufacturing in the Indian economy, which has hovered around 16 percent of GDP since 1980. China's share of manufacturing has stayed at around 33 percent since 1991. Similarly, the share of industry (that includes manufacturing) has grown slowly in India from around 25 percent in 1989 to around 28 percent today, while it has been around 46 percent in China ever since 1993. As a consequence, the GDP share of services has grown to well over 50 percent in India, more than 20 percent above the industry share, while in China it remains below the industry share at around 43 percent (Binswanger-Mkhize and d'Souza, 2011a). The poor development of industry in India, and of labor intensive manufacturing in particular, has led to adverse urban employment consequences.

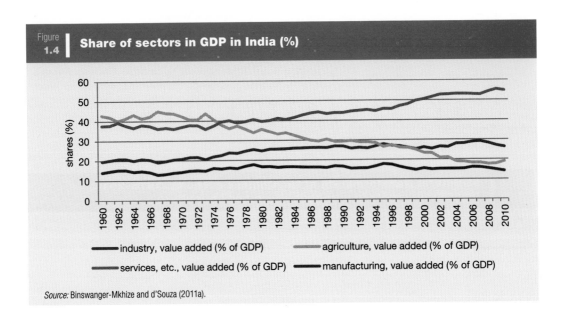

Figure 1.4 | Share of sectors in GDP in India (%)

— industry, value added (% of GDP)   — agriculture, value added (% of GDP)
— services, etc., value added (% of GDP)   — manufacturing, value added (% of GDP)

Source: Binswanger-Mkhize and d'Souza (2011a).

Agricultural growth and productivity growth

The 1980s were the golden years of Indian agriculture during which the growth of agriculture (3.3 percent), labor productivity (2.3 percent), and total factor productivity (TFP) growth (2.0 percent) were at their peak (Table 1.1). Much of this growth can be attributed to the spreading of the Green Revolution across most regions of India.

All these growth rates declined in the 1990s and 2000s. The decadal averages hide a deeper slump in agricultural production and productivity growth from the mid-1990s to the first half of the 2000s. A good illustration is the behavior of the annual total factor productivity (TFP) growth shown in Figure 1.5. While it hovered around 2 percent during the 1980s, it slowed to near zero

Table 1.1 | Growth of agriculture, agricultural productivity, and labor force

indicator	growth rates for decades in percentage or three year average cenetered on last shown					average growth rate of 2006–09
	1960–70	1971–80	1981–90	1991–2000	2001–09	
agricultural GDP growth	3.8	1.5	3.3	2.7	2.8	3.1
growth of agric. output/worker*	0.6	0.4	2.3	1.2	1.1	1.5
total factor productivity growth**	0	0.8	2.0	1.5	1.9**	
TFP growth in China	0	0	2.8	4.2	2.7**	
total population growth	2.1	2.3	2.2	2.0	1.6	NA
agricultural labor force growth	1.4	1.7	1.6	1.4	1.2	1.1
non-agricultural labor force growth	2.7	3.2	3.7	3.2	3.1	3.0

Notes: *Constant US$ of 2000, **Fuglie forthcoming, to 2007 only.

Source: Binswanger-Mkhize and d'Souza (2011a).

in 2001 only to rebound afterwards and to reach 3 percent and above in 2006 and 2007. Growth of agriculture also accelerated to slightly above 3 percent in the years since 2006, which explains the decadal growth of agriculture of 2.8 percent despite the poor performance during the early 2000s. However, the growth rate is still around 1 percent below the target rate of the Government of India for agriculture at 4 percent. Parikh et al., (2011) also show that rates of growth of agriculture in excess of 4 percent are needed if the economy grows at more than 8 percent per year, and imports of agricultural commodities are limited to less than 10 percent of domestic availability. The rising direct demand for food associated with rapid income growth, and the rising indirect demands for agriculture as intermediary inputs from other growing sectors of the economy lie behind this increased need in the required agricultural growth rate.

Table 1.1 also shows that since 1980 the TFP growth rate of agriculture in China has been persistently higher than in India, close to or exceeding 3 percent in all three decades since. As shown by the calculations of TFP growth of Fuglie (forthcoming), no other country in the world has ever shown such a prolonged period of high total factor productivity growth in agriculture. The faster structural transformation in China is therefore caused by its advantage in agricultural productivity growth.

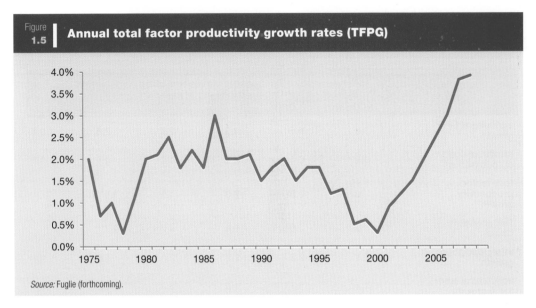

Figure 1.5 | **Annual total factor productivity growth rates (TFPG)**

Source: Fuglie (forthcoming).

Employment, unemployment, and wage trends

Rapid movement towards a structural transformation should show up in the Indian data by a tightening of the rural labor market and an increase in opportunities for rural-urban migration. This section shows that this is also not happening, and the following section instead shows that rural households are diversifying into the rural non-farm sector. The limited absorptive employment capacity of the urban economy has led the non-farm sector to become the main destination of growing rural labor forces. *While this is a structural transformation of sorts, it is a stunted one.*

Rural and urban employment trends

Table 1.1 shows that India's population growth rate has slowed down from a peak of 2.3 percent in the 1970s to 1.6 percent in the 2000s and is expected to slow to about 1 percent in the current decade. The growth of the labor force has accelerated, however. In urban areas it grew at 3.1 percent in the last decade, while in rural areas the growth rate was 1.2 percent (Table 1.1), for a total labor force growth rate of 2.8 percent. This is significantly larger than the population growth rate on account of the 'demographic dividend' associated with a slowdown in the population growth rate. Hazell et al. (2011) cite UN population projections that suggest that the rural population will peak at 900 million in 2022, and that the rural labor force may continue to grow until 2045. *Clearly, the Indian economy as a whole is facing an enormous employment generation challenge in both urban and rural areas for more than the next 30 years.*

Rural and urban males have always had fairly similar labor participation rates while the rates for rural females have been much lower, and even lower for urban females (Figure 1.6).

Figure 1.6 | **Trends in labor participation rates**

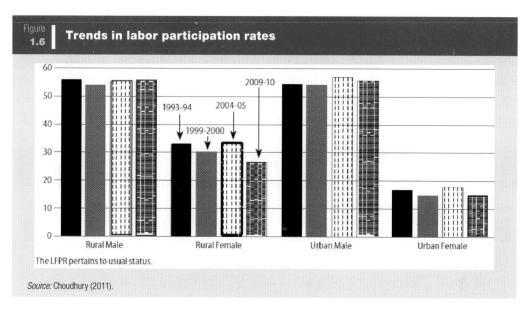

The LFPR pertains to usual status.

Source: Choudhury (2011).

Since 1973 there has been little discernible trend in rural male labor participation and only limited fluctuations. Female rural and urban participation rates fluctuated from 1977–78 to 2004–05. During the early years of the century there were significant increases in participation rates, especially for females, in both rural and urban areas. Since then labor participation rates have gone down for rural females to their lowest level over the entire period. Himanshu et al. (2011) interprets the movement of rural women into the labor force between 1999–2000 and 2004–05 as a response to the agrarian crisis of the period. The subsequent sharp drop in labor participation to 2009–10 is interpreted as a withdrawal from the labor markets as economic conditions improved again. Others have pointed to the very large increase in participation in education as a major reason for withdrawal of women from the labor market, but careful analysis by Choudhury (2012) suggests that this is not a good explanation.

Choudhury (2011) also shows that in both rural and urban areas there are some common trends: a slight decline in the manufacturing share of employment, which is consistent with the constancy of the manufacturing share in the Indian economy and its far slower growth in the past decade than planned; a decline in the share of agriculture and allied industries; a sharp increase in construction; and a large share of the labor force in urban areas in trade, hotels, and restaurants (much smaller in rural areas) and in both areas they have stayed fairly constant. As a consequence, rural non-farm sector employment has grown especially fast.

The employment data also reveal a significant trend towards the feminization of agriculture. Among rural workers, females have always been more likely to be engaged in the primary sectors, most of which is agriculture, than men, and, correspondingly, less in the secondary sectors. For example, in 1977–78, 88.1 percent of female workers were engaged in primary sectors compared to 80.6 percent of males (Table 1.2). By 2009–10, these percentages had gone down for both males and females as a consequence of the rise of the rural non-farm sector. However, for males, engagement in the primary sector had gone down to 62.8 percent, or by 25 percent, while for females they had gone down to 79.3 percent, or by only about 10 percent. On account of their higher labor participation rate there are still more men working in agriculture than women. Nevertheless, there is a clear trend towards the feminization of the agriculture labor force.

Table 1.2	Changes in the sectoral composition of the rural labor force					
NSS round	rural males			rural females		
	primary	secondary	tertiary	primary	secondary	tertiary
32 (July '77–June '78)	80.6	8.8	10.5	88.1	6.7	5.1
38 (Jan–Dec '83)	77.5	10.0	12.2	87.5	7.4	4.8
43 (July '87–June '88)	74.5	12.1	13.4	84.7	10.0	5.3
50 (July '93–June '94)	74.1	11.2	14.7	86.2	8.3	5.5
55 (July '99–June '00)	71.4	12.6	16.0	85.4	8.9	5.7
61 (July '04–June '05)	66.5	15.5	18.0	83.3	10.2	6.6
64 (July '07–June '08)	66.5	16.2	17.3	83.5	9.7	6.8
66 (July '09–June '10)	62.8	19.3	17.8	79.3	13.0	7.6

Source: Himanshu (2011).

As shown in Table 1.3, employment in India is very much concentrated in the informal sector. Between 1999–2000 and 2004–05 the proportion of workers in the formal sector declined from 8.8 percent to 7.5 percent. The National Commission for Employment in the Unorganized Sector (NCEUS) defines organized employment as employees who receive provident fund and social security benefits. Within the organized (formal) sector, the proportion of employees with informal contracts rose from 37.8 percent to 46.7 percent. The Indian labor market has shown a marked tendency to informalization of labor relationships, and only limited creation of high-quality jobs with secure contracts and pension and health benefits for urban workers as well as for migrants from rural areas. Employment in the rural non-farm sector has always been primarily informal, but

| Table 1.3 | Distribution of workers by type of employment and sector organization | | | | | II/1 |

sector	1999–2000			2004–2005		
	informal	formal	total	informal	formal	total
unorganized sector	341.28	1.36	342.64	393.47	1.43	394.90
	(99.60)	(0.40)	(100.0)	(99.64)	(0.36)	(100.0)
organized sector	20.46	33.67	54.12	29.14	33.42	62.57
	(37.80)	(62.20)	(100.0)	(46.58)	(53.42)	(100.0)
total	361.74	35.02	396.76	422.61	34.85	457.46
	(91.17)	(8.83)	(100.0)	(92.38)	(7.46)	(100.0)

Notes: 1. UPSS basis; 2. Figures in bracket indicate percentages; 3. Estimates by NCEUS.

Source: Government of India (2008), Table 4.7.

the small formal sector employment share has followed the trend to informalization as well (World Bank, 2010).

Urban employment growth, particularly in the manufacturing sector, has been inadequate to provide enough employment opportunities for workers from rural areas. The great informality of employment in the Indian economy and in the organized sector, and the deepening of urban poverty discussed in the next section sharply reduce the attractiveness of urban areas for rural migrants, especially for unskilled and semi-skilled ones. Urban areas remain a pole of attraction of highly-skilled workers. *Nevertheless, the poor employment prospects for low-skilled workers in urban areas means that male, and especially female workers, are stuck in rural areas.*

Agricultural employment, unemployment, and wages

Employment growth in Indian agriculture slowed down between the early 1990s to 2004–05 (World Bank, 2010). As discussed in Choudhury (2011), in 2009–10 the current daily status unemployment rates were the lowest for urban males at 5.5 percent, followed by rural males at 6.2 percent, 8 percent for rural females, and slightly over 9 percent for urban females. Unemployment rates were higher for 2004–05, with the growth of labor participation in the period preceding that year partly or fully driven by distress (World Bank, 2010; Himanshu, 2011). Urban unemployment rates, but not rural ones, today are also lower than in the 1990s. Nevertheless, the urban labor market is still very hostile for females, deterring rural-urban migration.

The growth rate of real agricultural wages declined between 1980 to the middle of the last decade, but has started to increase recently. As shown in Table 1.4, since then real wages in the entire economy have risen at a fairly rapid pace. The fastest real wage growth is observed for urban female salaried workers at 7.8 percent, followed by rural female casual workers at 6.2

percent and by urban male salaried workers. Since female participation rates fell, their faster rising wages are consistent with a voluntary withdrawal of females from labor markets, either as a consequence of growing family income and/or greater participation in education. Wages of casual male workers rose at 4.5 percent in rural areas and 4.2 percent for urban males, which in each case means a compound wage growth of close to 25 percent over the past five years. *There is no recent trend in divergence of unskilled wages between rural and urban areas.*

Table 1.4 | Average daily real wage rate for workers (in 2004–05 prices (Rs.))

year	rural		urban			
	male	female	male	female	secondary	tertiary
regular salaried					6.7	5.1
2004–05	145	86	203	153	7.4	4.8
2009–10	165	103	260	213	10.0	5.3
growth rate (%)	2.8	4.2	5.6	7.8	8.3	5.5
casual					8.9	5.7
2004–05	55	35	75	44	10.2	6.6
2009–10	67	46	91	53	9.7	6.8
growth rate (%)	4.5	6.2	4.2	4.1	13.0	7.6

Note: The wages for urban workers have been deflated by consumer price index (industrial workers) (CPI(IW)) and that of rural workers by consumer price index (agricultural labor) (CPI(AL)). This wage refers to the wage for casual workers engaged in work other than public work.

Source: Choudhury (2011).

How can one explain the recent rise in rural wages? First, after a sharp slowdown in agricultural growth from the early 1990s to the middle of the last decade (which reflected itself in the slowdown of agricultural employment growth), the agricultural GDP growth rate has accelerated again. Second, since the middle of the last decade, agricultural prices have increased significantly in real terms on account of rapid increases in procurement prices of major agricultural crops (Oxus, 2011), and perhaps under the influence of rising and high world market prices since 2008. Third, rural non-farm sector employment growth has also accelerated significantly over the past 20 years. Fourth, there has been a withdrawal of women from the rural labor force since 2004–05 with a shift of women to education.

A fifth explanation is the growth in public expenditures in rural areas that have increased rural purchasing power. Since before the beginning of the 11th Plan, public expenditures for the 13 flagship programs for agriculture, rural development, and social development have been increasing rapidly, and now amount to Rs. 186,539 crore, or approximately US$37 billion. Two-thirds of the expenditures are in programs that are only operating in rural areas. The rural component of the social programs will likely take the lion's share of these expenditures. The rural component of all programs therefore must reach or exceed 85 percent of the total expenditures, or about 158,000 crore, which is nearly 17 percent of agricultural GDP. Therefore, rural development, employment and social development programs, even if they encountered large leakages, are increasingly transferring purchasing power into the rural economy. These are likely to lead to increases in the demand for food and non-farm goods and services, generating multiplier effects on both agriculture and on rural non-farm incomes. In addition, a number of programs will also impact agriculture and rural development via their direct program impacts on output in these sectors. These direct and indirect impacts will be the drivers of the increase in real rural wages.

A highly visible component of the growing public rural expenditures has been the Mahatma Gandhi National Rural Employment Guarantee Act (MGNREGA). The program has been described by the Planning Commission as: "With a people-centered, demand-driven architecture, completely different from the earlier rural employment programmes, MGNREGA has directly led to the creation of 987 crore person-days of work since its inception in 2006–07. In financial year 2010–11, MGNREGA provided employment to 5.45 crore households generating 253.68 crore person-days. It has also successfully raised the negotiating power of agricultural labour, resulting in higher agricultural wages and improved economic outcomes leading to reduction in distress migration" (2011). The program has been widely seen as the major cause of rural wage rate rises. However, it is only a relatively small share of total rural government expenditures and of rural employment. It is likely that the other five factors discussed above together have been a more important driver of the recent real rural wage rate rises, and that rural wages would have increased even in the absence of MGNREGA.

Urban-rural differences in poverty, inequality, income, and consumption

The analysis presented so far raises a major puzzle: for the past few years economic growth has accelerated sharply to more than 8 percent. The inter-sectoral labor productivity differential has risen rapidly; agriculture grew fairly slowly in the period between 1990 and 2005; agricultural productivity growth also slumped in the same period; urban employment opportunities have grown fairly slowly, especially for lower-skilled workers and for women; and migration has been fairly slow. With these trends, one would expect a rising differential between urban and rural per capita incomes and consumptions, and a rising differential between urban and rural poverty rates. However, this has not been the case. As seen in Table 1.5, the rural poverty rate (using the old poverty line) declined from 50.1 percent in 1993–94 to 31.8 percent in 2004–05, or by 18.3

percent, while urban poverty declined from 41.8 percent to 25.7 percent, or by 6.1 percent.[4] In absolute terms the decline in rural areas is larger than in urban areas, but in relative terms, the rate of poverty decline in urban areas is slightly faster than rural areas. By 2004–05, in urban areas both the poverty gap and the squared poverty gap had become deeper, indicating a progressive urbanization of poverty (World Bank, 2010). The poverty data with the higher poverty line resulting from the Tendulkar Committee report show similar convergence for the period 1993–94 to 2004–05 (Table 1.5). *These trends are inconsistent with a growing divergence of rural and urban poverty.*

Table 1.5	Changes in rural and urban poverty rates			
percent of people below poverty line		rural	urban	difference
1993–94		50.1	31.8	18.3 = 45 %[1]
2004–05		41.8	25.7	16.1 = 48 %[1]

Note: [1] calculated with respect to the mean percentage.

Source: Tendulkar report (Planning Commission, 2009).

The ratio of urban to rural per capita income declined from 2.45 in 1970–71 to 2.30 during the 1980s and early 1990s. On the other hand, data on consumption shown in Table 1.6 suggest that the ratio of urban consumption to rural consumption increased from 1.54 in 1983 to around 1.70 in 2004–05 and 2009–10. Whether rural-urban disparities have increased is therefore dependent on the data used and the period considered. But neither data series suggest a sharp change in urban-rural disparities over the past 30 years.

Given the significant increases in non-agricultural to agricultural productivity differential and the agricultural trends discussed above, it is surprising that the urban-rural per capita income and consumption gaps have not increased sharply, and that the gap between the rural and urban headcount poverty rates has not increased sharply as well.

The drivers of rural poverty reduction

Ravallion and Datt (1996) show, in line with the international experience, prior to 1991 rural growth was the most important driver of poverty reduction and reduced rural poverty, national poverty, and even urban poverty. But urban growth only reduced urban poverty and had no impact on rural poverty or national poverty. In 2009, Datt and Ravallion updated their earlier work

---

4 Preliminary estimates of the national poverty rate prepared by Ravi and cited in Ahluwahlia (2011) suggest that the national poverty rate under the new Tendulkar Committee poverty line has declined further from 37.2 percent in 2004–05 to 37.2 percent in 2009–10, or at an accelerated rate of about 1 percent per year. The urban-rural poverty rates for 2009–10 have not yet become available.

Table 1.6	Consumption inequality, India				
	1983	1987–88	1993–94	2004–05	2009–10
Gini Coefficient of distribution of consumption					
rural	0.30	0.30	0.28	0.30	0.28
urban	0.30	0.35	0.34	0.37	0.37
urban-rural ratio of mean consumption (constant prices)*	1.54	1.44	1.64	1.72	1.69

Note: *Original shows urban-rural ratio.

Source: Ahluwahlia (2011), Table 6.

to 2004–05. They showed that rural growth remains significant for reducing rural poverty and national poverty. But since 1991, when economic growth started to accelerate, urban growth has become the major driver not only of urban poverty reduction, but for both national and rural poverty reduction. *Datt and Ravallion's new findings suggest that a spillover has emerged from more rapid urban growth to rural growth.*

Since agricultural growth had slowed down during the period 2004–05, the spillovers must have been felt primarily in the non-farm sector. In the past, the rural non-farm sector was viewed as driven primarily by agriculture (Hazell and Hagbladde, 1993), so to add an urban driver is a novelty. But there is more direct evidence of such a driver: Himanshu et al. (2010) show via a multiple regression using the within estimator in panel data of regions in India that higher non-farm employment by rural adults also significantly reduces rural poverty.

These results do not imply that agriculture has lost its impact on the rural non-farm sector, and more broadly on rural poverty. In Datt and Ravaillion's 2009 update, agricultural growth remains an important determinant of rural poverty reduction. This conclusion is reinforced by the same regression analysis of Himanshu et al. (2010) that showed that higher yields are associated with declining rural poverty, suggesting the impact of agricultural productivity growth on poverty remains high. In the same regression they also show a strong and negative impact of higher agricultural wage growth on rural poverty. This strong impact is not surprising as agricultural workers constitute about half of India's overall poverty population.

In conclusion, neither poverty nor per capita income and consumption show signs of rapid divergence between rural and urban areas as a consequence of the rising disparity of labor productivity between agricultural and non-agricultural sectors. Consumption inequality has recently increased in urban areas but stayed fairly constant in rural areas. While rural growth and agriculture were the main drivers of poverty reduction before 1991, since then urban growth has become a quantitatively more important driver of poverty reduction overall even in rural areas.

Nevertheless, growth in agriculture, in agricultural productivity (as measured by yields), and in agricultural wages remain important drivers of rural poverty reduction.

The rising importance of the rural non-farm sector

If urban areas are inhospitable to migrants from rural areas then where has the growing rural labor force found employment and opportunities for increasing their incomes? If there had been no such opportunities, undoubtedly rural poverty would not have improved as fast as urban poverty and rural-urban income and consumption parities would have declined. However, the rural non-farm sector has become much more dynamic than the farming sector, both in terms of GDP growth and employment generation. Between 1983–2004 rural non-farm GDP has grown at a rate of 7.1 percent, more than a percentage point faster than non-farm GDP, and 4.5 percent faster than agricultural GDP (Table 1.7). This faster growth of the non-farm sector started in the decade from 1983–1993. In the period 1993–2004, non-agricultural employment growth in rural areas accelerated from 3.5 percent to 4.8 percent. In the 1980s, 4 out of 10 rural jobs were in the non-farm sector, now it is 6 out of 10 (ibid). *Given the large size of the rural labor force these numbers mean that the rural non-farm sector has emerged as the largest source of new jobs in the Indian economy.*

Table 1.7 | Trends in non-farm employment and in national, rural non-farm, and agricultural GDP (annualized rates of growth, %)

year	non-farm employment	GDPN	non-farm GDP	agriculture GDP
1983–2004	3.3	5.8	7.1	2.6
1983–1993	3.5	5.2	6.4	2.9
1993–2004	4.8	6.0	7.2	1.8

Note: GDP at factor cost at 1993–94 prices. Agriculture GDP originating in agriculture, forestry, and fishing. Non-farm GDP defined as a residual.

Source: Himanshu et al. (2010), Table 3.

Growth in rural non-farm sector employment has occurred all over India, but has been highly uneven. It is highest in Kerala, West Bengal, and Tamil Nadu, and lowest in Chhattisgarh, Madhya Pradesh, followed by Uttarakhand, Karnataka, Gujarat, and Maharashtra (World Bank, 2010; Binswanger-Mkhize and d'Souza, 2011b).

Until 2004, the growth in non-farm jobs had come primarily from increases in services, transport, and construction. In 1983, close to 40 percent of rural non-farm jobs were in manufacturing. Despite continued growth of rural manufacturing, this share has declined to just a little above 30 percent in 2004–05. In 1983, social services and trade, transport, and communication both generated about 26 percent of non-farm jobs. Social services have since declined to about 18 percent of the jobs, while trade, transport, and communications have grown rapidly to about 33 percent. In 1983, construction was by far the smallest sector, with a share of only 10 percent. Since then it has grown the fastest and now generates close to 19 percent of the rural non-farm jobs. The high level of rural construction has visually transformed villages all over India, with much better village infrastructure and housing.

Foster and Rosenzweig (2005) show that non-farm enterprises producing tradable goods (the rural factory sector) locate in settings where reservation wages are lower. If the rural factory sector seeks out low-wage areas, factory growth will be largest in those areas that have not experienced local agricultural productivity growth. Thus, rural non-farm growth reduces spatial inequalities in economic opportunities and incomes. Nevertheless, the location of factories where wages are low has an equalizing impact on income distribution in rural areas.

Datt and Ravallion's 2009 analysis suggests that since 1992, urban growth has also fueled the rural non-farm sector. More direct evidence of spillovers comes from World Bank (2010), p 66: "During the two periods of analysis, 1983 to 1993–94 and 1993–94 to 2004–05, regression estimates suggest that non-farm employment increased more in regions where urban incomes also grew faster. Disaggregating the analysis by different types of non-farm employment, the results show that it is regular salaried jobs and self-employment activities that appear to be most strongly and positively correlated with urban growth; casual non-farm employment is uncorrelated with urban growth." Additional drivers of recent rural non-farm growth can be inferred from a closer look at the composition of employment growth (Box 1.1).

Box 1.1 | **Recent drivers of rural non-farm growth**

Between 1999–2000 and 2004–05, rural non-farm employment increased by 16 million by principal status, *of which eight million (nearly 50 percent) was in the form of self-employment, five million as casual employment, and three million as regular employment* (Himanshu, 2011). By industry, 5 million was accounted for by construction (equivalent to almost the entire increase in casual employment), 4 million by trade and hotels, 3.5 million by manufacturing, and 1.8 million by transport and communication. Within the large rural self-employment component that has been shown to be partly driven by urban growth, three industries account for nearly 60 percent of the increase: 2.2 million was accounted for by retail trade, 1.5 million by manufacture of wearing apparel, and 1 million by land transport. Another 25 percent of the increase was accounted for by 7 activity codes that include post and communications, where the largest increase was in the form of STD/PCO booths, maintenance and repair of motor vehicles, and hotels and restaurants (ibid). The STD/BCP booths and the maintanace and repairs of motor vehilces are fueled by technical change in communication, motorization of transport, and agricultural mechanization. Increases in hotels and restaurants reflect income growth that is partially driven by urban spillovers.

Who benefits from non-farm wage employment? It is primarily males in the age group of 18–26 years old who have some education and are moving out of agriculture into non-farm jobs (Eswaran et al., 2009). Women are barely transitioning into the non-farm wage employment sector. In growth terms, the number of rural men working off-farm doubled between 1983 and 2004–05; for women the increase was 73 percent. Individuals from scheduled castes and tribes are markedly more likely to be employed as agricultural laborers than in non-farm activities, even controlling for education and land. Even a small amount of education, such as achieving literacy, improves prospects of finding non-farm employment, and with higher levels of education the odds of employment in well-paid regular non-farm occupations rises. Finally, those in the non-farm sector own more land on average than agricultural laborers, except for those in casual non-farm employment (ibid).

The REDS data for 2007 show a significant differential between average farm and rural non-farm wages of 47 percent, and the premium has been stable since 1999 (Binswanger-Mkhize et al., 2011b). Eswaran et al. (2009) use NSS data to show that wage premia associated with education were growing over time. By 2004–05 NSS found these premia had increased to Rs. 86 for literate workers over illiterate ones, Rs. 197 for those who had attended middle school, and Rs. 696 for graduates. The authors conclude that if more middle school and high school graduates were available in 2004, they would have found employment in rural industry and services.

Until 2004–05, employment growth in the non-farm wage sector had accelerated while the growth in average earnings had decreased. These two trends have cancelled each other out, and for the last two decades, growth in total non-farm wage earnings has been constant (World Bank, 2010). In spite of the preponderance of non-farm jobs in rural employment generation, Eswaran et al. (2008) estimate *the contribution of the rural non-farm sector to rural wage growth to be only about 22 percent of the total growth, thereby confirming the importance of agricultural growth and productivity growth to rural wage growth.* In particular, the rural non-farm sector has not contributed to wage growth among the illiterate, but only among the more educated (Eswaran et al., 2009).

Box 1.2 shows that the rural labor market is significantly connected to the urban labor market, and that the farm and non-farm labor markets, while supporting a significant wage differential between them, are highly integrated.

As discussed in Box 1.1, a particularly dynamic development has been the growth in self-employment in the non-farm sector. The question has arisen whether such employment is a consequence of economic distress or of rising self-employment income opportunities (World Bank, 2010). In order to answer this question, data on income earned by rural non-farm self-employment is required that is not available in the standard NSS consumption, poverty, and employment surveys. Binswanger-Mkhize et al. (2011a) analyzed data from the 1999 and 2007 rounds of the Rural Economic and Demographic Surveys (REDS) of the National Council of Applied Economic

| Box 1.2 | **The behavior and impacts of farm, non-farm, and urban labor markets** |

Econometric results from Binswanger-Mkhize et al. (2011b) for the period of 1999–2007 are used to discuss labor market behavior. These come from the REDS national panel data set of over 5,000 households. A first finding in the table below is that in Indian villages the farm and non-farm labor markets are linked closely in a symmetric manner: the elasticity of the rural farm wage with respect to the predicted non-farm wage is close to 0.5 and the converse elasticity of the non-farm wage is almost the same size. A rise in the urban wage increases both these wages with an elasticity of around 0.17. Morover, the elasticity of the farm and non-farm wage to the aggregate agricultural price is almost identical at 0.04. Finally, a rise in either of the two wages leads to large reallocations of labor to the sector that has experienced the wage rate rise. The elasticities far exceed all other elasticities examined so far. The reason the two labor markets are so integrated is that the slightest change in their relative wage trends induces a lot of movement of the family labor to the other sector, quickly reducing the disparity.

Box 1.1 table 1: The responses of rural labor to changes in wages

	predicted	GDPN	non-farm GDP	agriculture GDP
labor force	0.020**	0.075**	-0.059**	0.036[1]
share of family labor in agriculture	3.262***	-5.571**	Na	-2.309[1]
share of labor in non-agriculture	-2.282**	4.944**	Na	2.662[1]
share of students	-.980[1]	.637[1]	--	-.353[1]
farm wage	--	0.484**	0.166**	
non-farm wage	0.488**	--	0.171**	

Notes: [1] Standard errors yet to be calculated; ** significant at 1 percent level.

A rise in the urban wage leads to a reduction in family labor force, which means that it induces rural-urban migration. The last column sums up the elasticities of the left hand variables with respect to the wages on the top. The resulting sum tells what would happen if the farm, the non-farm, and the rural wage were to rise by the same proportion. Such a rise of the national wage level would induce slightly more people to commit to work in rural areas. This suggests that people would prefer the rural areas if there was an overall income impact from higher wages. These preferences may well reflect their perception of the relatively hostile nature of the urban labor market discussed in section one. However, looking at the shares, the sums show that people would tend to move their work force from agriculture to non-agriculture, suggesting that while they prefer rural areas they would prefer to work in the non-farm sector. This supports the notion that people would rather move out of agriculture if they could.

research to fill this gap. REDS is a nationally representative panel of rural households that were originally selected in 1971 to study the Green Revolution with a slight tilt towards better agricultural areas.

Because of population growth and household subdivision, the sample grew from 4,690 households in 1999 to 5,759 households. Households have become smaller in size, contain a lower proportion of farm households, and, on average, own less land. (The decline in average owned and operational holding sizes, a consequence of rural population and labor force growth, has been a long-term trend in India since 1962 (Basole and Basu, 2011)). Despite these trends, per capita income grew from 8,498 Rs. in 1999 to 12,370 Rs. in 2007, i.e., by Rs. 3,881 (in Rs. of 1999), or at an annual rate of 5.7 percent, which is similar to urban per capita income growth. These data also show rising rural wages, but also a more than doubling of the prices of agricultural land between the two periods. Since the wealth of farmers is primarily in the form of land, they have experienced a significant increase in their real wealth.

Between 1999 and 2007, the number of households engaged in non-farm self-employment more than doubled from under 10 percent to nearly 20 percent. Unfortunately, the gender of the owners of these enterprises is not known, but given the growth of the rural self-help movement, it is possible that women participated significantly in this self-employment growth. While agricultural profits and agricultural labor incomes grew in absolute terms, it was the rural non-farm self-employment income component that grew the fastest: For households engaged in rural non-farm employment this component of income rose from Rs. 36,767 to Rs. 64,045, i.e., by 74 percent in only 8 years, or at a simple annual rate of 9.3 percent. Figure 1.7a and Figure 1.7b shows that for the sample as a whole, the shares of income shifted from agricultural profits and wages (-9.26 percent and -2.10 percent) towards non-farm self employment income (+12.19 percent). At the same time the share of non-farm wage income has stayed nearly constant at around 7.5 percent.

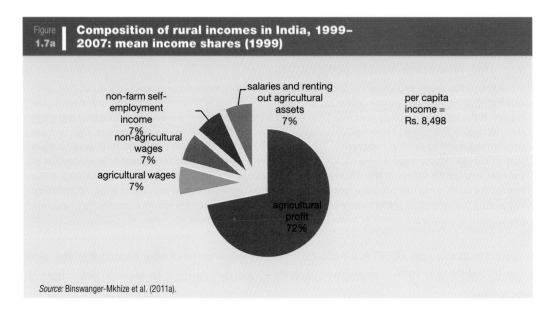

Figure 1.7a | **Composition of rural incomes in India, 1999–2007: mean income shares (1999)**

non-farm self-employment income 7%

non-agricultural wages 7%

agricultural wages 7%

salaries and renting out agricultural assets 7%

agricultural profit 72%

per capita income = Rs. 8,498

Source: Binswanger-Mkhize et al. (2011a).

The income data on the rural non-farm self-employment sector suggests that while it may contain some distress employment, this is not the main driver of its expansion, and that instead it has become the most dynamic source of income growth of rural households, including farmers.

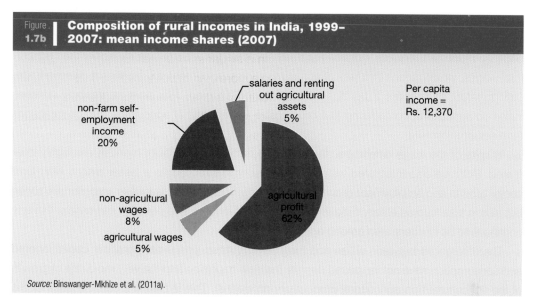

Figure 1.7b | **Composition of rural incomes in India, 1999–2007: mean income shares (2007)**

Source: Binswanger-Mkhize et al. (2011a).

What is observed among farms is not only diversification of agricultural production to higher-valued products, but also to more remunerative self-employment in the non-farm sector. There is therefore a marked tendency of agriculture to move to a productive and modern model of part-time farming.

Summary of employment and poverty trends

Urban and rural male labor participation have been around 55 percent for the past two decades. Meanwhile, female rural participation rates have fluctuated around 30 percent and female urban participation rates have been very low, fluctuating around 15 percent. Formal sector employment is a distressingly small component of employment in India and even in the formal sector there is a trend towards informal employment contracts. Between 1990–91 and 2004–05 the growth rate of agricultural employment (both wage and self-employment) has slowed down, while that of the rural non-farm sector has accelerated greatly, becoming the most significant source of the Indian economy. The share of women within the agricultural sector has been rising steadily, indicating a trend towards the feminization of agriculture.

Within the rural non-farm sector, self-employment accounts for as much of employment growth as wage employment, and diversification of farms into rural non-farm self-employment is now a reality for around 20 percent of farmers. Non-farm self-employment is driven by rapidly rising incomes in these enterprises, and therefore such employment cannot be regarded as distress employment. Indeed non-farm self-employment income has become the largest source of rural income growth.

Rural non-farm wage employment is accessed primarily by young males with some education, suggesting that females are at a disadvantage in obtaining such employment, perhaps because much of it requires mobility. Rural non-farm wages are significantly higher than agricultural wages, which means that lack of access to such employment is a significant disadvantage. *Unfortunately, this implies a significant impediment to women, who have increasingly concentrated on agriculture, contributing to a progressive feminization of agriculture, and on rural non-farm self-employment.*

In spite of the wage differential, the farm and non-farm rural labor markets are highly integrated. Both are also integrated with the urban labor market, but to a lesser extent. Non-farm sector growth and employment growth have not only happened in favorable agro-climate zones but also in less favored areas, mitigating inter-regional income and poverty differentials. Poverty continues to be concentrated among agricultural workers.

The differences between urban and rural poverty rates, and urban-rural per capita income and consumption have not increased despite the slow migration and the very rapid rate of growth of the agricultural/non-agricultural productivity differential. This is clearly a consequence of the growth of employment and income in the rural non-farm sector, and especially in rural non-farm self-employment and income. Since the early 1990s, this rural non-farm self-employment sector has become increasingly fueled by accelerating urban growth.

Implications for structural change

The new growing rural non-farm dynamic has led to a revision of the standard model of structural transformation that equates non-agriculture with urban areas. It now has to include the rural non-farm sector. *Structural transformation in the form of a decline in agricultural employment and in favor of non-agricultural employment is happening in India. However, the new form of structural transformation in India is a stunted one, because it primarily generates employment that is informal and/or insecure, and without the benefits of health and unemployment insurance and pensions.*

The structural transformation trends are in sharp contrast to the trends in China. Near zero population growth rates and rapid growth of labor-intensive manufacturing and in other urban sectors have led to a world record rural-urban migration that has left rural areas without young workers. Farms are increasingly operated by older farmers, many of whom are also women. Even in China, the rural non-farm sector has emerged as a dynamic sector, probably on account of spillovers from rapid urban growth to rural areas, as well as continued rapid agricultural growth. Urban and rural wages have started to grow very rapidly at around the same time as Indian rural wages, but the pace of real wage growth is much faster. It appears that China is on the way towards a normal structural transformation.

In spite of rapid economic growth, India's structural transformation is constrained by the weakness of employment growth in the urban economy, and most specifically in labor-intensive manufacturing. Most experts attribute the slow growth in labor-intensive manufacturing to restrictive labor legislation in India and to poor infrastructure for power, water, and transport. The dream of a structural transformation directly to a service economy with good and secure urban jobs has not been realized and is unlikely in the future as well. That there is nevertheless a structural transformation from agricultural production and employment towards non-agriculture appears to be a consequence of rising urban spillovers to rural non-farm self-employment, and this has prevented a greater divergence in poverty rates and per capita incomes and consumption. *Continued growth of high urban economic growth is therefore critical for rural income growth.* However, agricultural growth and higher productivity continue to be powerful drivers of rural poverty reduction, rural non-farm sector growth, and agricultural and rural wages. *An acceleration of agricultural growth and agricultural productivity growth to sustained higher levels than in the past two decades would therefore be highly beneficial for rural areas. As a consequence, agricultural and rural development policies, institutions, and programs remain important determinants of rural welfare.*

Vision of agriculture and rural poverty reduction

Using the results and insights from the past sections a vision for the agricultural sector and for rural poverty reduction over the next decades can be developed. The structural transformation in India and rural-urban migration will likely remain constrained by the slow growth of employment in urban areas, in industry, and especially in labor-intensive manufacturing. This is because it appears to be politically impossible to reform restrictive labor legislation, and because it will take a long time to overcome infrastructure bottlenecks. For most unskilled and semi-skilled workers, urban migration opportunities are likely to remain constrained, and limited to the informal sector, or to informal contracts in the formal sector. There is little chance that the urban economy will provide enough employment for the growing rural labor force to allow a large proportion to move to the urban economy. The rural labor force will therefore have to find a way to improve their incomes in rural areas.

Given the need to raise agricultural income and the economies of scale that mechanization and credit constraints bring to agriculture (Foster and Rosenzweig, 2011), it may appear paradoxical that farm sizes would continue to decline. However, this tendency is in line with past trends in India, where farm sizes have grown modestly only in Punjab, and declined everywhere else. This decline is in line with continued rises in rural populations and labor forces, and with the limited labor absorption potential of urban areas.[5] The rapidly rising prices of agricultural land will impart a portfolio motivation to hang on to land in the households owning land and remaining in the countryside. While land rental markets could lead to land consolidation, up to the latest data available, land renting has also continued to decline. To provide self-employment opportunities for family labor, and especially for women, most households will be reluctant to rent out or sell land in the future as well. With males having better opportunities in rural non-farm employment than females, agriculture will continue to feminize. With these trends agriculture will be dominated by even smaller part-time farm households; with a few full-time farmers at the top and a large majority of part-time marginal, small, and medium farmers.

Some of these trends are in contrast to China: absent or old owners of rural land rights are increasingly renting them out to relatives, to larger farmers, and to enterprises. Land rentals have risen from close to zero in the 1990s to around 20 percent of total agricultural land. As a consequence average operational holdings have started to rise, while the ownership distribution of land rights remains unaffected. These trends are also likely to continue.

Part-time farmers in India will get more income from non-agriculture than from agriculture. All types of farmers will focus much more on horticulture, milk, poultry, and eggs. Consumer demand will drive a trend towards traceability of agricultural output, quality control, and organic farming that will provide additional income opportunities. Farms will be much more capital-intensive, and use advanced biological and mechanical technology for crops, horticulture, livestock, and aquaculture. Water markets and other cooperative ways will be used to realize economies of scale. Depending on economy-wide growth, farmers will try to increase their agricultural incomes by adoption of modern technology, further diversification towards higher-valued crops, use of more machinery, and increasing reliance on family labor. The rural non-farm sector will continue to grow faster than agriculture, provide more income opportunities than agriculture, and produce an increased range of services and products, using progressively more modern technology. Declining farm size trends and the diversification of households into the non-farm sector will undoubtedly continue. *As a consequence, the emergence of a farm sector dominated by modern part-time farmers, many of them female, whose households will combine farming with non-farm employment of the men and/or self-employment in the non-farm sector is likely.*

While these trends are likely to continue under both very rapid and more moderate economic growth, and regardless of agricultural policies and programs, both an optimistic and a more pessimistic future are possible for agricultural and rural incomes, and for rural poverty reduction. The optimistic version is based on a combination of rapid economy-wide growth as well as rapid agricultural and rural non-farm growth; both partly driven by urban demand and technology

---

5 It is also consistent with trends in advanced economies that are dominated by small family farms, such as Japan, Taiwan, Korea, or European countries such as Italy, Spain, Switzerland, and Norway. However, in many of these countries the heavy subsidization of agriculture and constraints imposed on agricultural land markets have limited land consolidation via sales and rental markets.

spillovers. Agricultural growth will be driven by rapid technological change and productivity growth, improvements in water use efficiency and irrigation growth, and the diversification of agriculture. Both full-time and part-time farmers will have plenty of new technologies available and be able to adopt these technologies, and many remunerative diversification opportunities in agriculture and non-agriculture. This will result in the emergence of a highly modern part-time farming sector and rapid agricultural income growth, which will also spillover into more rapid rural non-farm growth. At the same time, the demand and technology spillovers from the urban economy will further accelerate rural non-farm sector growth. Non-farm opportunities will continue to be more accessible to young and educated males than to females, accelerating the feminization of agriculture. However, this may be associated with rising entrepreneurial opportunities for the female farmers. The combination of rising agricultural and rural non-farm incomes will support rapid income growth in rural areas, including rapid rural wage growth. Rural-urban incomes and consumption ratios will improve, or at least not deteriorate, and rural poverty will decline very rapidly, except in remote regions with poor agricultural endowments and poor prospects for rural non-agricultural development.

Rising incomes from agriculture and the non-farm sector will not only sharply reduce absolute poverty in rural areas but hunger as well, except perhaps in some tribal areas. Malnutrition may, however, continue to persist, as it has in the developed world, via the addition of obesity problems.

Under a pessimistic vision, economy-wide growth will be slower, and the slowdown in economy-wide growth will reduce the urban spillovers to higher agricultural and non-agricultural demand, and technology spillovers in the non-farm sector. Slow agricultural growth could not only result from reduced demand for food, but also if (i) technical change in agriculture remains slow; (ii) services for part-time smallholders are not scaled up and improved; (iii) technology adoption is limited more to the full-time farmers; and (iv) female farmers have limited entrepreneurial opportunities. The combination of relatively slow agricultural growth will reduce rural non-farm sector growth, which will also suffer from reduced urban spillovers. Rural income growth and wage growth will be lower. Rural-urban incomes and consumption ratios will deteriorate, and rural poverty will decline fairly slowly, even in better located and endowed rural areas.

The private sector is emerging as a key driver of many components of agricultural and rural development. All of the non-farm sector development and all of farm investment is a private sector activity. In addition, the private sector is transforming the marketing system from the farm to consumers; it has become a major source of new technology, including genetically modified organisms (GMOs), and seeds supplies; it has entered agricultural extension in a significant way, via contract farming and in input supply; the private sector is providing piped water in canal systems to irrigators; it has entered agricultural credit via contract farming and microfinance; and it is assisting in the administration of land record systems. NGOs have also entered agricultural extension, natural resources management, fostering of linkages of farmers to market opportuni-ties, and micro-finance. Opportunities for public-private partnerships and partnerships with the private sector and NGOs are growing significantly, and need to be mobilized much better.

Despite the constraints arising for rural welfare from slow urban employment growth, rapid agricultural growth can also contribute significantly. Over the coming decades, agriculture will continue to diversify rapidly towards high-value commodities, and deal with declining farm sizes, feminization of the labor force, increasing water stress, and climate change. These opportunities and challenges cannot be managed by small adjustments in existing institutions, policies and programs. In Centennial Group (2012), a full set of bold recommendations is spelled out that would help bring about an optimistic vision for agriculture—these recommendations are also discussed in Part I of this book. Box 1.3 further details how land reform, land administration, and land markets can support an optimistic vision.

An optimistic future for Indian agriculture—vision and policies

Over the next few decades, accelerating investment and productivity growth, and maintaining income parity in agriculture will require much accelerated technical change, further diversification into high-valued crops, continued growth of irrigation, and further diversification of farmers to the non-farm sector. Agricultural research will continue to be provided by both the public and the private sector, with the public sector having a particularly important role in upstream technolo-gies and in technologies with limited private appropriability such as open pollinated varieties or agronomic practices and soil conservation. The public sector will have to become more account-able to farmers and consumers more efficient. As the failure to document adverse side effects of transgenic crops becomes ever more apparent, biotechnology and transgenic crops will become more widely accepted, and competition among private sector providers will reduce the costs of biotech inputs. Transgenic crops will therefore become a major source of total factor productivity growth.

Agricultural extension will become much more pluralistic with rapid growth of extension by input suppliers and contractors of output, via scaling up of NGO extension efforts and of mobile applications for agricultural information on technologies and practices, inputs, and output mar-kets. At the same time, the public sector, via stronger support from the state levels, should be able to strengthen the Agriculture Technology Management Agency (ATMA) model of coordina-tion and provision of extension in much closer coordination with private sector providers. All extension providers will continue to struggle with the issue of how to provide extension to the many small and part-time farmers, and to the rising share of women farmers. The challenges of how best to provide extension in rain-fed farming, semi-arid, arid areas, and tribal areas will con-tinue to preoccupy the public and NGO sectors, which should find it useful to cooperate more.

Technical change, diversification, and continued irrigation growth may not be enough to maintain agricultural and rural incomes in line with rapidly growing urban incomes. Instead, significant financial support to farmers may be required. Current subsidies to fertilizer, electricity, water, and support to crop prices are tied to inputs and outputs. Subsidies are large but an inefficient means to transfer income to farmers, and they have adverse environmental impacts. Reformed, and

---

Box 1.3 | How land reform, land administration, and land markets can support an optimistic vision

Even though under our visions for the future of agriculture large-scale consolidation of land holdings is neither necessary nor likely, flexible and secure land transactions contribute in several ways to the realization of the vision, as demonstrated by a series of careful studies using the REDS data of the National Council of Applied Economic Research (NCAER):

- Land reform has not led to inefficient small holdings, but instead has led to higher asset accumulation in states that underwent more land reform, higher income growth, and higher educational attainments of children.

- Land rentals have steadily declined, and are unlikely to become a major avenue for the aggregation of large farms. Instead (i) land rental has been an important avenue for land access for poor, land scarce and landless households, and therefore has supported poverty reduction in an environment with limited rural-urban migration options. (ii) Those who rent land obtain higher returns to their labor than available in the casual labor market. (iii) State-level land rental restrictions reduce the ability of the poor to get access to land and their productivity.

- Land sales markets transferred land to more efficient producers who increased their incomes. However, village weather shocks encouraged distress sales by poor households. Where employment guarantee schemes were operating, they reduced such distress sales; MGNREGA will reinforce this mechanism. With such safeguards in place, constraints on land sales among land reform beneficiaries and in tribal areas can be safely eliminated.

- Amendments in the Hindu Succession Act that give equal rights to sons and daughters to inherit land significantly increased women's probability of inheriting land, although it did not bring about full gender equality. Girls raised by women who had inherited land had significantly higher levels of education than those raised by women not subject to the amended Act. In a feminizing agriculture, women's rights to inherit land is even more important.

- Computerizing registration of deeds and/or textual records is fully or partly completed in Andhra Pradesh, Gujarat, Karnataka, Maharashtra, Rajasthan, and Tamil Nadu. Computerization of textual records was facilitated through private sector contracting. In Maharashtra, computerizing registration of deeds has been associated with a 50 percent increase in the number of registered transfers. Stamp duty collected during the same period has more than doubled. Land transactions in sales and rental markets have been simplified and made more secure. Better land records will also make it easier to use small parcels of highly-valuable land as collateral for loans to finance investments in agriculture and in the non-agricultural sector.

- In tribal areas, individual or community land rights are neither recorded nor can they be transacted. There is also no system of land administration for traditionally 'marginal' lands. In tribal areas land administration should first focus on the registration of communal tenure, and eventually of individual tenure, if the communities decided in an open and transparent vote to move to private property. This is the approach that is now used in Mexico and other countries. Improved land

Box 1.3 | How land reform, land administration, and land markets can support an optimistic vision

administration would ensure greater security of tenure and facilitate rental and sales to enable tribal populations to obtain the same benefits associated with land ownership as other farmer groups.

The key recommendations resulting from these studies are as follows:

- Consider further provision of land to landless and land-poor people;

- Eliminate remaining constraints on land rental;

- Strengthen land inheritance rights for women;

- Clarify and record rights in marginal areas traditionally outside the system, and tribal areas, by recognizing and recording communal tenure, and by systematically resolving conflicts; and

- Further improve land administration in rural areas via computerization and spatial records.

Source: Deininger and Nagarajan (2011).

more efficient subsidies, will have to shift to broad cash transfers on a per farm basis rather than be linked to products, and include input vouchers favoring small farmers.

Existing constraints on agricultural marketing via regulated markets will have to be eliminated, and marketing and value chains will have to be modernized at an accelerated pace from the farm to the retail outlet. Intense competition in marketing will help constrain the markups in the value chain and therefore assist in combating food inflation. Although some small and part-time farmers may encounter greater marketing problems than larger and full-time farmers, all classes of farmers will be able to avail themselves of better marketing options thanks to the cell phone and rising incentives of retailers and processors to ensure themselves of high-quality outputs via contract farming.

Water management under canal irrigation will have shifted to more demand-driven modes of providing water in a timely and controlled manner, often via pumping and in pipes, and at much higher water use efficiencies. Groundwater irrigation as well as private pumping from canals and other water sources will continue to be the major source of irrigation growth. The problems of reliable electricity supply to both agricultural water users and rural consumers will be resolved in many states, but not without complex political problems and twists and turns. Groundwater depletion will remain a major threat in semi-arid, arid, and hard-rock areas, but solutions that are responsive to the aspirations of millions of irrigators, rather than of a command and control type, will emerge in many places. Water harvesting, groundwater recharge, and drip and sprinkler irrigation will help significantly in reducing depletion. Nevertheless, command and control interventions may be required in some of the most critical watersheds.

It is hard to see how the enormous challenges of agricultural growth, natural resource management, and social services for rural areas can be resolved without greater citizen empowerment and decentralization. Such reforms have been under discussion in India for a very long time but all initiatives so far have failed to bring them about. Reforms will have to be driven primarily by the states, but with support from strong incentives and perhaps further legislative interventions provided by the center as well. They will not come about without pressures from below. Therefore, support for transformative institutions such as self-help groups (SHGs), farmer associations, and organizations of the poor and marginalized will have to expand. More than in other areas the possibility of continued failure to reach these policy intentions is high.

Agricultural and rural development programs of the center will be consolidated from the hundreds of central and centrally-sponsored schemes to a sharply reduced set of block grants that will provide much more flexibility for implementers at state, district, block, and village levels. Many of them will also become much more empowering of the final beneficiaries who will take a much greater role in planning and implementing the schemes. Roles and accountabilities will be clarified and strengthened, along with monitoring, evaluation, and impact evaluation. As a consequence, implementation of agricultural and rural development programs could be significantly improved, become more transparent, and less a source of corruption.

Core literature on India 1960–2010: Structural change, the rural non-farm sector, and the prospects for agriculture

Timmer, C. Peter. 2009. *A world without agriculture: The structural transformation in historical perspective.* Washington, DC: American Enterprise Institute.

This paper reviews the theory of structural transformation elaborated in the 1950s and confronts it with data from recently industrialized and developing countries. It concludes that it has become more difficult to reach the point where the agricultural and the non-agricultural sectors start to converge in labor productivity.

Johnston, Bruce F. and John W. Mellor. 1961. The role of agriculture in economic development. American Economic Review 51(4): 566–93.

This classic paper develops the basic ideas and models that show the pathways by which agricultural growth can contribute to economic development and poverty reduction. Its ideas have been empirically tested many times and stood up very well.

Hazell, P., and S. Hagbladde. 1993. Farm-nonfarm growth linkages and the welfare of the poor. In M. Lipton, and J. van der Gaag, eds., *Including the poor* pp. 190–204. Washington, DC: World Bank.

The authors summarize the literature that emerged in the 1970s and 1980s on the composition and drivers of the rural non-farm sector. The drivers of non-farm growth are shown to be mostly the linkages with agriculture, via forward, backward and consumer demand linkages, but they have been much stronger in Asia than in Africa.

Datt, Gaurav and Martin Ravallion. 2009. *Has India's economic growth become more pro-poor in the wake of economic reforms?* Washington, DC: World Bank, Policy Research Working Paper 5103.

In this paper the authors update their analysis up to the early 1990s where they measured the contributions to urban, rural and national poverty reduction of urban and rural growth. They show that before 1991 rural growth was the main driver of urban, rural and national poverty reduction, which is consistent with the ideas of Johnston and Mellor in the second paper listed above. They then show that since the early 1990s, rural growth still contributes to rural poverty reduction, but that urban growth has emerged as a more important driver of urban and rural poverty reduction than rural growth.

Foster, Andrew D. and Mark R. Rosenzweig. 2003. *Agricultural development, industrialization and rural inequality.* Providence, RI: Brown University, Mimeo.

This paper shows that rural industrialization is particularly strong in areas that were bypassed by the Green Revolution, and therefore had slower wage growth and lower wages than the Green Revolution areas. This means that rural industrialization has sought out low wage environment and thereby has contributed to the inter-regional equalization of rural incomes.

Hazell, Peter B., Derek Headey, Alejandro Nin Pratt and Derek Byerlee. 2011. *Structural imbalances and farm and nonfarm employment prospects in rural South Asia.* Washington, DC: Report for the World Bank.

This paper updates paper number three in this lists with special reference to South Asia. It shows that non-farm employment and self employment has become a very important component of overall employment growth. It also warns that rural labor force growth in South Asia will continue for several decades and pose an enormous employment challenge for the countries.

Ahluwahlia, Montek. 2011. Prospects and policy challenges in the twelfth plan. *Economic and Political Weekly* 46(21): 88–105.

This paper reviews achievements of the Indian economy in the current five year plan, such as the accelerating poverty reduction, and uses this background to analyze the challenges that have to be addressed in the coming five year plan that will start in late 2012.

Chowdury, Subhanil. 2011. Employment in India: What does the latest data show? *Economic and Political Weekly* 46(32): 23–26.

The author reviews the disappointing employment trends in the Indian economy in the last decade and analyzes the reasons for it.

Eswaran, Mukesh, Ashok Kotwal, Bharat Ramaswami and Wilima Wadhwa. 2009. Sectoral labour flows and agricultural wages in India, 1983–2004: Has growth trickled down? *Economic and Political Weekly* 44(2): 46–55.

This paper analyzes the rapid growth of the rural nonfarm sector and the employment and wage growth that has been associated with it.

Himanshu, Peter Lanjouw, Abhiroop Mukhopadhyay and Rinku Murgai. 2011. *Non-farm diversification and rural poverty decline: A perspective from Indian sample survey and village study.* London, UK: Working Paper 44. Asia Research Centre, London School of Economics and Political Science.

This paper brings together a number of studies that analyze the drivers of rural non-farm growth and show how the rural non-farm sector has helped in reducing rural poverty in India.

Agricultural Diversification in India: Trends, Contribution to Growth, and Small Farmers' Participation

Chapter 2

Pratap S. Birthal, P.K. Joshi, and A.V. Narayanan

Introduction[1]

In the last quarter of the 20th century, Indian agriculture grew at an annual rate of about 3 percent. This made the country self-sufficient in cereals, improved food security, and contributed to the reduction in rural poverty. During the 2000s, agricultural growth, however, came under stress on account of several factors, like the deceleration of productivity growth, the decline in public investments, and increased weather uncertainty among others (Singh and Pal, 2010; GOI, 2010). An excessive emphasis on cereal production was also a reason for the slowdown in agricultural growth (Barghouti et al., 2005). Agricultural growth decelerated to 2.3 percent in the first half of the 2000s but recovered slightly to reach 2.6 percent by 2008–09. Agriculture, directly or indirectly, supports the livelihoods of about 60 percent of the country's population; and sluggish growth in the agricultural sector, if it persists for long, endangers national food security, leads to a rise in rural poverty, and reduces overall economic growth. Enhancing agricultural growth, therefore, is a major policy concern.

Indian agriculture is primarily small farm agriculture. Over 85 percent of the land holdings are less than or equal to 2 hectares (ha), with a mean holding size of 0.53ha (GOI, 2006). In this context, a crucial development question arises: should some small farmers exit agriculture, so as to improve the average landholding size and reduce employment pressures on agriculture, or should they continue in agriculture and enhance their incomes through increased intensification and/or diversification of their production portfolio towards activities, such as horticulture, dairy, and poultry production that have a strong potential for higher returns to land? Historical trends suggest that there are limited opportunities for a rapid transfer of labor from the farm to the non-farm sector, as the share of agriculture in the total workforce declined only slowly from 69 percent in 1983 to 58 percent in 2004–05. Further, the non-farm sector is not easily accessible to a majority of farm households due to financial, technological, and skill barriers (Coppard, 2001). In 2004–05, the non-farm sector engaged only 24 percent of the total rural workforce (Nagaraj, 2007).

1 Both Pratap S. Birthal and A.V. Narayanan are affiliated with the International Crops Research Institute for the Semi-Arid Tropics, Patancheru 502 324, Andhra Pradesh. P.K. Joshi is now Director, South Asia, International Food Policy Research Institute (IFPRI), New Delhi.

Nonetheless, there is evidence that the potential is high for agricultural diversification, from lower- to higher-value activities, to enhance income and employment for small farmers, at least in the short run (Barghouti et al., 2005; Joshi et al., 2004; Weinberger and Lumpkin, 2007). Many high-value agricultural activities, like vegetable cultivation, dairy, poultry, pig, and small ruminant husbandry, have short gestation periods, require low initial investments, and generate a stream of outputs that can be easily liquidated for cash. Besides, these activities are labor-intensive (Weinberger and Lumpkin, 2007; Joshi et al., 2006a). Weinberger and Lumpkin (2007) conclude that horticultural crops generate 1.2–6.0 times more income and require 1.6–5.4 times more labor than food grains. Small farmers have a larger endowment of labor in relation to land, and therefore, diversification towards high-value commodities can be an important opportunity for small farmers to utilize their surplus labor; with the possibility of substituting capital with human labor (e.g., manual weeding can replace use of herbicides; or mechanical harvesting, grading, and cleaning can be substituted with human labor).

On the demand side, there are significant opportunities for farmers to diversify their agriculture towards high-value food commodities. Rising per capita incomes and a fast-growing urban population[2] are fuelling rapid growth in their demand. Between 1990 and 2000, the per capita consumption of high-value food commodities (fruits, vegetables, milk, and meat) increased by 10–40 percent, against a decline of 5 percent in the consumption of cereals (Kumar et al. 2007). The factors underlying the demand for high-value food commodities have been robust in the recent past and are unlikely to subside in the near future, implying a further increase in their demand.

Agricultural diversification towards high-value commodities can enhance agricultural growth (Birthal et al., 2007; 2008) and make a significant contribution in reducing rural poverty. Empirical studies have shown that agricultural growth in India has contributed more to poverty reduction than the growth in other economic sectors (Ravallion and Datt, 1996; Warr, 2003). In this context, it should be mentioned that high-value food commodities (horticulture and animal products) account for close to half of agricultural incomes in India and have been growing faster compared to food grains (Birthal et al., 2008; Chand et al., 2009).

The question becomes whether small farmers benefit from the diversification-led growth? A commonly expressed fear is that they do not benefit from agricultural diversification since the production and marketing requirements of high-value crops are different than those for food grains. Small farmers may lack access to the capital, improved technologies, quality inputs, timely information, and support services that high-value agriculture needs. Further, most high-value commodities are perishable, and involve greater production and market risks, while small farmers are more risk-averse. Local rural markets are thin for these commodities, and the marketed surplus

---

2 Between 1990–91 and 2008–09, India's per capita income grew at an annual rate of above four percent; and the share of the urban population in total population increased from 25.7 percent in 1991 to 29.6 percent in 2009.

of individual producers is usually too small to be remuneratively traded in distant urban markets due to higher transportation and marketing costs. These constraints are more severe for farmers who live far from urban centers.

This chapter examines four important questions. First, is Indian agriculture diversifying from lower- to higher-value commodities? Second, does diversification contribute to agricultural growth, and if so, by how much? Third, is diversification-led growth inclusive? And fourth, what kind of technologies, policies, and institutions are required to foster agricultural diversification?

The chapter is organized into seven sections. The following section describes data and methodology. Section 3 discerns trends in agricultural diversification at national and regional levels. Section 4 discusses the growth effects of agricultural diversification. Section 5 investigates the distribution of benefits of agricultural diversification. Section 6 examines possibilities of linking small farmers with modern supply chains. Finally, Section 7 presents conclusions and policy implications.

Data and methodology

This chapter analyzes the dynamics of agricultural diversification and its contribution to agricultural growth for the period 1980–81 to 2007–08. This period is further divided into three sub-periods: 1980–81 to 1989–90, 1990–91 to 1999–2000 and 2000–01 to 2007–08. There are reasons to investigate the dynamics of diversification in these three periods separately: in the 1980s, the Green Revolution was at its peak and had spread throughout the country leading to broad-based growth of agriculture. In the early 1990s, the Government of India initiated a series of economic reforms that included the de-regulation of the agro-food processing industry, liberalization of agricultural markets, liberalized trade in agricultural commodities, and de-monopolized external trade from state control. Further, consumption patterns underwent a significant shift— out of staple cereals towards high-value food commodities. The program of economic reforms continued beyond the 1990s, but agricultural growth decelerated in the 2000s because of a number of factors, including the decline in public investment, slow growth in critical inputs, and increased weather uncertainty (GOI, 2010). These changes in consumption patterns, technologies, agricultural policies, institutions, and climatic conditions influenced the speed and contribution of diversification to agricultural growth.

India is a vast country with considerable heterogeneity in socio-cultural, economic, and agro-climatic conditions. This heterogeneity is likely to have influenced the nature, extent, and speed of agricultural diversification across regions. Therefore, the dynamics of agricultural diversification and its growth effects are investigated at the regional level. For the regional analysis, major Indian states, based on their geographical contiguity, were grouped into four broad regions: eastern, northern, western, and southern regions.[3]

The states within a region, by and large, are homogeneous in their agro-climatic characteristics, production patterns, and productivity levels. Most states in the eastern region have alluvial soils and high rainfall, and one-third of the total cropped area is irrigated. Rice is the main crop, but vegetables are widely grown. This region is heavily populated (its population density is almost twice the national average), has the lowest per capita incomes, and is the least urbanized. Northern states, except the hill states of Himachal Pradesh, Jammu & Kashmir, and Uttarakhand, have alluvial soils and a climate that ranges from semi-arid to humid. About three-fourths of the total cropped area is irrigated. Rice and wheat are the main crops in the plains; maize, wheat, fruits, and vegetables in the hills and mountains. Fertilizer consumption per unit of cropped area is among the highest of the regions. States in the western region have a semi-arid to arid climate and limited irrigation facilities. Agriculture is largely rain-fed; rice, wheat, sorghum, and pearl-millet are important crops. This region makes significant contributions to the total production of oilseeds and pulses. It is the least populated region, more urbanized, and has higher per capita incomes than any other region. The southern states, except Kerala, have a semi-arid climate. Agriculture is largely rain-dependent; the intensity of fertilizer use is the highest in this region. Rice is the dominant crop, followed by oilseeds, and pulses.

Data

The study uses data compiled from various published and unpublished sources and includes state-level data on the area, production, and yields of 42 crops were collected from various issues of the 'Indian Agricultural Statistics,' published by the Ministry of Agriculture, Government of India. Data on the value of output of these crops was compiled from published and unpublished sources of the Central Statistical Organization (CSO). Other important sources include the 'National Accounts Statistics' published by the CSO, and the 'Statistical Abstracts' of various states:

- cereals: rice, wheat, maize, sorghum, pearl millet, finger millet, barley, small millets;
- pulses: chickpea, pigeon pea, and other pulses;
- oilseeds: groundnut, sesame, rapeseed-mustard, soybean, linseed, sunflower, safflower, castor, and niger seed;
- fibers: cotton, jute, and sun hemp;
- spices: areca nut, cardamom, chilies, pepper, turmeric, ginger, garlic, and coriander;
- fruits: banana, cashew nut, and other fruits;

3 Eastern region comprises of Assam, Bihar, Jharkhand, Orissa, and West Bengal; northern region includes Haryana, Himachal Pradesh, Jammu & Kashmir, Punjab, Uttarakhand, and Uttar Pradesh; western region includes states of Chattishgarh, Gujarat, Madhya Pradesh, Maharashtra, and Rajasthan; and southern region consists of Andhra Pradesh, Karnataka, Kerala, and Tamil Nadu. Other states were not included in this classification because of the non-availability of time-series data.

- vegetables: potato, sweet potato, onion, tapioca, and other vegetables;
- beverages: tea and coffee; and
- sugarcane

The prices of different agricultural commodities were calculated by dividing their value output (at current prices) by their respective production levels. Data on output values of agricultural commodities at the state level for 2006–07 and 2007–08 were not available. These were generated by extrapolating commodity prices using their wholesale price indices. The output value of these crops was estimated by multiplying their production levels by their extrapolated prices. Current prices of different agricultural commodities were converted into real prices using the general wholesale price index of all commodities (1993–94 base) as a deflator. The data were de-trended applying the Hodrick-Prescott (HP) filter[4] with a modifying factor of 6.25.

The issue of small farmers' participation in high-value agriculture was examined using data extracted from a large-scale country-wide survey conducted in 2003 by the National Sample Survey Organization (NSSO) in its 59th round (GOI 2005). This survey had a specific focus on farming and farmers.

Decomposition of growth[5]

A change in the value of agricultural output or agricultural growth can come from any or all of the following sources: (i) changes in the total cropped area; (ii) changes in cropping patterns or diversification; (iii) changes in crop yields or technological change; and (iv) changes in the real prices of agricultural commodities. These sources are influenced by many policy and non-policy factors; hence their relative contribution to agricultural growth varies across time and space. For example, a change in the cropped area may occur because of changes in climatic conditions, population, urbanization, markets, infrastructure, and industrialization. Land reallocation decisions may be influenced by changes in domestic and world prices of different commodities, input requirements and their prices, domestic and international demand, service delivery, and extension programs. Changes in crop yields are a manifestation of the investment in agricultural research and development, technology, input use, irrigation, rainfall, mechanization, pest management, and agronomic practices. Likewise, the prices of agricultural commodities are influenced by domestic demand, international trade, domestic production, trade policy, world prices, taxes, and incentives.

To quantify the contributions of area, yield, prices, and land reallocation or diversification to agricultural growth, we followed the 'growth accounting approach' developed by Minot et al.

---

4 The Hodrick-Perscott filter is a data smoothening technique, commonly applied to remove short-term fluctuations from time series data. It generates a smoothened non-linear representation of a time series. The adjustment of the sensitivity of the trend to short-term fluctuations is achieved by applying a suitable adjustment factor.
5 This section draws heavily from Minot et al. (2006) and Joshi et al. (2006b).

(2006). Let A_i be the area under crop i, Y_i be its yield, and P_i be its price, then the gross revenue (R) from n crops ($i...n$) is:

$$R = \sum_{i=1}^{n} A_i Y_i P_i$$

(1)

Further, to quantify the effect of land reallocation or diversification, A_i is expressed as a share of crop i in the total cropped area as $a_i = A_i / \sum_i A_i$ and the Equation (1) can be re-written as:

$$R = \left(\sum_{i=1}^{n} a_i Y_i P_i \right) \sum_{i=1}^{n} A_i$$

(2)

The total derivative of Equation (2) provides changes in the gross value of output due to area, yield, prices, and land reallocation.

$$dR \cong \left(\sum_{i=1}^{n} a_i Y_i P_i \right) d\left(\sum_{i=1}^{n} A_i \right) + \left(\sum_{i=1}^{n} A_i \right) d\left(\sum_{i=1}^{n} a_i Y_i P_i \right)$$

(3)

The second term on the right-hand side of Equation (3) can be further decomposed from a change in sums to the sum of changes as:

$$dR \cong \left(\sum_{i=1}^{n} a_i Y_i P_i \right) d\left(\sum_{i=1}^{n} Ai \right) + \sum_{i=1}^{n} A_i \sum_{i=1}^{n} d\left(a_i Y_i P_i \right)$$

(4)

Further expansion of the term $\sum_{i=1}^{n} A_i \sum_{i=1}^{n} d(a_i Y_i P_i)$ in Equation (4) yields the following expression:

$$dR \cong \left(\sum_{i=1}^{n} a_i Y_i P_i \right) d\left(\sum_{i=1}^{n} Ai \right) + \sum_{i=1}^{n} A_i \sum_{i=1}^{n} a_i Y_i d\, P_i + \sum_{i=1}^{n} A_i \sum_{i=1}^{n} a_i P_i dY + \sum_{i=1}^{n} A_i \sum Y_i P_i da_i$$

(5)

Equation (5) decomposes growth due to changes in total cropped area, crop yields, prices, and diversification. Equation (5) is an approximation of the change in the gross revenue explained by area, yields, prices, and diversification as it does not contain 'interaction effect' of these variables. The first term on the right-hand side represents the change in gross revenues due to a change in the total cropped area. The expression $\sum_{i=1}^{n} a_i Y_i P$ is the weighted average of the gross revenue per hectare, the weights are the shares of each crop (a_i) in the total cropped area. The second term on the right-hand side captures the change in gross revenue due to a change in real commodity prices. The third term measures the change in gross revenue due to changes in crop yields or technology. The fourth term provides an estimate of the contribution of diversification to the change in gross revenue. Dividing both sides of equation (5) by the overall change in gross revenue (dR) provides us the proportionate share of each source in the overall change of gross revenue, or agricultural growth.

Trends in agricultural diversification

Table 2.1 tracks the changes in the composition of India's agricultural sector in the past three decades. The crop sub-sector constitutes over two-thirds of agricultural income and has continued to dominate the agricultural sector, even though its share has declined steadily over time. Livestock[6] is the next most important income source after crops, and its share in agricultural income has increased appreciably, from less than 17 percent in TE1982–83 to 25 percent in the 1990s and 2000s. Fisheries and forestry account for the rest of agricultural income. Fisheries gained marginally in terms of their income share in the 1990s, and forestry experienced a large decline. These changes in the agricultural sector, however, virtually stopped in the 2000s.

Changes in the sectoral composition of agriculture at the regional level, by and large, conform to the changes at the national (India as a whole) level (Table 2.1). Crops make up around two-thirds of the agricultural income in the eastern, northern, and southern regions, and as at the national level, their share in agricultural income has declined steadily. This trend was reversed in the western region, where the share of crops in agricultural income increased marginally from 68 percent in TE[7]1982–83 to 71 percent in TE2007–08. Livestock was the second largest source of income in all regions, but more so in the northern region where its contribution was nearly 29 percent. Its contribution was also higher in the southern region. Further, the share of livestock grew in all regions, but notably in the southern region where it increased from less than 14 percent in the early 1980s to close to 25 percent in the 2000s, primarily due to the remarkable growth of poultry production. The contribution of fisheries to agricultural income has been higher in the eastern and southern regions, given their proximity to the coast. The share of fisheries also increased, particularly in the eastern and northern regions because of an expansion of freshwater aquaculture.

Since the crop sub-sector has continued to dominate the agricultural sector, we closely examine the dynamics of agricultural diversification within the crop sub-sector. Table 2.2 shows the changes in the shares of crops (or crop groups) in the total cropped area and gross value of output. At the national level, the share of food grains (cereals and pulses) in the total cropped area declined steadily from 72 percent in TE1982–83 to 63 percent in TE2007–08, and in the gross value of output from 53 percent to 41 percent. The decline in the area and value shares of food grains was due to a decline in the shares of coarse cereals[8] and pulses, not of rice and

---

6 Separate estimates of Net Domestic Product (NSDP) from livestock at the state level are not available in the National Accounts Statistics. These are aggregated together with crops under the heading 'agriculture'. We have separated these in proportion to their value of production.

7 TE stands for triennium ending average.

8 Table 2.2 does not show estimated area and value shares of coarse cereals other than maize. These can be obtained deducting the shares of rice, wheat, and maize from the share of total cereals.

Table 2.1 | Changes in the composition of India's agricultural sector (% of net domestic product)

	crops	livestock	fisheries	forestry	total
India as a whole					6.7
TE1982–83	70.2	16.7	2.8	10.4	100
TE1992–93	69.7	22.0	3.1	5.1	100
TE2001–03	66.9	24.7	3.6	4.8	100
TE2007–08	67.4	24.5	3.5	4.6	100
eastern region	55	35	75	44	10.2
TE1982–83	72.3	15.8	4.6	7.3	100
TE1992–93	66.4	22.2	6.2	5.3	100
TE2001–03	67.9	20.6	6.8	4.7	100
TE2007–08	65.4	22.1	7.2	5.2	100
northern region					
TE1982–83	70.0	20.1	0.3	9.6	100
TE1992–93	71.3	25.3	0.7	2.7	100
TE2001–03	67.6	27.3	1.2	3.8	100
TE2007–08	65.5	28.9	1.4	4.1	100
western region					
TE1982–83	67.7	16.7	1.4	14.1	100
TE1992–93	68.6	22.4	2.0	7.0	100
TE2001–03	67.3	25.1	2.1	5.6	100
TE2007–08	70.9	22.6	1.7	4.8	100
southern region					
TE1982–83	71.5	13.7	5.4	9.4	100
TE1992–93	72.4	17.8	4.3	5.4	100
TE2001–03	64.7	24.9	5.4	5.0	100
TE2007–08	66.3	24.3	5.0	4.4	100

Note: Livestock also includes poultry.

Source: Ministry of Agriculture (2010).

wheat—the staple food crops. The share of coarse cereals, excluding maize, in the total cropped area declined from 19 percent in TE1982–83 to 10 percent in TE2007–08 and in the gross value of output from 7 percent to 3 percent. The share of oilseeds in both total cropped area and gross value of output increased drastically in the 1980s, but later, while their share in the total cropped area remained almost unchanged at 14 percent, their share in gross value of output declined from 14 percent in the early 1990s to around 10 percent in the 2000s. On the other hand, high-value crops (fruits, vegetables, and spices) consolidated their position, raising their share in gross value of output from 22 percent in TE1982–83 to 31 percent in TE2007–08, and in total cropped area from 4.2 percent to 7.6 percent. It is interesting to note that the value of output of horticultural crops is little more than the combined value of output of rice and wheat, and the value of livestock. These trends clearly demonstrate that Indian agriculture has been gradually diversifying towards horticulture and livestock (including poultry).

There are, however, considerable regional differences in the diversification patterns. Agriculture is most diversified in the southern region followed by the western region. In both regions, agriculture is rain-fed, and has been diversifying away from coarse cereals (sorghum, pearl millet, finger millet, and small millets) towards oilseeds, pulses, and horticultural crops. Oilseeds occupy about over one-fifth of the total cropped area in the southern and western regions. Maize has also gained in these regions, mainly because of its increasing demand as poultry feed. Incidentally, poultry production is concentrated in the southern and western regions which account for 40 and 25 percent, respectively, of India's gross value of poultry production. The area share of high-value crops increased from 7.7 percent in TE1982–83 to 10.8 percent in TE2007–08 in the southern region, and from a meager 1.6 percent to 4.7 percent in the western region. In both the southern and western regions, fruits have occupied a larger area compared to vegetables because of their comparatively low water requirements. In TE 2007–08, high-value crops contributed 37.8 percent to the gross value of output in the southern region and 25.5 percent in the western region.

In the northern region, crop composition is biased towards wheat and rice. Together, these crops have continued to occupy about half of the total cropped area in this region. Favorable pricing policies, assured procurement, the availability of high-yielding seeds, and better irrigation facilities were the main factors that motivated farmers to allocate more areas to rice and wheat. The area share of coarse cereals (excluding maize) declined from 10.0 percent in TE1982–83 to 6.9 percent in TE1992–93 and further to 5.4 percent in TE2007–08. There was also a drastic decline in the area share of pulses. Sugarcane, cotton, vegetables, and fruits are other important crops grown in this region. The high-value segment of agriculture expanded in this region; its share in the total cropped area almost doubled from 3.0 percent in TE 1982–83 to 5.9 percent in TE2007–08, and its share in gross value of output increased from 14.1 percent to 20.2 percent. The cultivation of high-value crops in this region is more prominent in the hill states of Himachal Pradesh, Jammu & Kashmir, and Uttarakhand. Other states, like Punjab, Haryana and Uttar Pradesh have a larger proportion of gross cropped area under irrigation, specialize in wheat and rice cultivation, have a high degree of mechanization in these crops, and diversify towards labor-intensive high-value crops. Wheat and rice are insulated from market risks given government procurement at assured minimum prices.

In the eastern region, agriculture is rice-based. Rice has occupied around 55 percent of the total cropped area in the past three decades. Vegetables are grown widely in this region, and their share in the total cropped area has almost doubled from 5.7 percent in TE1982–83 to 10.3

percent in TE2007–08. The area share of fruits increased from 1.7 percent to 2.7 percent in this period. It is interesting to note that high-value crops accounted for close to half of the total value of output of crops in TE2007–08, an increase from 36 percent in the early 1980s. A congenial climate and abundant labor supply explain the dominance of high-value agriculture in this region, which has the potential to emerge as an important hub for the cultivation of horticultural commodities. However, it remains below its potential due to poor infrastructure (mainly roads and markets).

In summary, Indian agriculture is steadily diversifying towards high-value food commodities, but not at the expense of staple food crops, rice, and wheat. Growth in the high-value commodities has been impressive in all regions—fuelled by increased domestic demand, and to some extent, supported by investment, incentives, and institutions that facilitated coordination between production, processing, and marketing. The National Horticultural Board was established in 1984 to promote the production, value addition, and exports of high-value crops, and the thrust continues with the National Horticulture Mission that was launched in 2005. The food processing industry was deregulated in 1991 to attract private investment, and the Agricultural Produce Marketing Committee (APMC) Act was amended to allow the establishment of private markets and direct procurement of farmers' produce by processing and marketing firms through contract farming or otherwise. However, the implementation of the amended Act by the state governments has been half-hearted. Restrictions on the inter-state movement of agricultural commodities were lifted in this period to encourage domestic trade. The rise in the number of supermarkets has facilitated the process of diversification by strengthening backward and forward linkages.

Contribution of diversification to growth

This section presents the decomposition of agricultural growth by crop and source—area under cultivation, prices, yields, and land reallocation or diversification—to quantify their contributions to growth. First, we look at trends in the value of output of different crops and their contributions to agricultural growth (at 1993–94 real prices). At the national level, agriculture grew at an annual rate of over 3 percent in the last three decades (Table 2.3). However, the trends for different crops are mixed. During the 2000s, growth in the output value of rice and wheat decelerated considerably; decelerated marginally for high-value crops; and improved for coarse cereals, pulses, oilseeds, sugarcane, and fibers. Nonetheless, the trends for high-value crops, oilseeds, and pulses were quite robust during this period, which provided a cushion to agriculture growth.

In the 1980s, agricultural growth was not centered on a few crops, but rather diverse crops. Cereals, oilseeds, and high-value crops each contributed 25–30 percent to overall agricultural growth. In the following decade, the contribution from high-value crops and cereals improved to 47 percent and 38 percent, respectively; oilseeds ceased to be a source of growth. During the 2000s, the contribution from cereals declined drastically to 15 percent and from high-value crops

| Table 2.2 | Changes in the area and value shares (%) of different crops in India |

period	rice area	rice VOP	wheat area	wheat VOP	maize area	maize VOP	total cereals area	total cereals VOP	pulses area	pulses VOP	oilseeds area	oilseeds VOP
India as a whole												
TE1982–83	22.6	23.4	13.1	13.2	3.4	2.2	57.6	45.5	14.3	7.2	9.4	9.9
TE1992–93	22.7	22.6	13.2	12.6	3.1	1.8	53.8	41.4	12.9	6.9	13.9	14.1
TE2002–03	23.3	20.9	14.5	13.9	3.6	1.9	53.0	39.7	11.9	6.0	13.7	10.3
TE2007–08	22.1	17.5	14.3	12.5	4.0	2.1	50.7	35.2	11.9	5.8	14.9	11.1
eastern region												
TE1982–83	55.9	36.1	7.2	5.4	3.6	1.8	69.0	44.0	11.6	6.0	5.5	4.3
TE1992–93	54.9	39.1	7.4	5.4	2.7	1.5	66.2	46.3	9.5	4.5	6.3	4.8
TE2002–03	56.1	35.2	8.4	5.1	2.7	1.4	67.8	41.9	6.0	3.1	4.7	2.9
TE2007–08	54.2	32.6	8.1	4.8	3.2	1.8	65.9	39.2	6.1	3.0	5.0	3.0
northern region												
TE1982–83	18.9	18.1	33.7	31.8	5.6	3.1	68.2	56.3	11.6	7.6	3.9	4.1
TE1992–93	20.7	20.6	35.4	30.4	4.5	2.2	67.5	55.4	9.2	6.4	4.7	5.4
TE2002–03	23.0	20.9	37.5	34.2	4.1	1.8	70.4	58.5	7.2	4.3	4.1	3.3
TE2007–08	23.2	18.9	37.3	31.5	3.9	1.7	69.8	53.9	6.6	3.5	4.7	3.7
western region												
TE1982–83	9.8	12.4	9.9	13.4	2.9	2.5	50.6	43.0	18.5	12.2	11.0	14.9
TE1992–93	10.2	10.2	9.4	11.5	3.1	2.1	46.6	34.1	17.3	12.7	17.6	22.5
TE2002–03	10.3	9.0	10.0	11.4	3.6	2.3	42.8	29.2	16.3	11.6	19.6	19.9
TE2007–08	9.4	7.0	10.4	10.5	3.7	1.8	40.1	25.1	16.2	10.3	21.3	20.5
southern region												
TE1982–83	24.0	29.2	1.0	0.4	1.5	1.2	49.8	38.0	10.9	2.7	16.2	15.7
TE1992–93	21.9	24.5	0.7	0.2	1.8	1.2	41.6	30.1	11.4	3.4	23.9	21.9
TE2002–03	21.6	22.5	0.8	0.2	3.8	2.1	39.7	27.8	13.5	4.0	21.3	14.1
TE2007–08	21.0	18.3	0.8	0.2	5.6	3.2	37.9	24.7	13.8	4.5	21.5	12.8

Note: VOP—value of output.

Source: Ministry of Agriculture (2010).

to 39 percent. In contrast, oilseeds, fiber crops (e.g., cotton), and sugarcane consolidated their shares during this period.

The trends and contributions of different crops to overall growth, however, vary across regions. In the eastern region, agriculture grew at a rate of 3.2 percent in the 1980s and 1990s, then decelerated to 2.9 percent in the 2000s. In the 1980s, rice had the largest share (42 percent) of growth, but was then gradually replaced by vegetables. In the 2000s, the share of vegetables in overall growth increased to 62 percent and the share of rice fell to 11 percent.

II/2

Table 2.2	Changes in the area and value shares (%) of different crops in India

period	fibers area	fibers VOP	beverages area	beverages VOP	spices area	spices VOP	fruits area	fruits VOP	vegetables area	vegetables VOP	sugarcane area	sugarcane VOP
India as a whole												
TE1982–83	5.2	4.5	0.3	1.3	1.0	2.3	1.3	8.8	2.0	11.0	1.8	8.3
TE1992–93	4.7	4.8	0.4	1.3	1.1	3.0	1.6	9.0	2.5	10.5	2.0	7.8
TE2002–03	5.2	3.9	0.5	1.6	1.2	4.0	2.1	12.7	3.3	12.7	2.3	7.8
TE2007–08	5.2	5.3	0.5	1.3	1.1	4.0	2.7	13.7	3.8	13.3	2.5	9.0
eastern region												
TE1982–83	3.2	2.4	1.0	3.6	0.8	2.2	1.7	7.6	5.7	26.8	0.8	2.6
TE1992–93	2.6	2.4	1.0	3.6	1.0	2.9	2.0	8.3	7.3	24.9	0.7	1.9
TE2002–03	3.0	2.5	1.3	4.6	1.0	3.5	2.5	10.3	9.0	29.5	0.5	1.3
TE2007–08	2.8	2.3	1.6	3.4	1.0	3.0	2.7	11.1	10.3	33.3	0.6	1.3
northern region												
TE1982–83	2.7	3.4	0.0	0.0	0.1	0.4	1.2	8.3	1.6	5.5	4.6	14.4
TE1992–93	2.9	4.9	0.0	0.0	0.1	0.3	1.8	7.0	2.2	6.5	5.1	13.9
TE2002–03	2.6	2.6	0.0	0.0	0.1	0.7	2.0	7.4	2.9	8.8	5.7	14.1
TE2007–08	2.6	3.4	0.0	0.0	0.2	1.1	2.2	8.3	3.6	10.8	5.9	15.1
western region												
TE1982–83	7.4	8.0	0.0	0.0	0.6	1.1	0.4	6.3	0.5	6.0	0.7	7.4
TE1992–93	6.5	7.3	0.0	0.0	0.8	1.9	0.6	8.4	0.7	6.3	0.8	6.2
TE2002–03	7.7	7.0	0.0	0.0	0.9	2.4	1.3	16.7	1.3	7.8	1.0	5.1
TE2007–08	7.6	9.7	0.0	0.0	0.9	2.3	2.1	15.7	1.7	7.4	1.4	8.6
southern region												
TE1982–83	5.3	3.7	0.9	2.1	2.8	5.6	2.9	12.9	2.0	8.2	1.5	7.7
TE1992–93	4.8	4.2	0.9	2.3	2.9	6.9	3.0	12.3	2.3	7.7	2.1	7.8
TE2002–03	5.2	3.3	1.3	2.6	3.3	10.1	3.8	16.4	2.9	8.9	2.6	8.7
TE2007–08	4.6	3.8	1.3	2.7	3.0	10.2	4.6	19.1	3.1	8.5	2.5	8.7

Note: VOP—value of output.

Source: Ministry of Agriculture (2010).

In the northern region, agriculture grew at a decelerated rate, from 3.2 percent in the 1980s to 2.6 percent in the 2000s. Rice and wheat accounted for a sizeable share of growth in the 1980s (57 percent) and 1990s (72 percent). Their share, however, dropped drastically to 16 percent in the 2000s, while high-value crops consolidated their position, as their share grew to 47 percent.

The southern region also experienced a decelerating trend in its agriculture, from over 3 percent in the 1980s and 1990s to 2.8 percent in the 2000s. In the 1980s, oilseeds were the fastest growing segment of agriculture, and contributed 44 percent to overall growth, yet their contribution subsequently fell drastically. Fruits, spices, and rice emerged as important sources of growth in the 1990s. In the 2000s, fruits and maize consolidated their share of growth. On the whole, the contribution of high-value crops to overall growth in agriculture almost doubled to 59 percent in the 1990s, but fell marginally to 52 percent in the 2000s.

Growth trends in the agricultural sector of the western region were the opposite of those in the other regions. Here, agriculture grew at an accelerated rate, from 3 percent in the 1980s and 1990s to 3.8 percent in the 2000s. In the 1980s, growth was fuelled largely by oilseeds (42 percent) and pulses (19 percent). In the subsequent decade, growth was propelled by fruits (32 percent) and wheat (15 percent). In the 2000s, growth sources diversified: 22 percent from oilseeds, 18 percent from sugarcane, 16 percent from cotton, and 15 percent from fruits. During this period, growth in the output value of fruits decelerated considerably: its aggregate share of high-value crops fell from 47 percent in the 1990s to 25 percent in the 2000s.

These results indicate that there are considerable regional differences in growth trends, where crops and their contribution to growth are concerned; and there have been significant changes in their patterns of growth and their relative shares of growth. These differences can be explained by the relative changes in area, yields, and prices of different crops. For example, rice and wheat accounted for a larger share of growth in the northern region in the 1980s and the 1990s, mainly because of widespread cultivation of improved varieties and the availability of good irrigation infrastructure. Likewise, oilseeds have, by and large, remained concentrated in the southern and western regions, and their higher contribution to growth in the 1980s can be attributed to favorable policies. In order to achieve self-sufficiency in edible oils, the Government of India launched a Technology Mission on Oilseeds (TMO) in 1984 that enhanced farmers' access to improved technologies, inputs, and support services. To harness its full potential, imports of edible oils were regulated through tariff and non-tariff means. The TMO lasted until the mid-1990s, and the country had largely succeeded in substituting edible oil imports with domestic production. Incidentally, the dismantling of the TMO coincided with India's opening up of its agricultural sector to external markets. The quantitative restrictions on imports of a number of agricultural commodities, including oilseeds and edible oils, were removed and tariffs were reduced. This resulted in an influx of imports of cheap edible oils, mainly palm oil. India now imports nearly half of its edible oil demand.

Table 2.4 presents the decomposition of agricultural growth in another way: by different sources, namely in terms of cultivated areas, yields, prices, and diversification. Yield improvements—a proxy for technological change—had been the main source of growth in Indian agriculture in the 1980s, when they contributed close to half to the overall growth. In the 1990s and 2000s the contribution of yields to agricultural growth declined to 37 percent.

The larger contribution of yields in the 1980s was an outcome of the investments in agricultural research and development during the Green Revolution. In the 1980s, there was a considerable

II/2

Table 2.3	Share of different crops to agricultural growth

	% annual growth in VOP			% share in overall VOP growth		
	1980s	1990s	2000s	1980s	1990s	2000s
India as a whole						
rice	2.9	2.6	0.4	21.3	18.1	1.9
wheat	2.3	4.6	1.9	10.4	19.0	6.8
maize	0.6	3.4	5.9	0.6	2.0	3.1
other cereals	-2.0	-1.0	3.9	-2.6	-1.2	3.0
pulses	4.0	1.5	3.5	8.5	2.3	5.2
oilseeds	7.3	-0.3	5.3	25.6	-1.5	13.7
fiber crops	2.4	2.1	10	3.9	2.5	11.3
sugarcane	1.7	3.3	6.8	3.8	8.2	14.4
beverages	4.8	4.8	0.0	2.0	2.1	0.0
spices	6.2	6.1	3.8	4.6	7.7	3.9
fruits	4.1	6.4	5.7	10.6	20.1	18.9
vegetables	3.3	5.5	4.9	10.7	18.9	16.2
other crops	1.1	4.9	5	0.6	1.7	1.6
all crops	3.2	3.2	3.1	100	100	100
eastern region						
rice	3.8	2.3	0.7	42.2	29.4	10.7
wheat	3.4	2.8	0.9	5.8	5.4	2.5
maize	-0.1	2.3	7.5	0.2	1.3	5.1
other cereals	-6.2	-5.0	-3.5	-0.9	-0.5	-0.2
pulses	1.9	-0.8	1.3	2.8	-2.7	1.6
oilseeds	8.0	-4.4	2.4	10.6	-7.5	3
fiber crops	3.4	3.4	0.8	3.0	2.5	0.7
sugarcane	0.3	-1.2	1.4	0.2	-1.3	0.5
beverages	4.9	3.8	-3.4	5.5	6.1	-4.9
spices	6.6	4.9	-1.1	4.7	4.4	0.8
fruits	4.1	4.2	3.9	9.2	14.6	17.3
vegetables	2.1	5.0	4.7	17.0	48.2	61.8
other crops	-1.9	1.3	8.4	-0.3	0.1	0.6
all crops	3.2	3.2	2.9	100	100	100
northern region						
rice	5.3	3.4	0.4	29.8	23.3	3.9
wheat	2.8	4.8	0.7	27.6	48.9	12.5
maize	0.3	1.5	0.6	0.3	1.0	0.5
other cereals	-2.2	1.1	3.4	-1.6	0.6	2.4
pulses	3.0	-1.4	-1.5	5.1	-2.9	-2.8
oilseeds	4.9	-2.4	4.7	6.6	-2.8	6.4
fiber crops	7.6	-4.8	7.8	8.0	-5.1	9.1

Table 2.3	Share of different crops to agricultural growth					

	% annual growth in VOP			% share in overall VOP growth		
	1980s	1990s	2000s	1980s	1990s	2000s
sugarcane	2.8	3.3	3.6	9.8	13.8	21.0
beverages	1.5	2.7	-1.8	0.0	0.0	0.0
spices	5.7	5.1	12.7	0.5	0.7	4.5
fruits	2.0	3.9	4.8	4.4	7.8	15.2
vegetables	5.4	6.1	6.6	8.6	13.9	27.4
other crops	26.8	10.1	-0.4	0.7	0.8	0.0
all crops	3.2	3.0	2.6	100	100	100
western region						
rice	1.0	1.7	2.2	5.0	5.1	2.5
wheat	0.8	4.6	5.5	5.5	15	8.5
maize	0.8	2.7	3.0	1.2	1.8	1.1
other cereals	-1.5	-1.5	0.6	-3.9	-4.3	0.4
pulses	4.7	2.8	4.9	19.0	8.8	7.8
oilseeds	7.3	2.1	7.9	42.2	11.7	21.9
fiber crops	-0.9	5.4	14.1	-0.1	10.2	16.1
sugarcane	-0.3	2.4	18.1	-0.2	4.1	17.7
beverages	0.0	0.0	0.0	0.0	0.0	0.0
spices	10.1	7.5	7.1	4.1	4.4	2.5
fruits	8.1	9.9	6.3	17.2	32.2	14.9
vegetables	5.6	5.4	6.3	10.3	10.5	7.1
other crops	-1.9	2.5	-6.1	-0.3	0.5	-0.4
all crops	3.0	3.1	3.7	100	100	100
southern region						
rice	0.9	2.5	-0.9	9.9	19.2	-6.7
wheat	-5.5	4.2	4.2	-0.4	0.3	0.3
maize	1.7	8.6	12.7	0.7	3.8	10.4
other cereals	-2.4	-0.4	2.6	-3.8	-0.5	2.4
pulses	6.8	3.6	5.5	6.4	4.1	8.0
oilseeds	7.8	-1.9	1.4	43.7	-10.2	5.3
fiber crops	2.6	2.0	5.9	4.1	2.6	7.3
sugarcane	1.9	4.9	3.2	4.0	13.0	9.4
beverages	4.8	6.0	4.1	3.3	4.0	3.2
spices	5.2	8.1	3.3	9.7	19.8	10.9
fruits	3.2	6.1	6.4	12.3	24.4	35.6
vegetables	3.1	6.2	2.2	8	14.5	5.2
other crops	1.6	5.2	6.5	2.1	5.1	8.7
all crops	3.1	3.0	2.8	100	100	100

Source: Ministry of Agriculture (2010).

II/2

| Table 2.4 | Contribution of diversification to agricultural growth (%) |

source	India as a whole			eastern region			northern region			western region			southern region		
	1980s	1990s	2000s	1980s	1990s	2000s	1980s	1990s	2000s	1980s	1990s	2000s	1980s	1990s	2000s
area	14.6	5.8	22.7	22.8	-18.1	15.8	8.1	19.3	13.1	11.6	14.1	24.1	24.7	-8.0	10.2
yield	48.5	36.5	36.9	29.8	46.9	48.3	61.3	27.2	39.2	49.7	17.7	25.1	43.4	47.7	45.1
price	9.5	23.1	10.7	14.3	34.1	4.6	3.7	33.8	22.5	8.1	10.8	11.0	9.4	16.3	3.2
diversification	27.0	33.4	28.3	33.5	35.6	40.8	26.6	18.7	25.0	29.5	56.9	39.8	20.5	42.1	38.6
interaction	0.4	1.2	1.4	-0.3	1.4	-9.4	0.4	1.0	0.3	1.1	0.5	0.0	2.0	1.8	2.9
total	100.0	100.0	100.0	100.0	100.0	100.0	100.0	100.0	100.0	100.0	100.0	100.0	100.0	100.0	100.0

Source: Ministry of Agriculture (2010).

increase in the use of modern inputs[9], such as improved seeds, chemical fertilizers, and electricity, that caused a rapid rise in crop yields. Rice and wheat yields grew at an annual rate of over three percent in the 1980s, but decelerated considerably in the 1990s (Annex 2.1). During the 2000s, growth in wheat yields decelerated further to 0.2 percent from 1.8 percent in the 1990s, while growth in rice yields remained almost stable at 1.6 percent. The deceleration in yield growth can be attributed to the slow growth of input use and irrigated areas.

Prices did not have any significant effect on agricultural growth, except in the 1990s when their contribution peaked at 23 percent, an increase from less than 10 percent in the 1980s. Their contribution in the 2000s was 10.7 percent.[10] Real prices of food and non-food commodities in the 1990s grew at 1.56 percent and 0.20 percent a year, respectively, as against 0.96 percent and 0.80 percent in the 1980s. During the 2000s, real prices of food as well as non-food commodities declined, only to recover after 2004–05.

Area expansion was not an important source of agricultural growth in the 1980s and 1990s. Its contribution to growth weakened in the 1990s, but improved in the 2000s. The higher contribution of area expansion to overall growth in the 2000s can be attributed to an increase in cropping intensities, which grew at an annual rate of 0.78 percent, against 0.49 percent in the 1980s and 0.42 percent in the 1990s.[11] It should be mentioned, however, that the scope to bring

---

9 In the 1980s, fertilizer and electricity consumption per ha increased at an annual rate of 7.5 percent and 14.5 percent, respectively. This decelerated considerably to 3.8 percent and 3.9 percent, respectively, in the 1990s. Growth in gross irrigated area accelerated marginally to 2.6 percent in the 1990s from 2.2 percent in the 1980s. In the 2000s, growth in electricity consumption and irrigated area decelerated to 2.2 percent and 1.5 percent respectively, and fertilizer consumption grew at an annual rate of 4.1 percent.
10 This chapter uses data up to 2007–08. In recent years, prices of food commodities have increased considerably and therefore the contribution of prices to overall growth is likely to be higher in the 2000s.
11 Cropping intensity was 123 percent in 1980–81, 129 percent in 1990–91, 133 percent in 2000–01, and 138 percent in 2007–08.

additional areas under cultivation is limited by the fixed supply of arable land. But the possibility exists to raise the intensity of land use, which is around 140 percent at present.

Diversification emerged as an important source of growth in the 1990s. Its contribution to overall growth increased from 27 percent in the 1980s to 33 percent in the 1990s, and remained at that level in the 2000s. Diversification occurred as farmers opted out of coarse cereals and pulses and moved towards vegetables, fruits, and spices. Together, the latter accounted for about two-thirds of the diversification-induced growth in the 1980s and 1990s, and 90 percent in the 2000s (Annex 2.2).

Growth sources, however, differ across regions, depending on cropping patterns, crop yields, and prices, as well as the trends associated with each of these factors. In the northern region, yield improvements or technology contributed over 61 percent to overall growth in the 1980s, then declined drastically to 27 percent in the 1990s, only to recover to some extent in the 2000s. A higher contribution of yields to agricultural growth in the 1980s was due to impressive gains in the yields of rice and wheat—the dominant crops in this region (Annex 2.1). Rice and wheat yield improvements alone contributed close to half of overall growth in agricultural production (Annex 2.2). In subsequent decades, yield growth of both rice and wheat decelerated considerably, and so did their contribution to overall growth.

In the 1990s, prices replaced yields as the main source of growth in the northern region. Their share of growth increased from 4 percent in the 1980s to 34 percent, due to a significant increase in the price of wheat, which grew at an annual rate of 1.8 percent, and contributed 17 percent to overall growth. Sugarcane, vegetables, fruits, and rice also contributed to the rising share of prices in overall growth. In the 2000s, the growth in wheat prices slowed to 0.4 percent and rice prices declined, lowering the share of prices in agricultural growth to 23 percent.

Diversification, with a share of around 25 percent, was another important source of agricultural growth in the 1980s and 2000s. The growth effect of area expansion was stronger in the 1990s—it contributed to roughly one-fifth of overall growth. Rice and wheat contributed over half of the growth due to area expansion, followed by sugarcane, fruits, and vegetables. These crops, with the exception of rice, also gained from land reallocation (diversification) during this period. In the 2000s, it was primarily vegetables that gained from diversification. Pulses, coarse cereals, and oilseeds lost area in the diversification process. High-value crops are labor-intensive and labor scarcity is one of the major constraints to their cultivation in the northern region.

The northern region has contributed significantly to national food security; yet, the continuous cropping of rice and wheat is degrading the region's soil and water resources. Future growth opportunities in this region will depend on a technological breakthrough in crop breeding for higher yields, conservation and a more judicious use of soil and water resources, and diversification of a production portfolio towards more remunerative crops.

Yield improvements contributed a larger share (50 percent) of growth in the western region in the 1980s. This was due to an increase in the yields of wheat, pulses, and oilseeds, which grew at an annual rate of two percent to three percent and together contributed one-third to overall growth. The growth effect of yield improvements, however, weakened considerably in the 1990s. During the 2000s, there was a significant improvement in the yields of cotton (due to *Bt* cotton), rice, and oilseeds, which enhanced the contribution of yields to overall growth to 25 percent.

In the 1990s, diversification emerged as the main source of growth in the western region. Its share of growth increased to 57 percent from 30 percent in the 1980s. During this period, land reallocation took place: mainly out of coarse cereals towards fruits and vegetables. Rapeseed-mustard, soybean and wheat benefited from land reallocation in the 1980s and 1990s. The speed of diversification slowed down in the 2000s, and its contribution to overall growth dropped to 40 percent. Nonetheless, fruits continued to occupy a larger share of diversification-caused growth. During the 2000s, close to one-fourth of agricultural growth came from area expansion, mainly under rice, wheat, oilseeds, pulses, and fruits. A number of factors could explain the higher levels of diversification towards fruits and vegetables in this region, but two important factors include the adoption of water-saving technologies (drip and sprinkler irrigation) and the emergence of producer cooperatives and associations for fruits that provided farmers with much needed access to markets (Birthal et al., 2007). The state of Maharashtra is at the forefront of these developments. For example, drip-irrigated areas in Maharashtra increased from a mere 500 ha in 1980 to over 0.4 million ha in 2008, about 60 percent of the country's total drip-irrigated area. The state took advantage of the central government's scheme of Export Processing Zones, particularly for grapes and onions.

In the eastern region, the yields of most crops have been lower compared to national averages. Nonetheless, there has been a continuous increase in the contribution of yield improvements to overall agricultural growth. In the 1980s, yield improvements accounted for 30 percent of growth, mostly from a significant increase in rice yields (Annex 2.1). The performance of other crops, like wheat, rapeseed-mustard, chickpeas, and pigeon pea also contributed, but at lower levels due to their smaller shares. The region accounts for nearly 50 percent of India's total vegetable area, but it experienced a declining trend in the mean yield of vegetables in the 1980s. The mean yield of fruit crops has also declined. Had the yield growth of these crops not decelerated, the growth effects of yield improvements would have been much stronger. The contribution of yields to overall growth improved in the 1990s and 2000s, most of which resulted from the yield increases of vegetables and fruits. As in the northern region, the contribution of prices to growth was very high in the 1990s—34 percent versus 14 percent in the 1980s and 5 percent in the 2000s; the larger contribution of prices in agriculture growth in the 1990s was associated with price rises of vegetables and pulses.

Diversification is emerging as an important source of growth in the eastern region. Its share in growth has steadily increased from one-third in the 1980s to 41 percent in the 2000s, mainly because of increasing emphasis on vegetables and fruits in the diversification process. Oilseeds and pulses were the main losers in this process. There are two important reasons for the higher share of horticultural crops in the overall growth of this region: a favorable climate (high rainfall and better seasonal distribution) conducive to the cultivation of vegetables and fruits; and the availability of sufficient human labor. Area expansion contributed 23 percent to overall growth in the 1980s. In the 1990s, however, the total cropped area declined at an annual rate of 0.45 percent, reducing growth by 18 percent. But, subsequently, its share in overall growth recovered.

Yields of most crops in the eastern region are low, and the deceleration in yield growth of important crops is a matter of concern. Agricultural development strategies should, therefore, focus on the introduction of improved technologies to narrow yield gaps, and on strengthening infrastructure and markets to sustain agricultural diversification and tap the growth potential of resources such as abundant labor and water.

In the southern region, technology has been an important source of growth, despite significant changes in the yield rates of different crops. The contribution of yield improvements to overall growth has remained around 45 percent in the past three decades. In the 1980s, there was a significant increase in the yields of rice, cotton, and spices, which accounted for most of the yield effect on agricultural growth. In subsequent decades, the growth of rice yields decelerated, and so did its contribution to growth. Improvements in the yields of coarse cereals, cotton, spices, fruits, and vegetables shared most of the yield effect in the 1990s and 2000s.

In the 1990s, diversification was an important source of growth in this region and remained influential in the 2000s. Its share of growth almost doubled from 21 percent in the 1980s to around 40 percent in the 1990s and following years. The nature of diversification, however, differed in these periods. In the 1980s, diversification was characterized by an area shift in favor of oilseeds, sugarcane, fruits, spices, and vegetables and away from cereals (including rice), cotton, and tobacco. In the 1990s and after, the tendency of diversification towards oilseeds and away from rice stopped; but the tendency to diversify into horticultural crops (mainly fruits) continued at the cost of coarse cereals (excluding maize, which gained in the process of diversification in the 2000s). Area expansion—in rice, oilseeds, and horticulture—contributed 25 percent to overall growth in the 1980s and 20 percent in the 2000s. The contribution of prices to agricultural growth, however, has been fairly small. An emphasis on the development of watersheds, markets, and institutions helped sustain crop yields and crop diversification towards more remunerative crops (Joshi et al., 2006b).

Diversification has emerged as an important source of growth in Indian agriculture in the past two decades. This can be attributed to the rapid growth in demand for high-value food commodities like fruits and vegetables that increased at an annual rate of 4.5 percent in the 1990s, compared to 3.0 percent in the 1980s (in the 2000s, it decelerated to 2.5 percent (FAOSTAT)). Demand-driven growth was supported by investments in public infrastructure (roads and markets) and favorable policies: (i) the food processing industry was deregulated in 1991; (ii) the limit to foreign direct investment (FDI) in food processing was gradually raised to 100 percent; (iii) agricultural markets were liberalized to allow private investment in markets; and (iv) it became easier for agro-processors to procure farm produce directly from producers outside the state-regulated agricultural markets, through institutions like contract farming or otherwise.

Each source of agricultural growth—from increases in areas, yields, prices, and diversification—is influenced by various policy and non-policy factors, and has separate implications for policy, technology, and institutions. If the area effect on growth is trivial, this indicates the limited scope of bringing additional areas under cultivation. There is, however, a possibility of enhancing total cropped area through the intensification of land use. For example, about 60 percent of 142 million ha of net cropped areas in India remains fallow in the post-rainy season; this land can be brought under cultivation through investments in irrigation, infrastructure, and agricultural research and development.

Prices are influenced by domestic and trade policies, and therefore, price-driven growth is neither sustainable, nor desirable because of its inflationary and distributional effects. In the case of India, government-set, 'minimum support' prices of rice and wheat have influenced agricultural growth more than any other commodity. Furthermore, small farmers do not benefit much from price increases of these crops since the benefits accrue in proportion of the marketed surplus, which is minimal with small farmers. On the other hand, the prices of high-value food commodities are market-determined. Nonetheless, given the crucial role of prices in land reallocation and in balancing supplies and demands for various agricultural commodities, it is important to have appropriate pricing policies and regulatory mechanisms in place.

In the long term, agricultural growth must emanate from technological change and diversification. Our results show that despite a decline in its contribution, technology has been the main source of growth in Indian agriculture. The diminishing role of technology on agricultural growth is a matter of concern that needs to be addressed. The slowdown in the contribution of technology could be due to a number of factors, such as the underinvestment in agricultural research, the inefficiency in delivery systems, slow growth in input use, the degradation of natural resources, and climate change, among others. Such factors need to be further studied to better understand their roles in yield growth. In the short term, there is an opportunity to accelerate yield growth and raise agricultural production by bridging the gap between actual and obtainable yields (Aggarwal et al., 2008; Murty et al., 2007) through improvements in input use efficiency, seed replacement rates, input delivery systems, etc. However, for a sustained shift in yield frontiers, the need for greater investments in agricultural research cannot be overstated. India spends about 0.5 percent of its agricultural gross domestic product on agricultural research, which is lower than the 0.7

percent average for developing countries and the 2–3 percent average for developed countries (von Braun et al., 2008).

The growth in high-value crops has resulted primarily from an increase in areas under their cultivation, not through yield increases. At the national level, the mean yields of fruits and vegetables during 1980–81 to 2007–08 grew at paltry annual rates of 0.44 percent and 0.06 percent, respectively. This implies the need to raise investments in research and for the development of horticultural crops. Apart from research, sustaining agricultural growth through increased diversification will require a host of factors, such as speeding up investments in public infrastructure (like roads and electricity, that generate widespread benefits) private investing in agribusiness-related activities, and encouraging institutions to remuneratively link farmers to markets. Fan et al. (2007) have identified investment in agricultural research and roads as important drivers of agricultural growth and poverty reduction in India.

Small farmer participation in high-value agriculture

The analysis of the sources of growth has demonstrated that diversification towards high-value commodities is an important and sustainable source of growth. Diversification towards these commodities is demand-driven, and is likely to remain so, as the underlying factors (income growth and urbanization) have been robust in the recent past and are unlikely to subside in the near future. The question then is: can small farmers benefit from diversification-led growth in agriculture?

Indian agriculture is dominated by small farmers, as mentioned at the outset of this chapter. Small land holdings (≤2.0 ha) comprise 86 percent of total land holdings and occupy 44 percent of the land area (Table 2.5). Their average size is small (0.53 ha) and is likely to further decline given the limited scope for land redistribution.

High-value agriculture is more remunerative and labor-intensive, but the capability of small farmers to diversify towards these is often doubted on several counts. First, the average size of small farmer landholdings may be too small to permit them to divert more land from staples towards high-value crops at the cost of their household food grain security. Second, poor access to capital, technology, inputs, and information may inhibit their participation in the production of high-value commodities (Birthal et al., 2007). Third, most high-value commodities are perishable, and therefore have greater production and market risks, which risk-averse small farmers are reluctant to bear. Fourth, their scale of production, hence their marketed surplus, may be too small to be remuneratively traded in the distant urban markets because of higher transportation and transaction costs (Birthal et al., 2005). Fifth, small farmers may be excluded from supply chains given their low volumes, the difficulty to comply to food safety standards, and the higher management costs associated with dealing with large numbers of small-scale farmers. Even though the agri-food marketing system in India is gradually changing from an open marketing

| Table 2.5 | Changes in the size and distribution of land holdings in India | | |

farm category	1980–81	1991–92	2002–03
% share in holdings			
small (≤2.0 ha)	75.3	80.6	86.0
medium (2.0-4.0 ha)	14.2	12.0	9.0
large (>4.0 ha)	10.5	7.4	5.0
total (million no.)	71.0	93.5	101.3
% share in area operated			
small (≤2.0 ha)	28.1	34.3	43.5
medium (2.0-4.0 ha)	23.6	24.1	22.5
large (>4.0 ha)	48.4	41.6	34.0
total (million ha)	118.6	125.1	107.7
% share of leased-in area	7.18	8.52	6.5
average size of holding (ha)			
small (≤2.0 ha)	0.62	0.57	0.53
medium (2.0-4.0 ha)	2.77	2.69	2.66
large (>4.0 ha)	7.69	7.53	7.23
all holdings	1.67	1.34	1.06

Source: GOI (2006).

system to a vertically coordinated system that will benefit farmers, the fear is that small farmers will be left out.

In order to comprehend the distributional impacts of diversification, we compare the cropping patterns for different categories of farmers and their contribution to the total production of fruits, vegetables, and spices. Table 2.6 shows the cropping patterns[12] for small, medium, and large farms. Three observations stand out prominently: (i) Small farmers allocate a larger proportion of their land to high-value crops. Fruits, vegetables, and spices together occupy 5.7 percent of the total cropped area, in contrast to 4.6 percent on medium farms and 3.0 percent on large farms. Using data from various rounds of the Agricultural Census, Birthal et al. (2008) report an increasing tendency among small farmers to diversify towards high-value crops. (ii) Small farmers appear to have a comparative advantage in the production of vegetables, rather than fruits and spices. This makes sense because vegetables generate quick and regular returns, require more labor and less capital—all factors that match their resource endowments. On the other hand, most fruit crops and certain spices (areca-nut and cardamom, for example) require more start-up capital

---

12 Area shares of different crops estimated using survey data from NSSO are suggestive of the cropping patterns on different categories of farms and may not necessarily be the same as obtained using data from other published sources.

and have longer gestation periods, which discourage small farmers to grow such crops. Birthal et al. (2007) found that small farms were significantly constrained by the limited access to credit in the cultivation of fruit crops. (iii) Even though small farmers allocate a higher share of their areas to high-value crops, they also allocate a larger proportion of their area to rice and wheat than do medium and large farmers. This indicates that small farmers do not risk their household food grain security when they decide to reallocate land to high-value crops. The cultivation of oilseeds and pulses is more prominent on large farms because these are not as labor-intensive as high-value crops and large farmers are relatively labor-constrained.

Table 2.6	Cropping pattern on different size farm categories in 2003 (% of gross cropped area)			
crop	small	medium	large	all
paddy	35.0	24.7	13.4	25.8
wheat	19.0	15.6	12.8	16.3
coarse cereals	16.5	19.4	18.3	17.7
pulses	8.0	11.0	14.6	10.8
oilseeds	7.2	12.0	14.0	10.4
fruits	1.2	1.4	0.9	1.2
vegetables	3.5	2.0	1.0	2.4
condiments and spices	1.0	1.2	1.0	1.1
sugarcane	2.6	3.5	2.5	2.8
cotton	2.1	4.8	15.4	6.9
other crops	3.9	4.4	6.0	4.7
total	100.0	100.0	100.0	100.0

Source: GOI (2005).

Small farmers also make a larger contribution to the production of high-value crops: they contribute to as much 70 percent of the total production of vegetables, 55 percent of fruits, and 49 percent of spices, as compared to their share of 44 percent in total land area. Their share in cereal production is also significant: 62 percent in rice, 54 percent in coarse cereals, 52 percent in wheat, and 69 percent in milk production. These figures indicate that small farmers, despite being more diversified towards high-value commodities, are also the custodians of India's food security.

The other issue that merits attention in the discussion of distributional impacts of diversification is the efficiency of small farms versus larger farms. A number of past studies concluded that

II/2

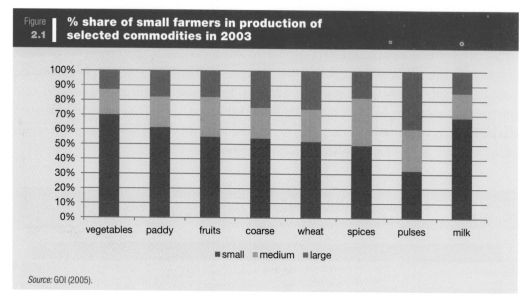

Figure 2.1 | **% share of small farmers in production of selected commodities in 2003**

■ small ■ medium ■ large

Source: GOI (2005).

small farmers are more efficient than large farmers, and attributed the efficiency gains to the labor advantage that small farmers have over large farmers (Bhardwaj, 1974; Bhalla, 1979; Bagi, 1981; Rao and Chotigeat, 1981). Fan and Chang-Kang (2005) reviewed the size-productivity relationships in a number of Asian countries and also found that small farms are more efficient than large farms. Most of these studies, however, have focused on cereals—mainly rice and wheat.

Table 2.7 compares gross returns[13] per unit of land for different crops across farm categories. On average, high-value crops generate more than Rs. 30,000 per ha, which is twice the gross revenue from rice and wheat, 2.5–3.0 times larger than from oilseeds, 4.0–4.5 times more than from pulses, and 5.0–6.0 times more than from coarse cereals. A further comparison across farm categories shows that—compared to large farms—the gross returns on small farms are 75 percent more for fruits, 36 percent more for spices and 12 percent more for vegetables. This shows that small farmers are more efficient in the production of high-value crops and coarse cereals, pulses, and oilseeds.

Gross returns from rice, wheat, sugarcane, and cotton are less on small farms than on medium and large farms, plausibly due to increased mechanization of agricultural operations with improved efficiency. On the other hand, high-value crops require more human and animal labor, making mechanization less important. Collins (1995) states that "many high-value crops are highly responsive to constant and careful monitoring of plant health; careful weeding, pruning and irrigation; harvesting based on assessments of when individual pieces of fruit and vegetables are ripe; and careful, efficient handling". These findings indicate that although small farmers might have

---

13 Gross returns are defined as the value of output of the main products and by-products.

Table 2.7	Gross returns on different size farm categories in 2003 (Rs./ha)			
	small farmers	medium farmers	large farmers	all farmers
paddy	14,751	14,940	16,972	15,163
wheat	16,798	17,939	16,908	17,068
coarse cereals	7,066	5,365	4,830	5,912
pulses	8,027	7,922	7,153	7,623
oilseeds	13,498	11,267	11,008	11,861
fruits	38,920	30,442	22,292	32,293
vegetables	31,381	27,647	28,072	30,230
spices	33,997	34,722	25,001	31,455
sugarcane	36,258	37,223	37,262	36,822
cotton	14,824	14,854	27,365	23,753

Source: GOI (2005).

lost their comparative advantage over large farmers in the production of certain crops like rice, wheat, sugarcane, and cotton, they are still more efficient in the production of high-value crops.

Diversification towards animal husbandry is another opportunity for small farmers to enhance their incomes and escape poverty. Livestock generates a continuous stream of outputs, which can be used by small farmers to earn cash to meet their daily food and other requirements. Animal husbandry requires little start-up capital, and as reproducible assets, livestock adds to the wealth and income of farm households. Livestock production requires more labor, allowing small farmers to utilize their available labor. Livestock income is more stable than crop income, and provides a cushion against the income shocks of crop failure. There is evidence that livestock could be an important pathway out of poverty for small farmers in developing countries (Datt et al., 2000; Dolberg, 2003; Kristjanson et al., 2004; Birthal and Taneja, 2006; Ojha, 2007).

More than two-thirds of all farm households in India are engaged in livestock production (Birthal et al. 2009); the share of livestock in agricultural income has increased from 17 percent in TE1980–82 to 25 percent in TE2007–08. The value of output of livestock is as high as the value of food grains. The demand for animal food products has also been growing rapidly (Kumar et al., 2007). This indicates the existence of ample income earning opportunities for farm households in the livestock sector.

Livestock is more equally distributed than land and is largely concentrated on small farms. In 2002–03, small farmers accounted for 74 percent of the country's total cattle, 71 percent of buffaloes, 78 percent of small ruminants, 89 percent of pigs, and 86 percent of rural poultry (Table 2.8). Interestingly, small farmers consolidated their position in livestock production. Between 1981–82 and 2002–03, their shares in different livestock species increased by 12–25 percentage points. Birthal (2008) concluded that small livestock producers are as efficient as—if not more than—large livestock producers.

Table 2.8	Share of small farmers in livestock population (%)		
	1981–82	1991–92	2002–03
cattle	51.0	71.3	74.2
buffaloes	50.6	57.5	70.8
small ruminants	55.3	65.5	77.7
pigs	76.4	70.3	88.6
poultry	66.7	73.9	80.6

Source: GOI (various years).

Despite the strong position of small farmers in the livestock sector, the scale of their production is smaller than that of large farmers. For example, in 2002–03 the average size of a dairy herd was 1.9 on small farms, 3.8 on medium farms, and 5.4 on large farms (Birthal et al., 2006). Birthal (2008) found that as dairy producers, 68 percent of marginal farmers (≤1.0 ha) and 61 percent of small farmers (≤2.0 ha) produced less than 1,000 liters of milk per household per annum. Production on such a small scale provides some nutritional benefits for small farm households, but cannot provide any surpluses for the market. Nonetheless, small farmers comprised 54 percent of farm households that produced more than 5,000 liters per household per annum. This suggests that small farmers have the potential to scale up dairy production if they can overcome particular production and marketing constraints.

Given the concentration of high-value commodities on small farms and a negative size-productivity relationship, small farmers do not seem to be at a disadvantage in the process of agricultural diversification towards these commodities; their labor and supervision cost advantages compensate for the disadvantages of higher marketing and transaction costs, and limited access to credit and information. Income and employment benefits of diversification-led growth are thus likely to be favorable for small farmers. These effects will be realized beyond the farm-gate, and will be larger if multiplier effects are considered. Labor demand of post-harvest activities, like transportation, packing, sorting, grading, and cleaning—all labor-intensive chores—increases (Weinberger and Genova, 2005), as does labor demand in agro-processing and agri-input industries.

Ravallion and Datt (1996) and Warr (2003) have shown that agricultural growth has a larger effect on poverty reduction than the growth in other economic sectors. The share of high-value commodities (horticultural and animal products) in the total output of India's agricultural pro-duction has been increasing consistently and at a rate faster than most other agricultural com-modities. These commodities account for about half of the value of agricultural output. Growth in the high-value segment of agriculture is likely to have contributed to rural poverty reduction. To substantiate this, we plot the relationships between agricultural productivity, the area share of fruits and vegetables, and the head-count poverty ratio at the district level (Figures 2.2a and 2.2b). The association between the area share of high-value crops and agricultural productivity is, as expected, positive (Figure 2.2a), which indicates that agricultural productivity is higher in the districts that have a larger share of the cropped area under cultivation of fruits and vegetables. Figure 2.2b plots the rural head-count poverty ratio against agricultural productivity; the associa-tion between the two is negative, indicating that diversification towards high-value crops has the potential to contribute to poverty reduction.

Growth in livestock production is also more pro-poor (Kristjanson et al., 200; Birthal and Taneja, 2006; Ojha, 2007). Birthal and Taneja (2006) plotted the growth rates of the rural head-count poverty ratio against the growth rates in the value of output of livestock and of crops for the major Indian states for the period 1984–1997 and observed that both crops and livestock had a favorable effect on poverty reduction. Poverty reduction, however, was more responsive to growth in livestock production. In a study of 600 households in Uttar Pradesh, about 23 percent of households were reported to have escaped poverty through the livestock route (Ojha, 2007). Mellor (2003) observes that the growth in livestock production has a larger effect on poverty reduction than similar growth in crop production. Faster growth in livestock production is likely

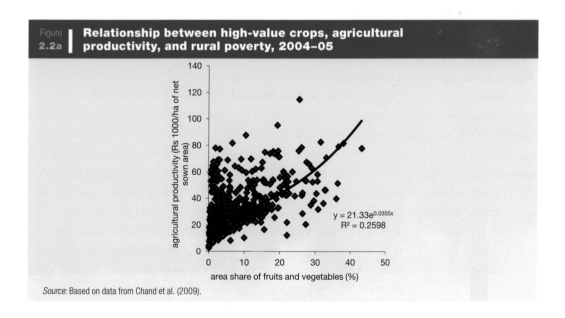

Figure 2.2a | **Relationship between high-value crops, agricultural productivity, and rural poverty, 2004–05**

$y = 21.33e^{0.0355x}$
$R^2 = 0.2598$

agricultural productivity (Rs 1000/ha of net sown area)

area share of fruits and vegetables (%)

Source: Based on data from Chand et al. (2009).

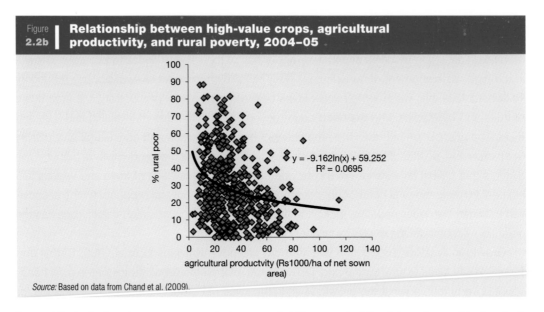

Figure 2.2b | Relationship between high-value crops, agricultural productivity, and rural poverty, 2004–05

$y = -9.162\ln(x) + 59.252$
$R^2 = 0.0695$

% rural poor

agricultural productvity (Rs1000/ha of net sown area)

Source: Based on data from Chand et al. (2009).

to contribute to female empowerment since nearly 70 percent of the labor demand in livestock production is met by women (Birthal and Taneja, 2006).

These results provide only an indication of the pro-poor potential of agricultural diversification. Besides diversification, there are other factors that contribute to poverty reduction; hence, the pro-poor effects of diversification need to be explored in more depth, using an appropriate econometric framework with due consideration given to the role of other factors.

Linking small farmers to markets

Our previous findings suggest that, despite scale limitations, small farmers do participate in the production of high-value commodities, are more efficient in their production, and make a significant contribution to their total production. Increased demand for high-value food commodities becomes an opportunity for small farmers to diversify towards these commodities. But many fear that small farmers may be displaced by large commercial producers in the marketplace. Small farmers usually have small marketed surpluses, rural local markets for high-value commodities are thin, and trading in distant urban markets may not be profitable due to high marketing and transaction costs. Although markets for agricultural commodities, including high-value commodities, are fairly well-developed in India, supply chains are long and inflate marketing and transaction costs for farmers, especially small farmers. Birthal et al. (2005) found that marketing and transaction costs are as high as 21 percent of the total cost of production milk and vegetables, and even 25 percent higher for small farmers. Gandhi and Namboodiri (2002) reported that farmers received one-third to one-half of the final prices of fruits and vegetables. Market and price risks are also greater for high-value commodities; prices of these commodities are volatile and fluctuate widely—sometimes on the same day depending upon the quantities and arrival times in the market. For commodities like broiler chickens, there is considerable seasonal price variability. The lack of assured and remunerative markets slows the speed of agricultural diversification.

With increasing dietary diversification and concerns for food safety, traditional marketing systems dominated by ad hoc transactions and intermediaries are gradually transforming into coordinated systems, like cooperatives, producers' associations, and contract farming. The corporate sector is entering the food retailing business. The new marketing systems are expected to improve market efficiency and accelerate the speed of agricultural diversification. But there is some doubt that small-scale producers will be included in these market-driven supply chains. Agribusiness or marketing firms often prefer a few large producers who can supply substantial volumes and comply with a firm's food safety and quality standards so as to avoid the higher transaction and supervision costs associated with contracting many small-scale producers (Dev and Rao, 2004; Kumar, 2006). It is argued that agribusiness firms may exploit small-scale producers because of their monopsony position in output markets and their monopoly in the input markets (Singh, 2002). The counterargument, however, states that reliance on a few large producers is potentially risky, especially when a processor does not have alternative supply sources. Contracting with a large number of small-scale producers distributes, and thus lessens the supply risk (Birthal et al., 2005).

There are a few examples of successful institutional innovations that remuneratively link small-scale producers to markets, such as dairy cooperatives. By providing an assured market to producers, they enhance farmers' incomes and contribute to growth in the dairy sectors of many states (Birthal 2008). Dairy cooperatives provide inputs, information, and technical support to producers, which eventually lead to improvements in production efficiency and a reduction in marketing and transaction costs. In a case study of dairy cooperatives in Punjab, members of dairy cooperatives realized 29 percent higher profits than those who sold their milk in the open market (Gupta et al. 2006). In 2009, dairy cooperatives procured 9.2 million tons of milk from 13.9 million farmer members.

With the liberalization of domestic markets, contract farming is emerging as an important means of linking farmers to markets. Birthal et al. (2005) quantified the benefits of contract farming in milk in Punjab and vegetables in peri-urban Delhi and found that contract farmers realized 78–100 percent more profits, faced about 90 percent less marketing and transaction costs, and received 4–8 percent higher prices over non-contract farmers. In the case of high-investment, high-risk commodities, like broilers, farmers use contracts as a means of shifting risk and accessing credit. Ramaswami et al. (2006) found that by using contracts, broiler producers in Andhra Pradesh could transfer as much as 88 percent of the market risk to the agribusiness firms. (It should be noted that in most these studies, yield rates were reported to be higher by 5–10 percent under contract farming). In the case of export-oriented grape production in Maharashtra, Roy and Thorat (2008) found that contract farmers realized 11 percent higher yields, 9 percent higher prices, and 60 percent higher profits than non-contract farmers. Studies on contract farming with a focus on commodities (e.g., potatoes in Haryana (Tripathi et al., 2005), hybrid cucumbers in Andhra Pradesh (Dev and Rao, 2004), tomatoes in Punjab (Rangi and Sindhu, 2000), Basmati rice in Punjab (Sharma, 2007) and cotton in Tamil Nadu (Agarwal et al., 2005)) reported that contract farmers realized significantly greater profits because of higher yields and higher prices.

This evidence suggests that farmers benefit from contract farming and cooperatives. The question becomes: are small farmers involved in contract farming? The evidence is mixed. Birthal et al. (2005) found significant involvement of smallholders in the contract farming of milk (56 percent) and vegetables (51 percent), but not in the case of broilers (32 percent). In gherkin contracts in Karnataka, Erappa (2006) reported that 50 percent of small farmers had contract arrangements. Roy and Thorat (2008) found that the distribution of contract grape producers was biased towards the lower end of land distribution. On the other hand, Kumar (2006) found that only 15 percent of contract farmers of various crops in Punjab were smallholders. In the case of contract farming of seed cereals in Haryana, about 30 percent of farmers were small farmers (Kumar et al., 2005). Deshingkar et al. (2003) concluded that small farmers are more involved in contract farming of labor-intensive crops.

Conclusions and implications

This chapter shows that Indian agriculture has been gradually diversifying out of staple crops towards horticultural and animal food products. Their share in the value of output of agricultural production (including animal husbandry and fisheries) is now close to 50 percent, 17 percentage points higher than in the early 1980s. This transformation of the production portfolio along with technological change has contributed to agricultural growth in recent decades. The share of horticultural crops in overall growth of the crop sub-sector increased from 26 percent in the 1980s to 47 percent in the 1990s but declined to 39 percent in the 2000s. Most of the increase in their share of growth came from diversification of land away from less profitable crops, mainly coarse cereals and pulses. As such, diversification contributed one-third to agricultural growth in the 1990s, marginally more than in the 1980s and the 2000s. There is, however, considerable regional variation in the trends and contribution of diversification to agricultural growth. In the western region, diversification was the main source of growth in the 1990s. Its share was also large in the southern region and improved over time. In the 2000s, diversification contributed about 40 percent to agricultural growth (except in the northern region). Technology contributed almost 50 percent to agricultural growth in the 1980s but fell to 39 percent in the 1990s and later. In the northern and southern regions, the contribution of technology to growth declined in the 1990s but recovered slightly afterwards. In contrast, technology emerged as an engine of growth in the eastern region, which lagged behind in agricultural development in the 1970s and 1980s.

Technology and diversification were more important as sources of growth than area expansion and prices.

This chapter also examines whether diversification-led growth was inclusive, and found that, contrary to common belief, small farmers do not benefit from diversification-led growth due to operational constraints—lack of access to markets, higher marketing and transaction costs associated with small marketable surpluses, and limited access to credit and information. However, they do not seem to be at a disadvantage in the process of agricultural diversification from lower- to higher-value commodities. Their participation rate in the production of high-value commodities is greater. Small farmers are more efficient in production, as their labor and supervision cost advantages compensate for the disadvantages of higher marketing and transaction costs, and limited access to credit and information. They contribute as much 70 percent to the total production of vegetables and milk, 55 percent to that of fruits, 62 percent to that of rice, and more than 50 percent to that of wheat and coarse cereals. This suggests that small farmers might augment their incomes with high-value crops but they do not undermine their household's food grain security. A review shows that farmers benefit from modern supply chains, driven by institutions like cooperatives and contract farming, in terms of higher yields, higher prices, lower production and marketing costs, and risk sharing. Evidence shows that small farmers are not altogether excluded from these supply chains. The income and nutritional outcomes of diversification-led growth are thus likely to be favorable for small farmers.

Technology remained the main source of growth in the past three decades but its declining share in growth presents a serious concern. The relative decline in its importance would be less alarming if the growth in yields were steady. However, the yield growth of most crops decelerated in recent years. This development could be due to a number of factors, such as the underinvestment in agricultural research, the reduced efficiency of research and extension, the degradation of natural resources, and climate change, among others. Given the wealth of studies on the high rates of return to investment in agricultural research and development, it is likely that underinvestment in agricultural research continues. Total factor productivity growth for most crops decelerated (Kumar et al., 2008). India spends only about 0.5 percent of its agricultural gross domestic product on agricultural research and extension, which is less than the average of 2–3 percent for developed countries and the average of 0.7 percent for developing countries (von Braun et al., 2008). This implies a need to: (i) increase investment in agricultural research and extension for an upward shift in yield frontiers; (ii) revisit the research agenda by examining the emerging opportunities and challenges—changes in consumption patterns, consumer preferences (quality, safety, and tastes), and food and feed grain demand—in different regions; and (iii) bridge the yield gap by enhancing farmers' access to improved seeds, quality inputs, crop management technologies/practices, and by improving the efficiency of delivery services. Research on horticultural crops merits special attention since most horticultural crops have experienced only modest yield increases. As small farmers proportionally devote larger areas to horticultural crops and are more efficient in their production, investments in horticultural research would have a larger effect on their incomes.

Diversification from lower- to higher-value commodities is an important avenue to foster rapid growth in agriculture, and also an opportunity for small farmers to enhance their incomes and escape poverty—especially in the regions that were somewhat neglected during the Green Revolution. The demand for high-value food commodities is likely to grow faster as the factors underlying demand growth (income growth and urbanization) had been quite robust in the recent past and are unlikely to subside in the near future. However, many studies indicate that the potential growth in high-value agriculture may be constrained due to the lack of investment in cold storage and refrigerated transportation, the poor flow of institutional credit to farmers and food industries, higher taxes and excise duties on processed foods, multiple laws governing the food industry, and the poor implementation of the new marketing act (Model Act, 2003) that permits transactions in food commodities outside the state-regulated markets (Birthal et al., 2005; Birthal et al., 2007; Birthal and Taneja, 2006; Landes, 2008). In the last few years, there has been some progress to reduce policy barriers to diversification. Harnessing the emerging opportunities in diversification would require: (i) increased investments in public infrastructure (roads, electricity, and communication) that reduce transportation and transaction costs and encourage the private sector to invest in agro-processing, cold storage, refrigerated transportation, and retail chains so as to effectively integrate markets for perishable commodities and minimize post-harvest losses; (ii) enhanced access of farmers to technology, credit, inputs, information, and services; and (iii) the implementation of appropriate policies that facilitate institutional arrangements like contract farming, producers' organizations, and cooperatives that provide farmers with easy access to output and input markets in order to share price risks and reduce marketing and transaction costs.

Growth in output prices served as an important source of growth in the past, but the price-led growth is not sustainable in the long term because of its potential inflationary and distributional effects. Some of the growth in prices is due to higher government-set prices (minimum support prices)—particularly of rice and wheat. Further, the benefits of increased prices are shared by farmers in proportion to their marketed surplus. Small farmers have reduced marketed surpluses and are unlikely to gain much from higher prices. Prices of high-value commodities are determined by market forces, and policy reforms and infrastructure can contribute to higher farm-gate prices or value realization.

Fourth, it is likely that area expansion will not continue to be a source of growth in the long term. The supply of land is fixed and the scope to bring additional areas under cultivation has been almost entirely exhausted. The only option that remains is to increase land use or cropping intensities, but they are already at about 140 percent at present. This option would require enhancing investments in irrigation infrastructure and providing incentives for the adoption of water-efficient technologies, such as drip and sprinkler irrigation.

In sum, the governments should ensure appropriate policies and a favorable investment climate to accelerate agricultural diversification and greater participation of the private sector in the supply/value chain of high-value commodities. These policies must be more stable and indicate a commitment towards the promotion of agriculture and agribusiness. The opportune time is now since the corporate sector is willing to invest in agribusiness to harness the emerging opportunities in domestic and global markets.

II/2

Table 2A.1	**Annual growth (%) in yields of important crops in India**

crop	India as a whole			eastern region			northern region		
	1980s	1990s	2000s	1980s	1990s	2000s	1980s	1990s	2000s
rice	3.15	1.26	1.62	3.85	1.58	1.56	4.13	1.17	1.15
wheat	3.21	1.83	0.20	2.20	1.82	-0.81	3.14	1.99	0.13
maize	2.59	1.97	2.38	4.44	3.02	-0.01	3.83	0.76	0.76
other cereals	0.88	1.03	2.75	1.55	-0.72	0.70	2.81	1.69	2.27
all cereals	3.05	1.99	1.50	3.68	1.74	1.21	3.82	1.80	0.61
chickpea	2.46	1.55	-0.12	1.75	-0.35	-0.16	2.16	0.70	-2.80
pigeon pea	-0.15	0.37	1.78	2.55	-3.41	1.31	-0.75	0.46	-4.45
other pulses	1.62	-0.82	0.17	0.90	-0.43	-0.66	-0.68	-1.77	-0.63
all pulses	1.59	0.54	0.64	1.07	-0.45	-0.09	0.87	-0.46	-2.09
groundnut	1.74	1.33	3.70	0.32	-3.35	4.78	2.20	-1.25	-1.94
rapeseed-mustard	3.04	0.33	1.84	3.76	0.16	-0.17	3.39	0.45	0.99
all oilseeds	1.70	1.72	0.52	3.60	1.88	-0.50	2.75	0.04	0.18
cotton	4.19	-1.41	13.84	3.67	7.52	5.39	6.10	-5.56	11.45
sugarcane	0.22	0.76	0.71	2.76	-0.21	-1.20	0.19	0.28	0.44
chilies	3.41	3.67	5.62	-1.28	0.75	3.25	0.51	-0.20	6.12
ginger	3.02	0.53	-1.38	3.27	2.21	0.03			
turmeric	5.76	0.79	4.37	3.91	0.77	1.09	5.40	4.29	3.27
spices	2.74	2.28	4.73	0.88	0.67	1.37	3.90	4.35	5.31
banana	3.56	3.27	0.69	0.00	3.33	-0.40	-	-	-
cashew nut	3.22	-0.67	3.74	16.58	-2.48	3.79	-	-	-
other fruits	-2.16	1.82	-1.10	-1.95	1.47	2.62	-2.40	-0.30	1.99
all fruits	-0.92	1.97	-0.71	-1.64	1.10	2.03	-2.40	-0.30	1.99
potato	2.25	1.55	-1.42	3.45	1.22	-2.45	1.73	2.27	-0.74
onion	0.23	-0.62	6.66	0.53	-0.37	12.49	-0.50	-1.57	5.29
other vegetables	-3.15	0.08	0.64	-5.36	0.58	1.71	1.69	0.25	2.06
all vegetables	-1.17	0.54	0.64	-3.23	0.86	1.00	1.59	1.45	0.40

Table 2A.1	Annual growth (%) in yields of important crops in India					
	western region			southern region		
crop	1980s	1990s	2000s	1980s	1990s	2000s
rice	1.17	-0.33	4.03	2.73	1.25	1.00
wheat	2.94	1.96	1.47	-0.99	1.91	2.43
maize	1.51	1.10	0.48	-0.28	2.02	3.61
other cereals	0.95	0.59	2.80	0.09	1.54	3.68
all cereals	1.74	1.72	2.84	2.17	2.17	2.83
chickpea	3.10	2.17	-0.41	-1.74	4.74	4.27
pigeon pea	0.89	0.87	2.28	-0.57	2.81	6.22
other pulses	2.37	-0.34	0.92	3.39	-0.66	0.20
all pulses	2.49	1.53	0.73	2.22	0.87	4.19
groundnut	1.29	2.81	5.48	1.75	0.83	1.92
rapeseed-mustard	2.17	0.16	2.25	-0.02	-0.26	6.01
all oilseeds	2.38	2.65	3.53	0.48	2.50	-1.54
cotton	1.35	1.15	16.83	6.02	-0.70	8.64
sugarcane	-0.55	-0.12	2.24	-0.30	1.84	0.14
chilies	3.07	1.39	0.24	5.19	4.82	7.32
ginger	1.29	3.04	-1.82	3.82	0.36	-2.45
turmeric	-3.13	0.01	0.81	5.44	0.14	5.09
spices	0.22	1.72	5.49	4.66	3.07	5.05
banana	3.16	4.66	-0.13	5.87	1.68	1.00
cashew nut	14.22	-1.91	7.61	1.48	-0.49	0.83
other fruits	-0.28	1.28	-7.51	-3.30	3.09	0.00
all fruits	0.71	0.60	-6.77	-0.81	3.71	1.00
potato	-0.39	0.82	2.96	5.57	-1.32	-7.25
onion	-0.04	-0.18	4.32	0.08	-0.82	5.56
other vegetables	2.69	-4.34	-4.34	-1.20	1.60	-0.38
all vegetables	1.89	-2.68	-0.25	0.34	0.85	1.28

II/2

Table 2A.2 | **Share of different crops by source in the overall increase in the value of output (million Rs./annum)**

crop	India as a whole					eastern region					northern region				
	A	Y	P	D	VOP	A	Y	P	D	VOP	A	Y	P	D	VOP
1980s															
rice	1,287	8,065	-1,474	249	8,121	659	3,235	-284	-151	3,451	171	2,443	-79	686	3,257
wheat	710	4,749	-1,618	191	3,987	101	330	-43	84	471	275	2,989	-1,014	783	3,016
maize	108	642	-233	-268	239	28	178	-77	-108	14	24	325	-64	-236	38
o. cereals	308	782	-932	-1,160	-1,009	9	20	-36	-69	-75	24	231	-127	-300	-178
pulses	415	1,225	2,231	-633	3,243	105	199	47	-115	227	67	125	827	-439	559
oilseeds	640	2,836	1,314	4,785	9,800	96	271	-20	496	862	38	391	8	245	721
fibers	237	1,944	55	-746	1,503	50	197	173	-173	247	36	563	84	166	875
sugarcane	434	9	-252	1,193	1,442	41	149	-89	-79	17	125	-41	380	593	1,071
beverages	80	256	317	97	762	76	114	218	38	453	0	1	0	-1	0
spices	134	697	459	338	1,651	45	25	144	159	380	3	16	63	-22	59
fruits	515	-1,132	2,067	2,611	4,065	145	-209	443	395	748	70	-630	160	901	479
vegetables	620	-1,935	1,828	3,668	4,084	494	-2,106	719	2,430	1,391	52	205	164	519	943
other crops	75	387	-139	7	324	12	28	-25	-175	-16	1	70	-3	4	73
total	5,562	18,527	3,624	10,332	38,212	1,861	2,431	1,171	2,734	8,171	886	6,688	399	2,898	10,912
1990s															
rice	619	4,670	2,058	1,167	8,579	-577	2,024	242	695	2,379	532	1,067	400	1,082	3,117
wheat	327	3,834	2,386	2,313	8,991	-79	295	-71	293	434	805	2,782	2,291	558	6,531
maize	48	575	8	316	950	-22	154	5	-26	103	56	82	91	-92	135
o. cereals	128	724	397	-1,788	-568	-5	-7	16	-44	-41	51	147	57	-171	82
pulses	190	462	1,641	-1,192	1,092	-72	-59	586	-638	-215	149	-57	310	-764	-381
oilseeds	415	1,790	-3,935	1,045	-702	-81	-22	-110	-411	-604	128	78	-522	-60	-371
fibers	152	-1,033	572	1,410	1,195	-34	102	-49	182	203	110	-825	193	-161	-684
sugarcane	219	756	892	1,984	3,902	-30	-30	72	-117	-106	363	110	871	476	1,839
beverages	32	262	238	450	989	-54		288	260	490	0	1	0	-1	0
spices	70	1,007	1,474	885	3,487	-38	49	249	109	363	8	9	39	32	92
fruits	232	2,546	2,706	3,914	9,544	-121	455	269	537	1,181	186	-200	356	708	1,036
vegetables	269	1,037	2,437	5,150	8,955	-346	852	1,213	2,194	3,896	183	390	429	839	1,851
other crops	37	662	79	157	954	-6	-12	47	-150	1	7	53	3	49	114
total	2,739	17,292	10,954	15,810	47,366	-1,465	3,795	2,757	2,882	8,085	2,577	3,636	4,518	2,495	13,360
2000s															
rice	3,066	5,612	-4,341	-2,960	1,391	440	1,938	-766	-713	901	313	1,198	-1,116	86	474
wheat	2,146	783	2,078	-103	5,023	63	-83	332	-112	210	508	408	685	-111	1,512
maize	346	975	111	822	2,309	22	12	214	174	432	27	81	61	-104	67
o. cereals	493	1,534	1,466	-1,271	2,228	1	2	3	-21	-15	25	196	182	-122	287
pulses	971	689	1,909	244	3,884	39	-4	85	25	137	63	-404	446	-436	-343
oilseeds	1,886	4,294	1,541	2,213	10,137	40	-52	51	216	252	56	127	300	288	776
fibers	812	9,588	-2,152	194	8,330	32	172	-46	-94	59	51	1,608	-512	8	1,100
sugarcane	1,402	1,139	5,090	2,629	10,671	17	-59	72	15	46	235	289	1,409	589	2,546
beverages	223	-229	-383	439	27	50	-264	-729	557	-414	0	2	-1	-2	0
spices	648	1,899	-151	511	2,911	38	134	-111	1	61	15	32	112	382	549
fruits	2,349	-1,141	971	11,780	13,961	154	996	529	615	1,456	129	778	486	445	1,848
vegetables	2,201	1,679	1,280	6,576	11,975	426	1,211	727	2,793	5,212	158	485	659	1,981	3,323
other crops	267	459	500	-108	1,151	6	73	25	-14	101	5	-51	15	23	-9
total	16,811	27,280	7,918	20,965	73,998	1,329	4,076	387	3,441	8,439	1,586	4,751	2,727	3,027	12,130

Note: A stands for area expansion; Y for yield expansion; P for price increases; D for diversification; and VOP for value of output.

Table 2A.2	Share of different crops by source in the overall increase in the value of output (million Rs./annum)

crop	western region					southern region				
	A	Y	P	D	VOP	A	Y	P	D	VOP
1980s										
rice	125	446	-214	132	485	617	2008	-770	-904	929
wheat	135	1,141	-437	-285	540	7	-6	-4	-37	-40
maize	25	159	-102	37	117	25	2	12	30	69
o. cereals	141	510	-599	-440	-381	143	58	-182	-372	-359
pulses	150	844	1,076	-194	1,862	70	163	260	93	594
oilseeds	217	1,056	560	2,185	4,137	411	940	704	1,963	4,080
fibers	75	376	-99	-370	-5	80	631	-102	-202	387
sugarcane	75	35	-397	294	-16	167	-126	-194	513	370
beverages	0	0	0	0	0	51	125	109	20	309
spices	13	29	230	28	303	132	564	-7	206	897
fruits	92	-77	476	1,190	1,686	341	-479	635	471	1,151
vegetables	76	322	286	308	1,008	190	-50	527	106	742
other crops	14	35	13	2	63	76	218	-115	29	204
total	1,138	4,875	794	2,888	9,799	2,310	4,048	874	1,917	9,334
1990s										
rice	201	-106	656	-66	678	-227	1386	864	367	2,404
wheat	205	889	103	761	1,992	-2	20	-1	18	34
maize	38	105	14	83	238	-16	88	-37	438	473
o. cereals	222	174	188	-1,137	-568	-32	290	160	-461	-58
pulses	239	517	554	-129	1,170	-33	131	231	186	518
oilseeds	444	943	-1,825	2,001	1,550	-160	717	-1,543	-266	-1,276
fibers	145	44	276	792	1,357	-42	-94	70	368	320
sugarcane	123	-116	-249	781	544	-87	559	211	917	1,626
beverages	0	0	0	0	0	-28	258	-52	313	495
spices	25	82	313	21	442	-91	688	867	987	2,478
fruits	112	556	1,170	2,376	4,276	-147	908	634	1,381	3,051
vegetables	103	-792	159	2,018	1,392	-92	617	571	833	1,817
other crops	19	57	71	53	204	-41	406	63	195	635
total	1,875	2,351	1,430	7,552	13,274	-997	5,974	2,039	5,275	12,516
2000s										
rice	718	1,681	-810	-600	944	204	814	-1,650	-366	-928
wheat	969	806	837	516	3,261	3	23	9	3	40
maize	187	127	9	96	420	58	428	5	913	1,449
o. cereals	509	951	-525	-757	143	39	481	417	-582	334
pulses	972	283	1,492	167	2,971	65	511	-26	507	1,111
oilseeds	1,963	3,520	1,294	1,491	8,374	185	350	178	-36	732
fibers	807	6,677	-1,513	274	6,159	48	1,250	-75	-253	1,011
sugarcane	622	768	2,018	2,830	6,770	106	11	1,021	79	1,309
beverages	0	0	0	0	0	39	21	345	43	446
spices	172	136	356	254	968	149	1,438	-123	127	1,546
fruits	1,568	-4,671	419	9,179	5,695	316	214	237	4,143	4,961
vegetables	691	-680	680	1,946	2,713	117	259	-334	653	727
other crops	66	2	-55	-144	-129	91	481	437	149	1,188
total	9,243	9,600	4,202	15,251	38,290	1,419	6,282	441	5,381	13,925

Note: A stands for area expansion; Y for yield expansion; P for price increases; D for diversification; and VOP for value of output.

Improving Water Use Efficiency: New Directions for Water Management in India

Chapter 3

Richard Ackermann

Introduction[1]

There are two big issues that continue to be carried forward from one Five Year Plan to the next: The aim to achieve 4 percent growth in agriculture, and the expansion of surface irrigation systems. Approaching the 12th Plan (2012–17), the 4 percent target is being carried over from the 10th Plan (2002–07)[2], and in turn from the 11th Plan (2007–12). In 2005, the Prime Minister suggested that a 10 percent growth rate was "eminently feasible, if . . . we manage to make a quantum leap in the growth rate of our agriculture . . ."[3] Yet agricultural growth has been consistently below that target (Table 3.1). As a result, GDP per agricultural worker is only about 75 percent higher in real terms than in 1950, against a 4-fold increase in overall real per capita GDP (Planning Commission, 2008).

Similarly, investment in irrigation—seen to be one of the principal factors affecting agricultural growth—has not yielded the desired results. The 11th Plan document notes that (Planning Commission 2008a, Vol. III, p. 43):

- Many major and medium irrigation (MMI) projects seem to remain under execution forever as they slip from one Plan to the other with enormous cost and time overruns.
- Owing to lack of maintenance, the capacity of the older systems seems to be going down.
- The gross irrigated area does not seem to be rising in a manner that it should be, given the investment in irrigation. The difference between potential created and area actually irrigated remains large.

Given the continued major influence of the monsoon on agricultural growth in India, the critical role of water for irrigation—representing as much as 90 percent of total water withdrawals—is obvious. Severe droughts continue to reduce GDP growth between two and five percent, despite a substantial decrease in the share of agriculture in overall GDP (Gadgil and Gadgil, 2006;

1 The author is especially grateful for valuable insights and suggestions from Tushaar Shah, N. Harshadeep, Hans P. Binswanger-Mkhize, Ramaswamy Iyer, Himanshu Thakkar, Mihir Shah, Upali Amarasinghe, Don Blackmore, Shashi Enart, Ashok Gulati, David Molden, Aditi Mukherji, Sachin Oza, Sanjay Pahuja, U.N. Panjiar, Kirit Parikh, Mohan Sharma, Vladimir Smakhtin, and Chander Vasudev. Of course, the author alone is responsible for misrepresentations and errors.
2 The 10th Plan aimed for a 3.97 percent (sic!) annual rise in agricultural GDP (Planning Commission, 2001, Vol.. II, p. 527).
3 Quoted in Amarasinghe et al. (2007b), p. 4.

Table 3.1	Compound average growth rates: gross domestic product at factor cost (constant 1999–2000 prices)	
	overall GDP (%)	GDP agriculture and allied activities (%)
1951–1960	3.9	3.0
1961–1970	3.7	2.3
1971–1980	3.1	1.5
1981–1990	5.4	3.4
1991–2000	5.6	2.7
2001–2007/8	7.7	3.2

Source: Reserve Bank of India (2009).

Virmani, 2004). At the same time, floods in parts of the country—often caused by the sudden release of water from reservoirs and/or the breach of increasingly unstable embankments, cause major damage to crops and people.

Approach

It is unfortunate that the discussion of water policy tends to be polarized along divergent lines: one perspective favoring a continuation of past approaches relying mostly on rehabilitation and further construction of surface irrigation structures and water storage infrastructure, pointing out that past policies have brought major benefits; another perspective highlighting the rapid rise of groundwater irrigation to compensate for poor surface water service delivery, the detrimental social and environmental impact of large surface storage, and the need for small-scale water harvesting. Seeing the debate in these terms misses the fundamental point: past discussion on water policy in India has often tended to focus on preconceived ideas of what would be *desirable,* rather than *what is actually happening* on the ground. In other words, the argument is over *design,* not *implementation.* And as with most other aspects of public policy in India, it is the shortfall in implementation that should ultimately inform what kind of water policy is most appropriate.[4]

Thus, advocates of more surface water irrigation rightly emphasize the economic loss caused by delayed completion of storage schemes and irrigation canals, but ignore the essential fact that these delays have been the hallmark of surface irrigation schemes for decades, and that no amount of exhortation will change this situation, because the supply-driven nature of the projects does not meet the demands of most farmers, nor is it responsive to the enormous constraints

---

4 Of course, it is no surprise that past major projects have fallen far short of their objectives, given that already at the design stage, project appraisals appear often not to have been completed prior to execution. One of the major recommendations of the report of the Working Group on Water Resources for the 11th Five Year Plan (MWR, 2006), written in 2006, is that "Project Appraisal should be made mandatory before execution of projects." (MWR, 2006, p. 5)

in terms of availability of good farmland both for those who must give up land for canals and for people displaced by large reservoirs. Further, with few remaining productive dam sites in India, the economics of additional construction of large water storage infrastructure are becoming tenuous, as judged even by international experts with very extensive experience in water storage construction, such as in the Murray-Darling River Basin in Australia.

This chapter takes a purely pragmatic approach to identifying practical and demand-driven ways to resolve some of the deep seated problems in the water sector. There is now a large enough body of irrefutable evidence of what works and what does not work in the Indian institutional environment, and above all, how farmers have responded (whether by design or by default), that one can make surprisingly robust recommendations going forward. Also, many of the alternative approaches to augmenting water resources that are in the public discourse are not new at all. For example, in 1949, the All India Congress Agrarian Reforms Committee already referred to soil and water conservation works and the need to capture rain-water "as near the place at which it fell" (AICC, 1949).

In the intervening 60 years, however, the social and institutional conditions have become far more complex, making all interventions in water resource development correspondingly more challenging. Changing food consumption patterns and farmers' desire to shift toward higher-value crops demand flexible irrigation water delivery and appropriate technological innovations so farmers can extract the most value from their rapidly diminishing plot sizes with limited on-farm labor (see, for example, Gupta, 2010).

The Government's *National Water Policy* (MWR, 2002)—essentially an amended version of the 1987 Policy—reflects the traditional preoccupation with 'water resource projects,' i.e., dams, reservoirs, and canal systems. 'Ecology' is listed in fourth place in terms of priorities, undermining a holistic notion of water management which would view ecology as an overarching theme to ensure the sustainability of the priorities that the Policy ranks higher: drinking water, irrigation, and hydro-power (not to mention flood control). R. Iyer, former Secretary of Water Resources, Government of India, laments that the National Water Policy 2002 "is a patchwork quilt uncharacterized by cogency or coherence, and uninformed by a philosophy or vision." (Iyer, 2002, p. 1705).

As the Ministry of Water Resources moves forward in preparing a new water policy, key observers have called for a 'new start' and a clear indication of what will be done differently from the past (Iyer, 2010; Shah, 2010b). The new policy should be based on a realistic assessment of implementation capacity and a concrete implementation pathway, and build on an interdisciplinary perspective that includes key actors outside the water sector—especially the power sector and the large Centrally Sponsored Schemes. Independent experts have also argued for clear recognition of: the difficulty of imposing direct water demand management systems (e.g., charges) on the highly informal system in India; the need to use scarce agricultural land as efficiently as possible; and the high value of using surface storage to replenish groundwater. At a minimum, the new water policy should insist on collection and timely dissemination of high quality data and on an effective performance management system for public water infrastructure (Shah, 2010b).

Water demand and supply

Total water demand in India was comprehensively estimated in 1999 by the National Commission for Integrated Water Resources Development (NCIWRD, 1999). In 2007, the International Water Management Institute (IWMI) made revised estimates as part of a long-term forecasting study based on calculations for each river basin (Amarasinghe et al., 2004; 2007b). Using IWMI's analysis (confirmed by FAO, 2010), total water demand in 2010 was about 761 billion cubic meters (bcm), or 68 percent of the total potentially utilizable water resources (PUWR) of 1,123 bcm (CWC, 2005).[5] By 2050, total water demand could reach 900 bcm—still less than the total potentially utilizable water resources (Amarasinghe et al., 2007b).

However, it hardly needs to be pointed out that these national aggregates hide wide variations in a country as complex as India, with 12 major and 46 smaller river basins providing some 690 bcm utilizable surface water potential, involving over 5,000 large dams (the third-largest number of large dams in the world)[6], and fed by rainfall varying between 100 mm per year in the westernmost part of the country, and 11,000 mm in the eastern-most part. Almost 50 percent of the annual precipitation occurs in just 15 days each year, and the amount of monsoon rains can vary substantially in each location from year to year. Himalayan rivers (Ganga, Brahmaputra, Meghna) are snow-fed and perennial, peninsular rivers depend on the monsoon (i.e., most of their flows are concentrated in about 4 months).[7]

Underlying this system of surface waters is a ground water system which can be thought of as an enormous—though complex—water storage basin with an annually replenishable (dynamic) net capacity of about 431 bcm (396 bcm taking into account natural discharge. This augments India's dam storage capacity of about 250 bcm—289 bcm after completion of ongoing projects; 397 bcm if all projects on the drawing board were completed (FAO, 2010; CWC, 2005; CGWB, 2011a).[8]

From a purely technical perspective, groundwater storage is considered superior to surface water storage in all respects except for the slower infiltration (Keller et al. 2000): it loses less to

---

5 1 bcm = 1 billion cubic meters = 1 km³ = 1 Gigaliter.
6 The International Commission on Large Dams (ICOLD) defines large dams as 'those having a height of 15 meters from the foundation or, if the height is between 5 to 15 meters, having a reservoir capacity of more than 3 million cubic meters' (ICOLD 2003). According to the latest *National Register of Large Dams in India*, issued in February 2009, there were 4,711 completed large dams and another 390 under construction. Two-thirds of all large dams in India are in Maharashtra (1,821), Madhya Pradesh (906), and Gujarat (666) (CWC 2009).
7 This situation is expected to hold well into the 22nd Century, even if climate change predictions are correct.
8 The Central Ground Water Board (CGWB) refers to a feasible groundwater storage capacity of 214 bcm for purposes of its evolving country-wide Artificial Recharge Master Plan, of which 160 BCM is considered 'retrievable' (CGWB, 2005). The (revised) master plan envisages raising post-monsoon groundwater levels to 8 meters (originally 3 meters) below ground level. Most resources (Rs. 1,800 crores) are now flowing to the Dugwell Recharge Scheme that would create 41.1 bcm potential increased recharge in the 7 groundwater-stressed hard-rock states (Shah, 2008b; http://cgwb.gov.in/Groundwater/Artificial_Recharge.htm accessed July 18, 2010).

evaporation (especially the exceptionally high rates of potential evapo-transpiration which, on average, can reach 10 times the amount of rainfall in the areas of heavy monsoon rainfall and which make surface storage uneconomical—CGWB, 2005); is available where and when needed; has better water quality; and does not silt up. Each year, India loses between 1.3 bcm and 2 bcm of surface storage capacity as a result of poor management of irrigation systems and siltation from natural erosion and deforestation (NCIWRD, 1999). The irreversible reduction in the useful life of major dams is imposing a cost of more than Rs. 2,000 crore per year, which is what it would cost to create the equivalent new storage capacity. The greater efficiency of groundwater use for irrigation is evident from the fact that 220 bcm of groundwater abstractions irrigate more than double the net area irrigated by 400 bcm of surface water withdrawals (FAO, 2010). In other words, each unit of groundwater irrigates almost four times as much land as does surface water, and it uses one-tenth the storage per hectare (ha) of net irrigated area compared to surface reservoirs.[9]

Thus, to achieve a given level of agricultural productivity, one bcm of groundwater is worth far more—up to 10 times more, depending on the crop—than one bcm of surface water. That is why in the US, for example, 71 percent of the irrigated land of the 16,000 farms that each sold over US$1 million of agricultural products in 2008 depended on groundwater (USDA, 2010). More generally, even though the US dam storage capacity is more than double that of India's, 66 percent of total irrigated area in the US is irrigated by groundwater—in percentage terms about the same as in India today (Table 3.3)—using only 42 percent of total irrigation water (USDA, 2010).

However, it is not a question of surface water versus groundwater. Each source has its appropriate place in a continuum of water storage options—from deep or shallow aquifers to natural wetlands (lakes, swamps), ponds and tanks, practices to retain soil moisture, to small and large constructed reservoirs (IWMI, 2009). The challenge going forward will be to find the economically and environmentally most suitable ways to employ these different types of water storage options, and above all, to promote institutions that can provide a reliable service to farmers, from a combination of surface and groundwater sources.

In projecting future water demand, while the NCIWRD assumed the ratio of surface water to groundwater to be 55:45, IWMI's more recent scenarios more realistically assume a ratio of 40:60. Despite greater irrigation efficiency, this use of groundwater is still likely to lead to significant water shortages at the regional level in peninsular India. The crucial point to be made here is that further surface water development through construction of additional large storage is not going to reduce the demand for groundwater—no more in India than in the US or, say, Spain. As mentioned above, the debate should therefore not be over what might be desirable, depending

---

9 For this calculation, net irrigated area was divided by gross water abstractions to estimate the number of hectares irrigated by 1 bcm. Exact figures are not available for gross irrigated area, but it is likely to be larger for groundwater than for surface water. These numbers say nothing about the sustainability of groundwater use at any particular location.

on one's point of view, but over how the growing demand for 'just-in-time water' (now mostly supplied by groundwater) can be accommodated in a sustainable manner. There is a range of instruments for achieving this, as discussed later in this chapter: water recharge at the basin level (e.g., through conjunctive use involving tank, canal, or river seepage) or at the local level (e.g., artificial recharge through dug-wells); increasing crop productivity per unit of water; shifting water use to higher-value activities, including outside of agriculture; and using the supply of electricity to manage water abstraction levels.

In contrast, the Indian government's overwhelming financial and human resources emphasis in the water sector continues to be on the traditional management of surface waters for their own sake. In this view, there remains substantial scope for constructing large surface storage infra-structure to capture 'wasted' water that goes unutilized to the sea.[10]

To what extent this is the case and what should be done about it is, in fact, open to analysis and should not be a matter of opinion. *Peninsular Rivers* are mostly fully regulated, and at least two of the largest river basins (Krishna and Kaveri) have reached full or partial closure (Figure 3.1 shows the projected degree of development[11] of water resources in the different river basins). For example, in the Krishna Basin, India's fourth largest river basin covering parts of Karnataka, Andhra Pradesh, and Maharashtra and home to 80 million inhabitants, the storage capacity of the major (8) and medium reservoirs has reached total water yield (Venot et al., 2007). As a result, in dry years, virtually no water reaches the sea. On the other hand, the capture of so much water within the Basin and the evaporation of an additional 36 bcm of water has changed the regional climate, increasing humidity and changing temperature regimes, aggravating saline groundwater intrusion, and putting at risk the delicate wetland and estuarine ecology which is important not only for aquatic habitats and fisheries, but also for preventing shore erosion. The lack of adequate environmental flows in the Krishna River has significantly aggravated water pollution problems from cities, since domestic and industrial effluents can no longer be sufficiently diluted by flowing water.

As a whole, experts point out that there is little scope for economically viable additional large water storage infrastructure for irrigation in peninsular India.[12] New projects under consideration or construction at this time are already being situated in relatively flat topography involving dis-proportionate areas to be flooded and correspondingly large numbers of people to be resettled, and exacerbating inter-state disputes (an example is the proposed Polavaram Dam in Andhra

---

10 The concept of water storage per capita (or also water storage as a fraction of annual runoff) is often invoked to argue that India is miss-ing out on a lot more water that could be productively used. However, as Iyer (2008) points out, this is a fallacious concept, since the water storage per capita depends on the topography of a country. A flat and featureless country, by definition, will have lower water storage per capita than a hilly country. Thus the numbers would differ substantially among, say, India, Nepal, and Bangladesh. The concept also does not take into account aquifer storage or even small water harvesting structures.

11 The degree of development is defined as the ratio of water withdrawn for the first time (e.g., upstream in a watershed) to PUWR. PUWR = potentially usable water resources.

12 There may be some additional opportunities for hydropower development, which would merit a separate discussion.

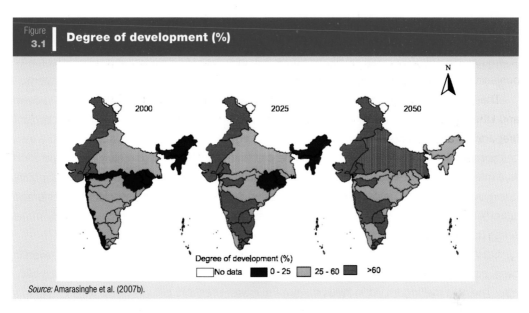

Figure 3.1 | Degree of development (%)

Degree of development (%)
No data | 0 - 25 | 25 - 60 | >60

Source: Amarasinghe et al. (2007b).

Pradesh, to which both Orissa and Chhattisgarh are strongly opposed). This explains why delays in completing—or even initiating—major and medium surface water storage projects are inevitably getting much longer, and why the cost "had already reached the fairly mind-boggling figure of Rs. 1,42,662 per ha by the end of the Ninth Plan" (GOI, 2006, p. 31).

The existing storage infrastructure in peninsular rivers is mostly designed to smooth out the southwest monsoon flows in, say, 9 out of 10 years. There may still be the 1 in 10 year flood, for which, however, there is no economic justification to invest in substantial additional infrastructure. Instead, better weather and flood forecasting is required, along with flood insurance and possibly the designation of flood diversion areas, whereby farmers are asked to temporarily (and against compensation) set aside embanked land to accommodate flood overflow.

Water flow in the *Himalayan Rivers*, particularly the Ganges, is far greater than in peninsular Rivers. However, there remain few options for surface water storage except to a relatively minor extent in the Himalayan Foothills. Farther upstream, the area is too mountainous and involves very high silt loads from the highly erosive Himalayas; downstream of the foothills, within India, the area is too flat to construct additional storage capacity. For the Ganges system, out of 250 bcm of potentially utilizable water, about 37 bcm are presently captured, and a total of at most 50 bcm would be captured if all possible dams under consideration were to be built. These would add little in the way of irrigation or flood prevention benefits (Blackmore, 2010):[13] Tributaries at risk are already fully embanked, and floods have occurred not because water has flown over the embankments, but because embankments have been repeatedly breached as a result of poor

---

13 This analysis and the conclusions are based, in part, on work carried out by Don Blackmore, former Chief Executive of the Murray Darling Basin Commission in Australia.

maintenance (e.g., Kosi in Bihar) or inappropriate dam management (e.g., Hirakud in Orissa). Within-year water storage on the Ganges could augment low-flow, but this would be of questionable economic merit, since groundwater tables are quite high and there is substantial potential for conjunctive use (a prerequisite for which is much better monitoring of water flow in the system).

There is significant additional hydropower potential in states such as Himachal Pradesh and Uttarakhand. However, these projects have also suffered from delays due to poor project preparation and appraisal, unresolved social and environmental concerns (especially the loss of important forest reserves in erosion-prone catchment areas), and ultimately, poor operation and maintenance of existing facilities, with electricity generation per MW of installed capacity falling over the years. The complexity of addressing the many problems associated with storage schemes has therefore shifted the focus increasingly to run-of-river projects which only make sense upstream of existing reservoirs.

On the *Indus River*, as well, the topography and the dominance of groundwater use—even in canal command areas—means that additional surface water development will yield limited benefits. The next major dam in the Indus system—at a cost of US$12 billion—would yield less than a 1.5 percent increase in regulated flow (Blackmore, 2010). This is an area where far more attention would need to be given in the future to better connect surface water development to groundwater use and to strategically utilize surface storage to replenish the rapidly depleting groundwater tables (in Australia, for example, 25 percent of all groundwater use will be sourced from induced stream flow leakage). Existing reservoirs in India are doing this to some extent, though generally not by design.

The Bhakra Dam and Reservoir in the Indus River Basin is often used to justify other large dam projects since it is seen as an icon in India's history—as a project that contributed in a major way towards achieving the country's food grain self-sufficiency. It is the second highest dam in Asia and the second largest reservoir in India. The structure has attracted extensive studies demonstrating its high indirect economic benefits (Bhatia et al., eds., 2008), but also questioning whether the dam deserves to be held up as a model of historic proportions (Dharmadhikary et al., 2005). On balance, the evidence seems to suggest that Bhakra is a rather typical project, with all the positive and negative aspects of other large dams and the related surface water infrastructure. As one analyst points out: "Bhakra has had both multiplier and divider effects."

Recent experience

Rather than get into the debate about the benefits of the 50-year old Bhakra project, it is easier to look at the recent experience with traditional supply-side water resource development projects. Two large flagship programs which have been under intense public scrutiny and which some have called 'best practice' serve as typical examples: the Accelerated Irrigation Benefits Programme (AIBP) and the Sardar-Sarovar Project (SSP) on the Narmada River (which has benefitted from AIBP funds since AIBP's inception). In the first half of 2010, the Government's Comptroller and Auditor General (CAG) submitted a comprehensive report on the AIBP as well as a separate report on the *Nigam* (company) tasked with managing the SSP.

Accelerated Irrigation Benefits Programme (AIBP)

Established in 1996–97, AIBP was originally intended to provide central assistance to complete large irrigation schemes and accelerate the creation of additional irrigation potential. In 2004, CAG concluded that "only 11 per cent [of the envisaged irrigation potential to be created under the AIBP] could be utilized. The poor programme performance was also reflected in high Development Cost per hectare" (CAG, 2004). AIBP was nevertheless continued with no significant modifications. Six years later, the most recent CAG Performance Audit Report presented in Parliament on May 7, 2010 states that (CAG, 2010a, p. 10):

> Most of the deficiencies pointed out in the earlier Audit Report continued to persist, and AIBP had still not achieved its targeted objective of accelerating completion of large irrigation projects and delivering the benefits of irrigation water to the farmers.

Central loan and grant assistance from the Government of India (GOI) for AIBP has been almost Rs. 40,000 crore.

It is worth briefly placing the 2010 CAG findings in context: when AIBP was launched in 1996–97, net area irrigated from canals and tanks in India amounted to 19.9 million ha.[14] By 2007–08, net area irrigated from canals and tanks had declined to 18.5 million ha, a decrease of over 7 percent (MOA, 2009). In fact, the net irrigated area covered by surface canals and tanks today is the same as it was in 1980–81 (when it was also about equal to that covered by wells and tube-wells), despite rapidly escalating expenditures as depicted in Figure 3.3, which shows only expenditures on major and medium irrigation projects (CWC, 2010). As Thomas and Ballabh (2008, p. 100) put it: "These projects have thus become a sort of 'sink', which keeps on drawing money, without being able to provide returns on the large capital invested."

Meanwhile, net irrigated area served by groundwater has shot up to at least double (and possibly as much as 6 times[15]) the area covered by surface irrigation schemes (Figure 3.2 and Table 3.3), driven, above all, by 20–25 million tube-wells used by small farmers who could not get access to government-supplied surface water—either at all, or in the necessary amounts at the appropriate time.

---

14 *Net Irrigated Area* is the area irrigated through any source once in a year for a particular crop.
Total/Gross Irrigated Area is the total area under crops, irrigated once and/or more than once in a year. It is counted as many times as the number of times the areas are irrigated and cropped in a year.
Cropping Intensity is the ratio of Total Cropped Area (counted as many times as there are sowings) to Net Area Sown (counted once).
15 The higher number is derived from the findings of the IWMI report analyzing satellite images of irrigated area in India (Thenkabail et al. 2009). This report suggests that as much as 113 million ha—80 percent of total net area sown—is irrigated at some point in the year, leaving only some 28 million ha as purely rain-fed lands.

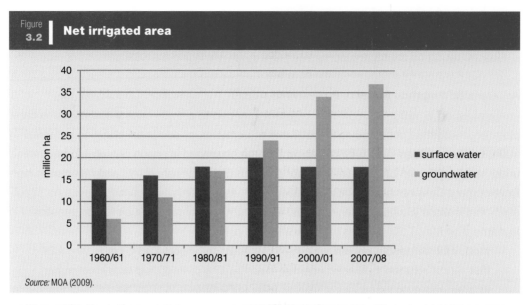

Figure 3.2 | **Net irrigated area**

Source: MOA (2009).

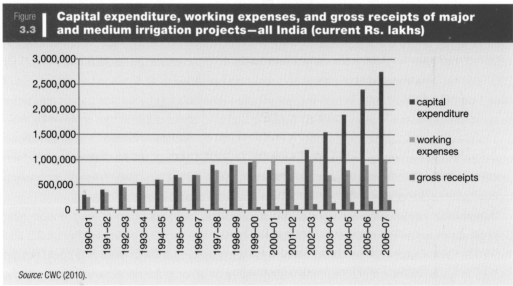

Figure 3.3 | **Capital expenditure, working expenses, and gross receipts of major and medium irrigation projects—all India (current Rs. lakhs)**

Source: CWC (2010).

Groundwater abstraction is taking place *within* virtually all surface irrigation command areas. In Punjab, for example, most irrigated areas are irrigated by groundwater (73 percent according to GOP, 2009; up to 96 percent in the *rabi* season according to NSSO, 2005b)—even within the command area of the Bhakra Dam (GOP, 2009).

Sardar Sarovar Project (SSP)

This multi-purpose (irrigation/power) 139-meter high dam and canal system on the Narmada River is the tallest of a series of 30 large dams and over 3,000 smaller dams being built along the Narmada Valley which stretches from Madhya Pradesh through Maharashtra to Gujarat. Now over 10 years behind schedule and 10 times over cost by 2012 (MWR, 2006), SSP was meant to expand irrigation potential in Gujarat and Rajasthan by about 1.8 million hectares. The main SSP canal is now more or less complete, and water has reached Rajasthan. However, the CAG report, tabled in the Gujarat State Assembly at the end of March 2010, states that utilized Cultivable Command Area (CCA) in Gujarat is only 6.56 per cent of the envisaged CCA[16], created in water-fed and not in water scarce zones (CAG, 2010b), and suggests that the large investment in the canal network has remained 'largely unfruitful.'[17] Despite major changes in the ground reality in which SSP's irrigation distribution network is supposed to be constructed—including 39,000 pending legal cases against land acquisition[18], and the fact that 80 percent of net irrigated area is already served by wells and tube-wells (MOA, 2009)—

> [t]he Detailed Project Report originally prepared (January 1980) . . . remained unrevised. Though the deadline of 2000 was fixed for achievement of full irrigation potential, no detailed plan to execute the project was prepared. . . . No data was maintained . . . on the impact of providing irrigation facility on agricultural productivity or agricultural pattern in the SSP command area. . . . The Company has not framed a comprehensive long term policy. . . . [R]epairs and maintenance was not done.
>
> (CAG, 2010b, p. x–xi)

Part of the plan for the distribution of SSP water was to rely on Water Users Associations (WUAs) to construct distribution systems in the village service areas. As the Tata Institute of Social Sciences report "Performance and Development Effectiveness of the Sardar Sarovar Project" spells out, "[a]lthough 1,186 WUAs were registered in 2006, only 10 percent were active and none of them constructed a distribution system. . . .Farmers are investing in diesel pumps and pipes to lift water from the main canal" (TISS, 2008; Parasuraman et al., 2010).

And why would WUAs actively engage in construction of the distribution network, if they are being asked to give up valuable farmland in return for uncertain water deliveries controlled by the Irrigation Department, and they need to continue to rely on groundwater anyway? Indeed, even in

---

16 The Cultivable Command Area in Gujarat using water from the SSP is estimated to have reached 10 percent of potential by the end of 2011.

17 "Project Woes—CAG raps SSNNL: Rs. 18Kcr fund unfruitful." *The Times Of India Ahmedabad,* Mar 31, 2010.
In addition, the CAG report on the AIBP "found substantial diversion of funds and other financial irregularities in the AIBP components of the Sardar Sarovar Project (SSP), Gujarat." (CAG, 2010a, p. xi) A series of CAG reports over the years on the company responsible for managing the SSP have pointed to massive financial irregularities. "In the period 2001-06, . . . almost 53 percent of the expenditure . . . was related to debt repayment by the [company]." SSP, which started construction in 1987, has displaced close to 50,000 families in 245 villages in Gujarat, Maharashtra, and Madhya Pradesh (TISS, 2008).

18 Rajiv Shah, "Narmada canal network: SSNNL to rope in big players." *Times of India,* May 12, 2010.

Gujarat's existing canal commands, "[t]here is not a single *taluka* . . . which is irrigated exclusively by gravity flow from canals" (Shah et al., 2009, p. 52).

Meanwhile, a proposal to promote the use of buried distribution pipelines to circumvent the problems of land alienation and to build on the farmer initiative of lifting water from the main and branch canals has been turned down by the government[19], even though such pipelines are among the most commonly used methods of tube-well water distribution in Gujarat. Using buried pipeline networks could have saved over 100,000 ha of land with a market value of Rs. 20,000 crore (Shah et al., 2010).

Ironically, despite the absence of irrigation water from the SSP, Gujarat in the period 1999–2000 through 2007–08 has posted the highest agricultural growth rate of any state in India: 9.4 percent (against the India-wide average of 2.8 percent)[20] (MOSPI, 2010), largely driven by increases in productivity and gross cropped area under *Bt* cotton, and to a lesser extent, rabi wheat, in the dry districts of Saurashtra and North Gujarat, where no SSP water—or any other canal irrigation—has been available (even though that was one of the main justifications for the SSP). The growing of cotton and *rabi* wheat was made possible by the fact that groundwater tables have actually been rising in parts of Saurashtra, Kachchh, and North Gujarat after recent monsoons (Figure 3.8). Groundwater tables in Saurashtra—the driest part of Gujarat—benefitted from a spontaneous large-scale water harvesting movement that involved the construction of some 300,000 recharge wells and 100,000 check dams in Saurashtra (Shah et al., 2009).

Institutional Incentives

In this context, it is not so much a matter of blaming poor service delivery on a 'weak state,' and of trying to identify ways to bring the three 'Fs'—funds, functions, and functionaries—in alignment (see, for example, World Bank, 2006 and World Bank, 2007, for an extensive treatment from this perspective). Instead, it is simply argued here that there is a massive mismatch between supply and demand for the kinds of irrigation services required by farmers, especially small farmers. Where these services meet demand, a vibrant market can ensue. Where the government has failed to respond to the needs of farmers, they have simply taken matters into their own hands and largely ignored (or undermined) what's on offer. Seeing the ground reality as it is—not as some would like to see it—would be the first step in finding solutions.

The next step is to focus on the incentives and disincentives both within the government as well as between the government and its 'clients.' Recent literature (Shah, 2008a) drawing on the "New Institutional Economics" (North, 1990; 2005; Williamson, 2000) convincingly shows that measures most likely to succeed are those that involve low transaction costs but offer high pay-offs to an individual (or sometimes a group) that is directly engaged and accountable for

---

19 Rajiv Shah, "Expert panel rejects Narmada pipeline plan." *The Times Of India Ahmedabad,* May 29, 2010.
20 Overall State GDP growth in the same period, at 8.7 percent, was actually lower than agricultural growth (MOSPI 2010).

the outcome of an initiative (i.e., in many cases this will mean someone who has some form of property rights). With the right incentives, this can also encourage 'self-generated' initiatives and campaigns *(swayambhu)* with far-reaching benefits.

Within the government, the proliferation of agencies leads to competing mandates and a general tendency of designing programs that meet internal bureaucratic demands. Thus, the 11th Five Year Plan contains separate supply-side strategies for surface irrigation and ground-water development. This is no surprise since the mandate to plan for future needs is entrusted to silo-type large government agencies dealing independently with surface and groundwater, each of which will naturally want to maximize the projects and corresponding resources it can obtain. Just as there is internal pressure in government agencies to promote large surface irrigation schemes, there is now a growing tendency to advocate large groundwater programs. In the process, the vital connection between surface and groundwater development goes lost, as does the urgency of using groundwater recharge strategically to save electricity: "every one metre rise in the depth from which farmers pump ground water saves the country 131 crore units (kWh) of electricity at generating stations" (Shah, 2008b, p. 43).

There are formidable bureaucratic obstacles, for example, to the notion that the purpose of surface water schemes might need to be redefined to serve the needs of groundwater users. As Iyer points out (Iyer, 2001, p. 1119),

> there is a dispersal of different components or aspects such as major/medium projects; minor irrigation; command area development; ground water; watershed development; rainwater-harvesting; water management; and so on. Different divisions/departments/agencies tend to deal with these matters with little coordination, much less integration. . . . The Irrigation Acts vest the management and control of waters in the hands of the state, and project planning and implementation are largely internal activities of the state.

Reforming water bureaucracies in this situation is a herculean task, partly because the supply-side emphasis precludes the establishment of clear yardsticks to measure service performance vis-à-vis clients at the end of the water distribution chain, partly because of the crushing legacy of attitudes from the past, and the accumulated staffing: In Haryana, for example, 83 percent of the allocation for irrigation Operation and Maintenance (O&M) goes to paying salaries; Uttar Pradesh's Irrigation Department employs no less than 110,000 people, only 5 percent of whom are professionals (Briscoe and Malik, 2006). However, there has been no new recruitment for many years, and most of the professional irrigation engineers will have retired in the next 5 to 10 years. "[I]rrigation departments are highly centralised and function with a top-down approach failing to establish any linkages with the farmer-users" (Raju and Gulati, 2008, p. 87).[21]

---

21 Attempts to 'privatize' (parts of) irrigation departments, as in the case of the *Nigam* managing the SSP, have evidently failed to deliver the desired results in terms of service and accountability.

For irrigation engineers and, more generally, irrigation departments, the incentives are rather clear: design and start new projects with the goal of achieving maximum irrigation potential *created*, with little regard to irrigation potential actually *utilized*. Of course, the latter depends heavily on complex social and political factors at the outlet of branch canals (including questions of property rights), and is one reason why irrigation departments are quite happy if this task can be delegated to WUAs (which, in turn, are reluctant to take on controversial tasks that may not be in their interest). Unsurprisingly, bureaucratic plans tend to be overly optimistic, and the discrepancy between potential created and utilized (a factor of 5 for India as a whole) has been getting greater as population densities, social conditions, and the proliferation of groundwater use have increased in prospective command areas. Four IIM reports (IIM, 2008) amounting to 983 pages analyzing this discrepancy essentially concluded that this is a matter of poor planning by irrigation departments (Shah, 2010a).

In this paradigm, there is little incentive to carry out operation and maintenance, because irrigation department staff are not directly accountable and are neither punished nor rewarded for long-term performance of the system. As Gulati et al. (2005, p. 20) point out, "there is no structural link between farmer payments and irrigation agency budgets. O&M (and capital) expenditures depend on government budget allocations and are not tied to cost recovery rates." The experience with AIBP demonstrates clearly that financial constraints are not the reason for poor performance. Hence raising water charges will not address the fundamental problem of poor irrigation services (ironically, states receiving AIBP funds were required to increase their irrigation charges). Nor will shifting responsibility for oversight and management (O&M) and its associated costs to water users associations improve matters, as they have little control over the quality of water delivery and therefore will generally be reluctant to invest in upkeep of a system over which they feel no ownership.

Some states have established financially "autonomous" irrigation agencies (Nigam), ostensibly to improve irrigation service performance. In some cases, these Nigams have managed to raise substantial capital from private sources to complete projects faster—though ongoing maintenance continues to be neglected. As was seen from the CAG audit of the *Nigam* responsible for the Sardar Sarovar Project, the top-down corporate mentality of agency staff appears not to have changed, and there have been massive abuses in mobilizing funds at very high interest rates, with the expectation (which has been proven correct) that the state (and indeed the Central

Government) as guarantor will make up for the financial inefficiency. The experience has been much the same in several States.[22]

Vis-à-vis the farmer, government departments and officials are often asked to meet unrealistic objectives and perform high-transaction cost tasks which are neither in the interest of the government official (especially the lowest-paid *Chowkidar,* the *Talatis,* or the irrigation engineer whose focus is only on technical issues) nor that of the farmer.[23] To overcome the lack of progress at the ground level, the temptation arises to introduce laws, regulations, or economic instruments (copied from industrialized countries) which are just as unlikely to be followed.

Nor is a radical change in the human resources profile of irrigation departments (assuming this would even be possible) likely to change much, unless the overall mandate and incentives are drastically changed as well—a process which is primarily political in nature. As explained later, the most straightforward approach might be to simply curtail the irrigation department's mandate to the delivery and regulation of bulk water, extending the evolving experience in Maharashtra.

Because of its supply-side focus, the government is forced into inefficient compromise solutions such as 'free' electric power or free irrigation water to make up for: (i) the fact that farmers actually have to pay significant amounts anyway for maintenance of their equipment which constantly breaks down because of poor quality power supply, and (ii) the large benefits preempted by wealthy farmers who control large quantities of virtually free water in the head reaches of canal commands, leaving farmers in the tail reaches with little (and unreliable) water or no water. In Andhra Pradesh, for example, marginal farmers pay up to 64 percent of their gross farm income on the costs associated with groundwater irrigation (annualized fixed costs, motor winding, and pump maintenance), even if electricity is essentially free (World Bank, 2001). In this environment, asking politicians to charge higher rates (let alone marginal cost) for electricity or water is akin to asking them to perform *hara-kiri* (Shah, 2008a). And even if the rates were increased, collecting them would be another story.

Dealing with these supposedly 'misguided' policies requires a careful understanding of the local conditions and tailor-made approaches that offer incentives that are appropriate to the needs of large numbers of independent water extractors. Abruptly imposing formal solutions recycled from industrialized countries—such as water pricing, withdrawal permits, and water rights in the framework of river basin agencies (a 'blue-print approach' sometimes described

---

22 Gulati et al. (2005) argue that reforms should be based on the following six principles:
(i) the irrigation agency must be financially autonomous; (ii) irrigation staff salaries must come from the fees charged for irrigation water; (iii) the irrigation agency must be accountable to user groups; (iv) third-party intervention in the form of an Independent Regulatory Commission for Canal Irrigation (IRCCI) may be necessary to prevent a deadlock between the irrigation agency and farmers when it comes to costs and incentives; (v) the primary tasks of the IRCCI should be to ensure transparency in contracts, obtain technical help, and act as a dispute settlement body; and (vi) the pricing of water should be related to consumption to keep costs low.
23 Thomas and Ballabh (2008) provide an interesting case study of the Mahi Right Bank Canal in Gujarat, detailing the incentives faced by the different stakeholders in the process of recovering irrigation charges—both those in the irrigation and revenue departments, and by farmers.

by international institutions as 'Integrated Water Resource Management')—is bound to fail in an informal system such as the one in India (Shah and van Koppen, 2006; Shah, 2008c). Boxes 3.1 and 3.2 illuminate this point.

Participatory Irrigation Management (PIM) and Water Users Associations (WUA): a tally

Many policymakers see Water Users Associations as the grand solution that will overcome the growing difficulty faced by irrigation departments to deal with the complex social issues in canal water delivery to the farmer, with deteriorating cost recovery and poor operation and maintenance.[24] Ironically, this is simply a revival of a solution already attempted in colonial times. As described, for example, by Mosse (2003) in a painstaking anthropological study of water policy in south India, an increasingly centralized government in British India essentially imposed the idea of a traditional local community "separate from the state, to which the state could devolve resource management responsibility" (Mosse, 2003, p. 270). And just as it did under the colonial government in the 19th Century, irrigation management transfer (IMT) today usually fails because the government's rights to water remain unchallenged, i.e., the focus on changing individual behavior obscures the fact that property rights and state policy have not changed. "Those familiar with field realities sense that PIM threatens irrigation officials" (Singh, 2000, pp. 376–377). A later section in this chapter points out that an equivalent argument applies to watershed management, where the state versus community dichotomy and the continued de facto control by state institutions inhibits Panchayats from taking charge of their own destiny, despite elaborate new institutional structures.

There is now extensive evidence that participatory approaches to canal irrigation management rarely work except where non-governmental organizations (NGOs) have provided sustained assistance over periods of 10–15 years and essentially acted as intermediaries between farmers and the irrigation department (though without *locus standi* vis-à-vis government).[25] WUAs are either co-opted by the irrigation department (e.g., in states where an irrigation engineer, called 'competent officer,' actually heads the WUA), or are called on to make "potentially conflict-ridden decisions" (Parthasarathy, 2006, p. 637) that wedge the WUA in an uncomfortable position between farmer and irrigation department. Examples include fee collection (for uncertain service levels) and alienation of land to build canals. Throughout this process, the irrigation department rarely gives up power, continuing to extract rents wherever possible (incoming new officers often

---

24 Indeed, strengthening WUAs is a major aspect of the centrally sponsored Command Area Development and Water Management Programme whose purpose since 1974–75 has been to support irrigation departments link main canal construction (such as under AIBP) with on-farm water use, and hence close the gap between irrigation potential created and utilized.
25 In the example of one state visited by the author, the NGO translated the documents produced by the irrigation department and discussed the proposed projects with affected farmers. The presence of the NGO relieved the irrigation department of the last vestige of having to address social issues in irrigation planning.

try to claw back whatever fee collection powers were transferred). Added to this, WUAs often discriminate against small farmers who either are kept out of the loop of decision-making or are relegated to performing manual tasks.

NGOs with extensive experience coaching and supporting WUAs confirm that acquisition of land for canal construction, capricious behavior of government departments, political interference, and lack of long-term vision (including how to integrate groundwater and surface water use) remain critical challenges that beg the question how sustainable WUAs can be, even after many years of nurturing (WALMI/DSC, 2010).

Independent observers go further: Pant (2008, p. 36) points out that despite having been tried for over 30 years in India, PIM "has yet to achieve even a semblance of acceptability and replicability, not to talk of scaling up." Mollinga et al. (2004) make much the same case with regard to the 'big bang' reform in Andhra Pradesh (AP): In one part of the state, irrigation reform was captured by the local political elite in cooperation with the Irrigation Department. In most parts, however, "water distribution has not been taken up by [WUAs] and Distributory Committees, the power of [WUAs] to collect fees/water rates has remained unutilized, and head-tail issues have not been addressed" (Mollinga et al., 2004, p. 255). Thus, ten years after the creation of 10,800 WUAs in Andhra Pradesh, net canal irrigated land has decreased by 40,000 ha. At the same time, without the involvement of WUAs, the net irrigated land covered by tube-wells in AP has shot up by a staggering 1.2 million ha (MOA, 2010).

An exhaustive review of 108 cases of irrigation management transfer in Asia (Mukherji et al., 2010) concludes that PIM had not produced fruitful results in developing countries. In most of Asia, it "has neither significantly improved productivity, operation and maintenance, nor has it produced other net benefits" (Mukherji et al., 2009, p. 18).[26] The review concludes that lack of success is due to a conceptual failure, not an implementation failure: PIM as a policy is not participatory.

Other detailed studies carry the debate to the next level. Parthasarathy (2006) compares the discrepancy between objectives and outcomes of PIM programs to 'an open pair of scissors:' while targets of area to be covered by PIM go up, achievements go down. He goes on to raise the question whether government agencies should continue to retail water in light of the failure of PIM. Rather, this calls for new approaches that concentrate responsibility and accountability to manage water distribution. Unfortunately, PIM has tended to undermine spontaneous, self-creating local institutions such as private irrigation service providers that have emerged in numerous locations.

---

26 A PIM intervention was defined "as successful when there is a marked improvement after transfer, or transferred systems fare better than non-transferred ones because *users receive adequate and reliable supply of water at reasonable and affordable costs over a sufficiently long period of time enabling them to increase their crop production, productivity and incomes*" (Mukherji et al., 2010).

This last point may be the most important in understanding how best to move forward. Would WUAs do better if they were genuinely given full autonomy, including over choice of technology, management systems, fee collection, implementation of upgrading programs, and all the money that can be collected from selling water—i.e., if irrigation departments changed their stripes and became fully cooperative? The answer in most cases would be 'no,' because WUAs by their nature are diffuse entities, with no clearly identifiable individual or corporate entity that is accountable to meet service standards and that can enforce contracts for water delivery and payment. In other words, the high transaction costs involved in reaching decisions tend to impede successful outcomes, especially in the complex social conditions of Indian villages.

In contrast, PIM can be successful in the few instances where the local social conditions (caste, ethnicity, education, income) are relatively homogeneous and/or where farmers have large land holdings, grow high-value crops, and may be tied into the international market. These conditions can be found, for example, in South Africa[27] or in Latin America, but rarely in India.

In India, some WUAs are known to have hired an individual on a contract basis to carry the responsibility for representing the WUA vis-à-vis the government. This comes closer to a private sector model that does not try to artificially re-create an idealistic community structure. As the next section elaborates (Shah and van Koppen, 2006, p. 3420),

> [t]he rise of a class of intermediaries between users and natural sources of water—in the shape
> of water service providers—is a precondition to meaningful demand management.

Private irrigation service providers—an alternative paradigm to WUAs

In virtually all infrastructure projects in India, land acquisition is one of the single biggest factors causing delays. However, unlike most roads, factories, or power plants, many irrigation systems can be run underground at a cost that today is not unreasonable considering the time, land, water, and energy savings. Already in 2000–01, several million hectares were irrigated with underground conveyance systems according to the 3rd Minor Irrigation Census. Retrofitting canal irrigation by piped systems is being done in several countries now (even in countries not suffering from land shortages, such as the US, Canada, and Spain), and evidence suggests that land values and farmer demand would justify such retrofitting in India as well. Indeed, this is the best way to use surface water to provide the pressurized water required for micro-irrigation (the alternative today is groundwater pumping, as required for participation in the National Mission on Micro-Irrigation). If there is a demand for piped water (including for drinking purposes), then a market is likely to develop if it is allowed to—all the more so as piped water allows volumetric pricing.

Indeed, in the context of the stalled Sardar Sarovar Project distribution system, it was seen that farmers had begun to lift water from the minors as soon as water began to flow, and were

---

27 An example of a successful WUA was mentioned at the WALMI/DSC meeting in Gujarat (WALMI/DSC, 2010): a 50-mile canal system in South Africa, serving 20,000 hectares owned by only 60 farmers.

conveying it in piped networks to their fields. The designers of the SSP consider this contrary to the plans and deem it 'illegal.' However, a slight shift in paradigm could turn this situation into an opportunity, as has happened on quite a broad scale in Maharashtra's Upper Krishna Basin (and as is happening, in different circumstances, in China—Box 3.1).

In Maharashtra, the pressure to use unutilized water from incomplete irrigation projects led the irrigation department to enlist the active support of private operators (through bank credit and electricity connections) to lift water from a series of storages created by diversion weirs along tributaries of the Krishna River. As described by Shah (2010), the irrigation department between December and June releases water on a 15-day schedule to fill up the dykes, starting with the lowest one first and thus protecting tail-enders. There are no canals, and the water is conveyed by the private operators through an extensive network of plastic pipes (70 percent of which are buried underground) up to a distance of 30 kilometers (km), thus avoiding the need for costly land acquisition, and reaching an area larger than the original design command. These schemes—of which there may be as many as 100,000 in Maharashtra—work very reliably and can offer irriga-tion-on-demand that is similar to tube-well irrigation. Many are owned and operated by farmer groups and cooperatives who have invested around Rs. 5,000 crore for a service that comes very close to what would be obtained from on-demand groundwater irrigation. As Shah (2010) points out, these lift irrigation schemes probably employ some 100,000 workers as water managers.

This is a perfect example of supply meeting genuine demand in the irrigation business. Irrigation departments have become bulk water providers in a system they are capable of manag-ing, leaving the distribution in a simple way to private and/or cooperative operators who serve as efficient and competent intermediaries with the farmer. Since there is no distribution canal sys-tem, these operators have full control over the bulk water once it has been delivered to the dykes along the river, main or branch canal and can therefore operate with the necessary reliability to be able to be held accountable to those who purchase the water. The use of pipes, rather than open canals, significantly enhances the reliability and precision with which irrigation services can be provided, not to mention the fact that their use reduces water consumption (as a result of lower evaporation), saves valuable agricultural land, and avoids frustrating land acquisition disputes common in the development of open canal irrigation systems.

Groundwater—the great equalizer

Declining farm size is a critical factor for understanding India's water economy:[28] The need to maximize returns to scarce land—not water—has led farmers to seek groundwater without which it would be difficult, if not impossible, to obtain multiple high-value crops year-round, as doing so requires reliable, on-demand, just-in-time small amounts of irrigation water in a decentralized

---

28 As earlier discussed, the dependability of, and easy access to, groundwater alone is enough to make it the preferred choice for higher-value agriculture, even in the US, where farm size is generally not a constraint.

manner. Small farmers also tend to be in the tail reaches of canal commands (with unreliable or no water) or altogether outside such commands, so groundwater for irrigation is their only choice.

A recent study of four villages in Andhra Pradesh is illuminating. Two of these villages have existing canal infrastructure; the other two rely entirely on groundwater but are supposed to become beneficiaries of the Polavaram mega irrigation project about to be constructed.[29] In the villages with canal infrastructure and WUAs, authorities allowed unauthorized water withdrawal in the upper reaches of canal irrigated areas ("the [irrigation] department is unwilling even to go on field inspections if they anticipate trouble." Sharma et al., 2008 p. 222) and insisted on dealing only with WUAs, even though these were known to be corrupt (according to the villagers, "the situation was better when there was no WUA, because the Irrigation Department officials were somewhat more accountable" (Sharma et al., 2008, p. 222).

Not surprisingly, 73 percent of large landholders (belonging to dominant caste groups) are in the head reach, against only seven percent of smallholders (Sharma et al., 2008, pp. 235–6):

> richer, more powerful farming communities have been observed to move into new irrigation areas, buying out small pockets of scattered lands belonging to the poor. For these communities who are connected to influential networks, irrigation plans and designs are known in advance, leading to a significant amount of land trading and consolidation even before the water flows through the canals. . . . Studies of the irrigation maps . . . show that at the macro level, irrigation plans have followed conscious political designs, and at micro levels, canal pathways are defined by elite interests and needs.

Further, it is revealing that the village with the largest amount of canal irrigation has the least amount of crop diversification, while the village relying 100 percent on groundwater had the most dynamic farming system (Sharma et al., 2008).

By favoring wealthier farmers in head reaches, canal irrigation is bucking the much broader trend toward increasingly smaller plots of land dependent on ground water. By 2002–03, the system of partible inheritance had led to some 107 million operational holdings and reduced average area operated per holding to 1.06 ha (Figure 3.4), split up, on average, into 2.3 separate parcels per holding. As Table 3.2 shows, marginal holdings of one hectare or less now constitute 70 percent of all operational holdings (up from 39 percent in 1960–61); farms between 1 ha and 4 ha make up another 25 percent. The percentage of large holdings (greater than 10 ha) declined the most—from 4.5 percent in 1960–61 to 0.8 percent in 2002–03, and large farms now operate 12 percent of the total farming area in India, down from 29 percent in 1960–61 (NSSO, 2006a). In other words, all farm sizes have, on average, been affected by the trend of decreasing size.

---

29 The Rs. 9,000+ crore Polavaram Project is expected to displace some 145,000 people in Andhar Pradesh, Chhattisgarh, and Orissa, and submerge several archeological sites. The project is affecting especially tribal populations who have barely been included by the state in the decision-making process (Bondla and Rao, 2010).

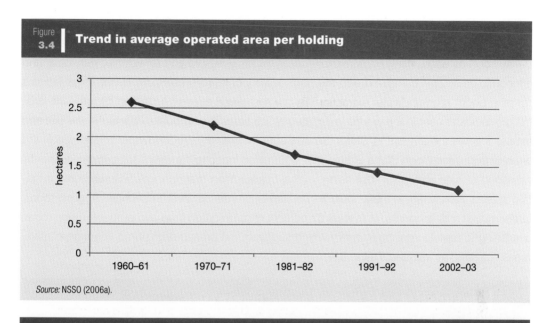

Figure 3.4 | Trend in average operated area per holding

Source: NSSO (2006a).

Table 3.2 | Changes in the size distribution of operational holdings

category of holdings	percentage of operational holdings				
	1960–61	1970–71	1981–82	1991–92	2002–03
marginal (<1 ha)	39	46	56	63	70
small (1–2 ha)	23	22	19	18	16
semi-medium (2–4 ha)	20	18	14	12	9
medium (4–10 ha)	14	11	9	6	4
large (>10 ha)	4	3	2	1	1
all sizes	100	100	100	100	100

Source: NSSO (2006a).

These country-wide trends mask considerable differences among states. For example, in the agriculturally most developed states of Punjab and Haryana, land concentration has increased, so that now two percent of Punjab's largest farmers operate 19 percent of the land while two-thirds of the farmers have marginal land holdings (in Haryana, the equivalent numbers are one percent operating 15 percent of the land).

The pressure on small and marginal farmers to maximize the return to land is further increased by relatively high (unofficial) tenancy rates, at least in some states. Although formally outlawed in most parts of India (and therefore under-reported in National Sample Surveys), land lease markets are quite strong and driven by the value-added that irrigation can provide. In western states, this has brought about much greater flexibility in the types of tenancy contracts, while in the north-eastern states, the classical 50:50 crop share remains the most common form of tenancy.

As soon as water lifting equipment became widely available and affordable, the social and economic pressures, combined with the poor delivery of irrigation services by the state, led to a massive boom in groundwater extraction. By the mid-1980s already, India had become the largest groundwater user in the world—estimated at 2.5 times the amount extracted in the US; and almost 3 times the amount in China (Shah, 2009), where most groundwater is pumped in the densely populated North China Plains (over 300 million inhabitants) with a temperate continental monsoon climate and lower (but rising) evapo-transpiration than in India.[30] Small pump irrigation spread everywhere in India, including in canal commands, and it continues to grow today. Groundwater now accounts for at least 67 percent of net irrigated area and probably much more, according to remote sensing imagery which suggests the presence of pump irrigation in many areas classified as 'rain-fed' (Table 3.3).

Table 3.3	**Net irrigated area: average annual growth rates**				
year	surface water canals and tanks		groundwater wells		of which: tube-wells
	million ha	Average growth for preceding period (%)	million ha	average growth for preceding period (%)	average growth for preceding period (%)
1960/61	14.9	2.3	7.3	2.0	
1970/71	17.0	1.3	11.9	5.0	
1980/81	18.5	0.9	17.7	4.0	
1990/91	20.4	1.0	24.7	3.4	4.1
2000/01	18.4	-1.0	33.8	3.2	4.7
2007/08	18.5	0.0	37.8	1.6	2.1

Notes: Satellite imagery (Thenkabail et al., 2006; 2009) suggests the following:

net irrigated area: 113 million ha (i.e., 80 percent of area sown is irrigated at some point)—of which 70 million ha involve groundwater use outside major command areas; 43 million ha may involve some conjunctive use.

net area sown: 140 million ha.

total cultivable land: 182 million ha.

annual water withdrawals for irrigation: 400 bcm surface water; 221 bcm groundwater.

Source: MOA (2009).

---

30 As explained in Box 7.3, there are further reasons in addition to the more temperate climate (and the relatively small area planted to paddy) why less groundwater is used in the North China Plains—the area where most groundwater is used in China—even though farm sizes are even smaller than in India. (i) Foremost, electricity is metered, fees are collected, and deep well pumping becomes very expensive; (ii) in many areas, groundwater permits have been rather strictly enforced by village party leaders; and (iii) By and large, hard budget constraints and accountability of local government to provide reliable services implies better surface irrigation services as well (Shah et al., 2004), though surface water is becoming increasingly scarce as competition grows from urban and industrial sources. River basins in the North China Plains are effectively closed.

For the vast number of small farmers, the benefits from tapping India's groundwater resources could not possibly have been achieved with canals and tanks which traditionally favored larger farmers. The comparison with industrialized countries highlights this particularly well: surface irrigation, as practiced for example in the Murray Darling Basin in Australia, is a viable solution only where there are relatively few customers, where the needs are relatively homogeneous, and where formal regulations can be enforced to ensure discipline in the system (though even in Australia's Murray Darling system, the government has been unable—indeed unwilling—to curb excessive surface water abstraction, despite deteriorating environmental conditions (Connell, 2006)). In India, the stories of tail-enders not getting water are legend: in Haryana's major irrigation systems, up to 84 percent of tail-enders are water-deprived; in Orissa, up to 72 percent are affected (Shah, 2009). Fewer than 10 percent of smallholders benefit from canal irrigation. Meanwhile, farmers at the head reaches typically planted water-intensive crops and used irrigation water not for protective irrigation in accordance with the original system design, but for full irrigation.

So farmers have voted with their feet (and their wallets). From the 1970s onward, as the state increasingly broke down and O&M of existing facilities fell into disarray, a fundamental paradigm shift took place: Public investment gave way to private investment on a massive scale, by adding as many as 25 million pumps and boreholes at a cost of well over Rs. 60,000 crore (Pearce, 2007). What had been a supply-driven system has now become a demand-driven one. Between 1980–81 (when net area irrigated by surface water and by ground water was about the same) and 2007–08, farmers added 20 million hectares of net groundwater irrigated land (Table 3.3), mostly at their own cost, but benefiting, to some extent, from subsidies for electricity and/or diesel, as well as for drilling equipment and pumps. Already 25 years ago, 70–80 percent of the value of irrigated production was estimated to be based on groundwater, contributing almost 10 percent of India's GDP (Dains and Pawar, 1987; World Bank, 1998a). Moreover, as mentioned earlier, groundwater is being used far more efficiently—even if still not sustainably—than surface water: 400 bcm of surface water is irrigating a net area of about 18.5 million ha, while 220 bcm of groundwater is irrigating more than double the net area.

To get to this point, small farmers jumped on the pump irrigation bandwagon faster than larger farmers: between 1970–71 and 1995–96, pump irrigation by marginal/small farmers (under 2 ha) shot up by 325 percent, while for large farmers (larger than 10 ha) it increased 55 percent. Small farmers have ended up owning about half the wells and pumps (Shah, 2009). Now they have direct access to water, and it is available on demand and just-in-time, assuring twice the yield that would have been available from canal irrigation. In most areas, the link between irrigation and canal/tank command areas has been broken. As mentioned earlier, Punjab is a good example, where the old command areas have become largely irrelevant—partly a historical holdover from times when farmers lost trust in the canal system which had become heavily politicized. In Punjab today, gross area irrigated is about the same inside and outside command areas (NSSO, 2005b).

India-wide, the groundwater revolution has probably been the single biggest factor leading to poverty reduction, for four inter-related reasons:

- It is a more 'equitable' technology than public surface canal irrigation, and it does not discriminate against small farmers the way government subsidies do (MSP, fertilizer subsidies, etc.);

- Just-in-time irrigation drought-proofs crops far more by reducing the risk of failure if the monsoon fails (by making irrigation possible during peak moisture stress);

- By providing year-round water, pump irrigation allows diversification and intensification to higher-value foods (vegetables and fruits require dozens of small quantity waterings a year) and dairying. Gross water productivity for vegetables and fruits, such as tomatoes, is 19 times that for wheat, and more than 10 times that for rice. Milk has overtaken rice in gross value of output—78 percent of landholders with less than 0.01 hectares are engaged in dairy production (NSSO, 2005b);

- At least in areas served by free or flat-rate electricity, an active water market allowed those too poor to afford their own irrigation pumps (or those who chose not to purchase pumps) to partake in the benefits of on-demand water, because pump owners could pump and sell extra water at little or no marginal cost.

In many ways, "the land-starved smallholder is the actor-in-chief in the current irrigation drama" (Shah, 2009, p. 232).

Water markets

By 2000, at least one in six farm holdings owned a water extraction mechanism. Today, the ratio is about one in four. This has led to vibrant water markets in which pump owners share their pumps with neighbors who cover the variable costs and contribute to overhead. A survey in western Uttar Pradesh highlights the 'egalitarian' effect of pump irrigation: more than half of large farmers purchased pump irrigation services, while 52 percent of small and marginal farmers were water sellers (Shah, 2009). Recent evidence from UP shows that caste hierarchies are breaking down (Pant, 2004). In other parts of the country, pump irrigation has tended to favor wealthier farmers who could afford the initial investment in pump sets, and then sell water at elevated prices to poorer farmers.

Where farmers have access to electricity for irrigation, the practice of State Electricity Boards (SEBs) to charge farmers a flat fee or no fee (originally intended to save on metering and billing costs) provides a boost to water markets because electricity costs simply represent overhead. Millions of smallholders who would never have access to canal water and could not afford their own pump sets can now obtain irrigation water.

Table 3.4 | Geographic distribution of electric and diesel irrigation pumps (2003)

state	percentage of farmer households irrigating land using	
	diesel pumps (%)	electric pumps (%)
eastern India		
Assam	87	4
West Bengal	87	13
Bihar	97	2
Orissa	61	38
Jharkand	81	2
Uttar Pradesh	84	16
western and southern India		
Andhra Pradesh	20	78
Chhattisgarh	28	63
Gujarat	35	63
Haryana	53	47
Karnataka	7	89
Kerala	15	85
Madhya Pradesh	34	65
Maharashtra	12	87
Punjab	29	71
Rajasthan	61	34
Tamil Nadu	27	72
all India	66	

Source: NSSO (2005b) and Shah (2009).

Of course, this dynamic is claimed to be at the heart of the problems faced by State Electricity Boards, which have been running up large deficits, especially in peninsular states where water tables are dropping and water extraction costs are rapidly increasing. More than 70 percent of the over Rs. 30,000 crore farm power subsidies go to these hard-rock areas (Table 3.6), where a one meter drop in aquifer level corresponds to Rs. 1,100 crore worth of power subsidies a year (Planning Commission, 2007).

Having said that, as much as one-third of what is termed 'agricultural consumption' of electricity may be due to theft by commercial and industrial users: "State Electricity Boards and their employees may have benefited from the ability to hide inefficient functioning and what may be collusion in theft behind agricultural use" (Dubash, 2007, p. 48). To the extent that SEBs (and/or their successor unbundled companies) are being forced under the new regulatory regime in India to account more transparently for system losses, one might expect the published numbers for aggregate technical and commercial losses in some states to rise in the short term.

The situation is more complex in those regions (mostly eastern India) where farmers do not have access to electricity and therefore must use diesel or kerosene to power their pumps.[31] India-wide, 66 percent of farmer households irrigate land using diesel pumps (Table 3.4), which are only suited to pump shallow groundwater down to about 10 meters depth (in contrast to electric pumps which can go much deeper). For these farmers, diesel prices have increased substantially, despite subsidies (and may increase even more with price deregulation), and the cost of irrigation is much higher than for electric pump users (Table 3.5).

Table 3.5	Cost of irrigating sugarcane—diesel vs. electric pump	
	diesel pump (Rs. per acre)	electric pump (Rs. per acre)
own irrigation source	1,620	37
purchased pump irrigation	3,780	1,080

Source: Shah (2007b)—case study of Akataha village (Deoria, Eastern Uttar Pradesh).

Thus, in 1990, 1 kilogram (kg) of rice could buy 1 liter of diesel (Figure 3.5); by 2007, 6 kg of rice were required for 1 liter of diesel (Shah, 2007b). The effect of these price increases is borne more heavily by the poor who in the past purchased irrigation from monopolistic pump owners who inflated their pump rental rates every time diesel prices increased, even though the real price of pump sets decreased significantly over the last two decades (and pump owners can in any case avail themselves of subsidies). As a result, farmers who can afford it are adopting water saving crops and irrigation technologies, while many of the poorest farmers have reverted to dryland farming, which has caused a decline in pump irrigation in the eastern Gangetic Plain—a region where groundwater pumping often has a beneficial effect to reduce water-logging and flooding (Shah, T. et al., 2009).

In short, to the extent that farmers have substituted private capital and institutions for public capital, the electricity and diesel subsidies have not been entirely unjustified. Far more than the fruitless capital investments in canal irrigation, or food and fertilizer subsidies which have mostly benefited wealthier farmers in a few states, the electricity and diesel subsidies have been more

---

31 The absence of electricity for irrigation in large parts of the Eastern Ganga Basin (eastern Uttar Pradesh (UP) and Bihar) arises from the fact that SEBs could no longer afford the power subsidies. In eastern UP, they let the electricity system atrophy in favor of continuing to serve a politically better connected constituency in western UP (where the Food Corporation of India has traditionally made its MSP purchases). By contrast, Orissa introduced metering, which forced farmers to switch to diesel.

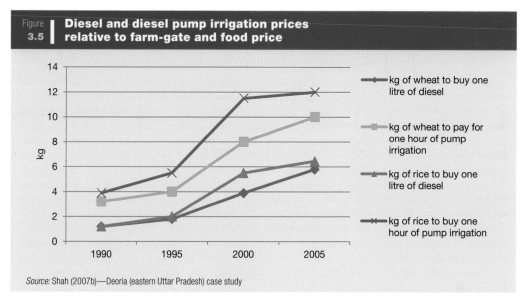

Figure 3.5 | Diesel and diesel pump irrigation prices relative to farm-gate and food price

- kg of wheat to buy one litre of diesel
- kg of wheat to pay for one hour of pump irrigation
- kg of rice to buy one litre of diesel
- kg of rice to buy one hour of pump irrigation

Source: Shah (2007b)—Deoria (eastern Uttar Pradesh) case study

equitable and have contributed substantially to poverty reduction and higher farmer incomes—although the growing divergence between electricity and diesel prices is increasingly pricing poorer farmers out of the market. Going forward, the big question is how to manage this irrigation-power nexus. What kind of strategy can support and sustain these private investments? To start, the economics of pump irrigation are more favorable than those of flow irrigation, with capital costs per irrigated hectare only 20–25 percent of new (and about the same as rehabilitating) major and medium canal irrigation. Moreover, as pointed out earlier, pump irrigation consumes at least 20 percent less water per ha than canal irrigation (Shah, 2009).

Making groundwater use sustainable

To understand the groundwater-power nexus and how it might be managed, this chapter examines four broad themes:

- Groundwater depletion and the hydro-geologic setting
- The role of electricity as a mechanism to control groundwater (and ways to improve fiscal sustainability)
- Implications for public investment policy
- The comparative advantage of different regions in India for various types of agriculture (considering groundwater depletion and energy costs, labor costs, etc.)

Groundwater depletion and the hydro-geologic setting

At the risk of oversimplification, there are two types of hydro-geological conditions in India which lead to different institutional dynamics of groundwater use (Figure 3.6 and Table 3.6): (i)

Box
3.1 | **How different are China and India?**

Both the similarities and the differences between the two giant countries are instructive.

China and India both have large agricultural populations with high population densities. Both have a high proportion of agriculture under irrigation, and a similar reliance on groundwater, (70 percent of net irrigated area is served by groundwater in the North China Plains, home to 200 million people and breadbasket of China) which, in turn, has spawned lively water markets. Just as in India, aquifer depletion in the North China Plains is very location-specific, implying that there are no one-size-fits-all solutions. The government is unable—or unwilling—to control excessive groundwater pumping despite the existence of numerous laws and regulations for this purpose (Wang et al., 2007). Further, in many parts of China, surface irrigation services are at best mediocre, irrigation charges are correspondingly low, and studies show that farmers would not be willing to pay more—let alone would be able to pay more—to cover the full water supply costs. Some already feel pushed against the wall for having to pay 'unofficial fees' (Liao et al., 2008). From the Indian experience, it also sounds somewhat familiar when we learn that farmers hardly participate in decisions related to irrigation management. However, there are also examples of well-managed surface irrigation systems in which the local government enters into long-term contracts with service providers who must maintain specific standards and also compete to retain the loyalty of groundwater users.

Interesting—though not surprising—is the observation by institutional experts that China's water sector suffers from an excessive supply focus (Xie et al., 2009), which has led to numerous poorly conceived irrigation (and other) projects with dubious economic justification.

More revealing are the differences between China and India, which merit some reflection. First, a simple yet stark difference: electricity—including for agricultural uses—is metered (both at the point of use and at the transformer) and charged throughout China. If nothing else, this sets a natural limit to the amount most farmers are willing to spend in their search for deep groundwater. A part-time farmer-electrician functions as a commission agent of the township electricity bureau to collect the fees.

It is somewhat harder to describe the structure and workings of government, and in particular the water sector institutions, at the six different government levels in China. The most prominent feature is the substantial authority of the local government at the township and/or village level. This level must support itself entirely out of local taxes and is therefore directly accountable to the local population for the quality of services it provides. It forms a kind of anchor for the enormous, extremely fragmented national bureaucracy. Even by Indian standards, the water sector, for example, is large—with some 40,000 staff deployed in water bureaus throughout each province, not counting locally-funded village staff. The vertical and horizontal fragmentation would be worse if it weren't for a bottom-up initiative coming from villages and townships to consolidate all water related agencies into *water resource bureaus,* then broadening their roles by renaming them *water affairs bureaus* (Shah et al., 2004). This is certainly an effective first reform step which is expected to yield much greater benefits once the functions of the Ministry of Water Resources at the top have been consolidated and the vertical fragmentation is reduced, if not removed. Purists who compare the Chinese system with that in some industrialized countries gripe at the lack of organization into river basin agencies. However, moving in that direction at this stage of Chinese development would take the only effective part of the government—namely the local government—out of the picture. The 'nine dragons' (meaning many masters) will no doubt continue to manage China's water for quite some time. However, the process set in motion—particularly from the village/townships upward—implies a genuine and overdue reorientation of the large water bureaucracy from water resources *development* to water resources *management.* In pursuing this gradual reform, the authorities decided to maintain the size of the bureaucracy but imbue it increasingly with a service-oriented spirit.

aquifers in alluvial and coastal/floodplain sediments that are found especially in the Indo-Gangetic Plain—so-called unconsolidated formations of porous sand or clay that have a high capacity to store rainwater and therefore a very large renewable groundwater potential[32], and (ii) hard rock aquifers (in consolidated geologic formations) of peninsular India, with much more fragile replenishable groundwater resources, often with low storage capacity. On average, about two-thirds of the annual replenishment of groundwater occurs through rainwater during the *kharif* season; the remaining third comes from canal seepage, return flow from irrigation, seepage from water bodies, etc. (CGWB, 2006). There are also much deeper aquifers, mostly below the alluvial areas of the Indo-Gangetic Basin, which get recharged only very slowly or not at all. The quantity of water stored here is estimated to be some 25 times as much as the amount of annually recharged groundwater at shallower levels—about 10,800 bcm. The Central Ground Water Board has suggested that part of this water could be beneficially used without harm to the aquifer (Planning Commission, 2007).

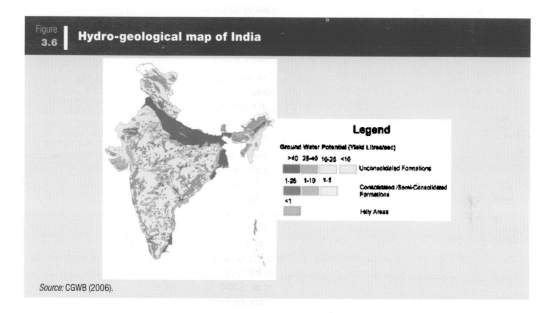

Figure
3.6 | **Hydro-geological map of India**

Source: CGWB (2006).

Most of the debate about groundwater depletion refers to the hard-rock areas of peninsular India and to parts of the arid alluvial plains in northwestern India. In just six states (Gujarat, Haryana, Maharashtra, Punjab, Rajasthan, and Tamil Nadu), 53 percent of assessment units (Blocks, Mandals, or Talukas) assessed jointly by State Ground Water Departments and the Central Ground Water Board in 2009 are semi-critical, critical, overexploited, or saline (against

---

32 For example, the annual replenishable ground water potential in the Ganga basin (171 bcm) is more than four times the amount in the next most prolific river basin, the Godavari basin (41 bcm) (Rao 2004).

Table 3.6 | Hydro-geologic settings and social behavior

major alluvial plains			
arid	upper and trans-Gangetic Plains	Decreasing total agricultural productivity; secondary salinization (freshwater lens on top of saline deep ground water of marine origin; soil salinization and sodic lands); micronutrient deficiency; ground water depletion.	Farmers get together to raise enough capital to maintain large tube wells capable of reaching increasingly greater depths. Water is distributed according to land holding.
humid	lower and middle Gangetic plains	Area prone to water-logging and flooding due to improper drainage. Low seed replacement rate. Best potential for agriculture; ground water abstraction helps reduce post-monsoon water-logging and flooding. Farmers mostly use diesel pumps.	Since these are heavily recharged aquifers, there is little incentive to cooperate to ensure sustainability of ground water. Each user operates in his own interest.
hard-rock areas	inland peninsualr India	Resource depletion and geogenic contamination of ground water: increased levels of fluoride, arsenic and iron affecting drinking water supply. Depletion of aquifers requires increasingly deeper, more powerful and expensive tube-well pumps that must run on electricity.	Depending on the extent of the local aquifer and the number of potential users, ground water extractors could cooperate (small aquifer, few users) or compete with each other (larger aquifer, many "anonymous" users).

Source: Shah (2009), Planning Commission (2008).

a national average of 27 percent), meaning that the annual groundwater draft was greater than 70 percent of the net annual groundwater availability (Figure 3.7). Where groundwater is overexploited, draft is greater than 100 percent of net annual availability, i.e., water is just being mined (Planning Commission, 2007).

However, ground water levels can change rather significantly over time. There are often also separate geological strata with differing aquifer properties. This makes it all the more urgent for the Ground Water Board to have a broad data dissemination program and to rapidly make public the quarterly monitoring results from its over 15,000 ground water observation wells.[33] Kulkarni and Shankar (2009) have recommended a National Ground Water Management Programme which would include aquifer mapping and characterization for the entire country at the scale of

---

33 The tardiness of disseminating current ground water information may be leading to substantial mis-investments in aquifer recharge as a result of States' reliance on outdated data. UP is already known to have complained.

Figure 3.7 | **Groundwater status (March 31, 2009)**

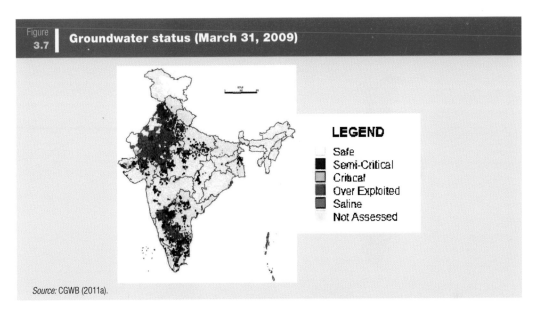

Source: CGWB (2011a).

watersheds of the order of 1,000 to 2,000 hectares. As they point out, "the exercise of resource mapping must empower the community to understand the resource and develop effective strategies of its protection" (p. 15). In a similar vein, the report of the Planning Commission (2007) recommends that the Ground Water Board should increasingly serve as a facilitator in working with states—and at lower levels—to design responses to changing ground water conditions.

Figure 3.8 compares the water level fluctuation in January 2011 against the average water level in the preceding ten-year period (2001–2010). This is an important comparison, as it shows that a significant rise in water levels has occurred in many parts of the country, especially in patches of Andhra Pradesh (108 Mandals have recorded increased water levels due to higher rainfall and more efficient water use practices); Tamil Nadu (55 Blocks—14 percent—show a reduction in groundwater draft and increased recharge, though 36 percent of Blocks remain overexploited); Western Rajasthan (13 Blocks show an increase in groundwater recharge); and Gujarat (increased water levels in 13 Taluks are due to good rainfall, rainwater harvesting and artificial recharge, and canal seepage). The map also shows where groundwater is consistently overexploited even though replenishable resources are mostly abundant: especially in Haryana, Punjab, and the Delhi metropolitan region; but also in parts of Rajasthan and southwestern Uttar Pradesh (CGWB 2011a, CGWB 2011b).

Since rainfall contributes about 68 percent of total annual replenishable resources, groundwater level fluctuations (Figure 3.8) are heavily influenced by rainfall levels. These have been more or less normal in the period 2004–2009—even above average in parts of peninsular and western India. That said, this leaves almost one-third of the annual replenishable resources to be determined by explicit or implicit 'policy' choices such as more efficient water use practices—water

II/3

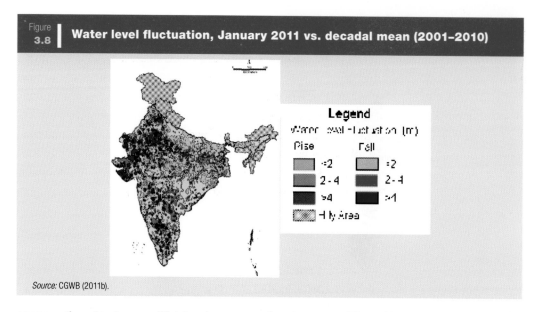

Figure 3.8 | **Water level fluctuation, January 2011 vs. decadal mean (2001–2010)**

Source: CGWB (2011b).

conservation structures, artificial recharge, canal seepage, etc. These kinds of measures take on much greater importance before and during drought years, when groundwater is crucial to make up for scarce rain and surface water resources.

Some of the most severe ground water depletion and salinization has occurred in those states where Food Corporation of India procurement traditionally took place and where the Minimum Support Price (MSP) subsidy encouraged double cropping with the help of tube-well irrigation on most of the sown areas. For example, in Punjab, 90 percent (for *kharif*) to 96 percent (for *rabi*) of cropped area is irrigated by tube-wells using groundwater (NSSO, 2005b). In that state, only 23 of 138 blocks were deemed safe in 2009; 110 were over-exploited, and another 5 were in critical or semi-critical condition (CGWB, 2011a). To address the problem of groundwater depletion, Punjab has suggested extending the MSP to additional crops to wean farmers away from water-intensive rice cultivation, and to encourage contract farming for horticulture (e.g., chick-pea). Other measures contemplated include large-scale artificial groundwater recharge, control of electricity supply, and micro-irrigation (Planning Commission, 2007). The Punjab Preservation of Sub-Soil Water Act 2009 is intended to reduce annual groundwater draft by 7 percent by requiring a delay in paddy transplanting until after June 10 as a way of avoiding the high evapo-transpiration in the early summer. IWMI has estimated that this will save about 175 million kWh of electricity that would have been used for groundwater pumping each year (IWMI, 2010).

Again, it is important to mind the details. Shergill (2007) has pointed out that the annual water consumption requirement for the wheat-rice rotation, which is blamed for excessive groundwater withdrawals in Punjab, is less than for sugar cane, moong-sunflower, moong-winter maize, and it is about the same as for the cotton-wheat rotation. He also suggests that the fall in the water table has not "crossed the danger mark" (Shergill, 2007, p. 81) given the observed fluctuations in the well depths. This proposition will need to be tested with the most recent data, comparing water levels both before and after the monsoon, and accepting that conditions could vary significantly from block to block.

The groundwater situation in northwestern India merits rather close scrutiny, given the large contribution these states make to agricultural production and government food purchases in India. Punjab's notion of extending the MSP to additional crops suggests that not only could this incentive be used to influence crop choices in Punjab, it could also drive a gradual shift of the cereal agriculture toward the Gangetic Basin where ground water is generally more easily available—indeed, where additional groundwater abstraction could stem water-logging and reduce flooding.

Long-term scenarios

Assuming a continuation of current practices, Amarasinghe et al. (2007b) have projected groundwater depletion for 2025 and 2050 (Figure 3.9).[34] The dominant use of groundwater for irrigation (up to 4 times more efficient than surface water)[35], changing food consumption patterns (more animal products, vegetables, milk), declining demographic growth in key peninsular states,

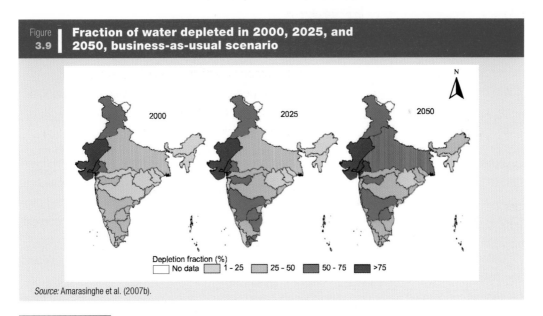

Figure 3.9 | **Fraction of water depleted in 2000, 2025, and 2050, business-as-usual scenario**

Depletion fraction (%)
☐ No data ☐ 1 - 25 ☐ 25 - 50 ■ 50 - 75 ■ >75

Source: Amarasinghe et al. (2007b).

---

34 The Depletion Fraction is defined as the ratio of evaporation to potentially utilizable water resources. Evaporation consists of process evaporation (evapo-transpiration from irrigation and transpiration from domestic and industrial sectors) and non-process evaporation (from swamps, reservoir and canal surfaces).

35 Chambers (1988) demonstrated more than 20 years ago that groundwater irrigation provided much higher yields per hectare than canal or tank irrigation. For example, average food grain yields in Tamil Nadu in 1977–79 were 6.53 tons per net groundwater irrigated hectare (t/ ha), against 2.60 t/ha using canal irrigation.

and higher irrigation efficiencies are expected to decrease water demand for irrigation by 2050, even from a business-as-usual perspective and taking into account environmental flow require- ments. By 2050, 85 percent of the *additional* water demand will be for industrial and domestic use, and there will be a growing amount of water trade between irrigation and urban users (as is already happening in Chennai, where farmers are selling their water to the city—not always with beneficial results, as it often means they must give up farming without getting alternative livelihood opportunities). Figure 3.10 shows the current status of water use for irrigation, and for domestic and industrial needs.

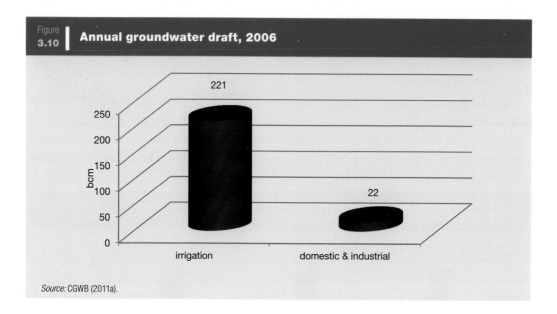

Figure 3.10 | **Annual groundwater draft, 2006**

Source: CGWB (2011a).

Amarasinghe et al. (2007b) calculate that groundwater availability would be sufficient, *on aver- age,* to meet most of the future food demand. However, the business-as-usual model predicts significant *regional* water crises by 2050, with depleted aquifers and significantly reduced river flows in specific areas. In several river basins, essentially all available water would be consumed, possibly leading to growing inter-state conflicts where these river basins cross state boundaries.

Unfortunately, the mechanisms to resolve such inter-state conflicts are extremely weak and are unlikely to be improved anytime soon: water and irrigation management are state matters, and the centrally appointed water disputes tribunals have taken decades to reach decisions that have subsequently not been enforced by the states, or have been appealed to the Supreme Court (Gujja et al., 2010). Worse, Maharashtra, for example, has built a series of worthless storage facili- ties just for the purpose of enhancing its claim to water from the Narmada River. All of this rein- forces states' tendency to tacitly promote groundwater use, which circumvents questions about inter-state water use. The fact that river-sharing agreements (such as for the Narmada Basin) make no reference to groundwater means that declining stream flows due to heavy groundwater abstraction may in the future require a reassessment of the allocations decreed by the Water Disputes Tribunal (Ranade, 2005).

Regulating groundwater use

Could a better legal approach help protect groundwater resources? The rules governing water use are defined by the Easement Act of 1882, according to which access to groundwater is part of an easement attached to land ownership, while surface water is state property. According to Indian law, it is up to the states to develop appropriate incentives to protect aquifers. However, when ground water is extracted beyond its replenishable level, the Central Ground Water Board can, in principle, declare the area as being under environmental threat, and demand that the state regulate further exploitation under the Environment Act of 1986.

In practice, the weak institutional capacity of the state and corresponding absence of credible enforcement, and an inefficient court system mean that a formal regulatory approach cannot succeed—all the more, as millions of groundwater users are involved. In this sense, international experience from countries with well-developed water institutions is not very helpful. Even in industrialized countries, there are lower limits governing the size of landholdings subject to groundwater regulation: in Australia, groundwater diversion for up to 2 hectares is exempt from regulation. In India, this would imply that more than 86 per cent of farm holdings would be exempt even under Australian law (Table 3.2 and Table 3.7). In the US, California does not have the power to limit groundwater withdrawals, and groundwater is generally not metered (Box 3.2).

Table 3.7	Groundwater use in Australia and India		
	groundwater irrigated area	amount of groundwater abstracted	number of abstractors
Australia	5.5 percent	5 bcm per year	a few thousand
India	over 67 percent	221 bcm per year	25 million

Source: Shah (2008c).

That said, there is scope for meaningful laws and regulations in *urban* areas where water pricing and tradable water rights among agricultural, urban, and industrial users could usefully be developed. The urgency to address urban groundwater abstraction is highlighted by CGWB data from 2007 (Figure 3.7) as well as recent news reports, such as from Bangalore, which is reported

Box 3.2 | Why laws, regulations, and water charges will not solves India's groundwater problems—lessons from the US, Spain, and Mexico

Groundwater regulation is a difficult matter not just in India; it has proven to be intractable in many other countries as well—even in countries with a comparatively small number of users.

United States

There are a surprising number of similarities between the US and India with respect to groundwater, despite an enormous difference in the number of users. First, the crucial difference: while India has as many as 39 million irrigated farms (NSSO, 2005b; NSSO, 2006a) extracting 220 bcm a year using some 25 million groundwater extraction devices for irrigation, the US has fewer than 98,000 groundwater irrigated farms extracting 74 bcm of water a year from 408,000 irrigation wells.

However, some aspects of groundwater extraction are remarkably similar in the two countries. Of the 22.2 million ha of irrigated land in the US, 66 percent is served by groundwater (in India, at least 67 percent of irrigated land is served by groundwater). In the US, groundwater irrigated farms typically represent the most profitable farms. Aquifers have been deteriorating for years, and despite countless measures, there has been little progress in arresting the problem (average well depth has been increasing to 74 meters— USDA, 2010). States have formed actively managed groundwater districts, bought out water rights from users at great expense, supplied surface waters to make up for curtailed ground waterpumping, and restricted exploitation, especially by ordering irrigated land to be taken out of cultivation. Some aquifers have responded positively, while others continue to deteriorate because legally mandated recharge has not kept up with withdrawals. A comprehensive report on the issue declared that "data on ground-water levels and rates of change are 'not adequate for national reporting'" (USGS, 2003; Reilly et al., 2008).

The state of California is a good, albeit somewhat extreme, example. Even though the state has one of the most elaborate systems of surface water conveyance in the world, about the same amount of land is irrigated by groundwater as by surface water (USDA, 2009b). Half of all vegetables, fruits, and nuts in the US are grown in California, where 24,000 farms take water from 61,000 wells (USDA, 2009a; 2009b; 2010).

In California, groundwater is regulated by a patchwork of local laws, though most local governments (with some notable exceptions) are unable to seriously control overdraft despite legislation in 1992 which provided the authority to develop groundwater management plans. The state can limit withdrawal from surface reservoirs but not from aquifers. As a result, 11 out of 42 ground waterbasins are critically overdrawn (mostly in the agriculture-rich Central Valley), which has also led to substantial land subsidence. 25 basins are operating in accordance with court settlements—a time-consuming and expensive process because there is no statutory guidance on how to apportion groundwater. "Groundwater users have been able to prevent any significant change in California's groundwater law" (Ashley and Smith, 1999, p. 49). In part, the government's hands are tied because farmers have shifted from row crops to fruit orchards, which cannot be left dry regardless of the level of drought. In 2005, 2006, and 2007, the Governor of the state vetoed legislation to measure groundwater usage, saying that a state-run system would be too expensive and cumbersome (Barringer, 2009).

But the tide may be slowly turning, with the state starting to collect uniform data following an emergency drought declaration in 2009, though there remains considerable resistance. In the 2010–11 budget, there are proposals before the legislature to introduce mandatory comprehensive monitoring (failing which the state would deny grants and loans), introduce 'Active Management Areas,' and modify the law to reflect the physical interconnectedness of surface and groundwater (Legislative Analyst's Office (LAO), 2010). However, an attempt to require permitting of groundwater abstraction state-wide is likely to fail (again), even though control of implementation would be left to the local level.

| Box 3.2 | Why laws, regulations, and water charges will not solves India's groundwater problems—lessons from the US, Spain, and Mexico |

Spain

In Spain, the Water Act of 1985, plus later amendments, took groundwater rights away from individuals and conferred ownership of the resource on the state. The Act made river basin authorities responsible for issuing permits and restricting groundwater extraction in overexploited basins. In order to get a handle on groundwater abstraction in overexploited basins, the authorities required well owners to join groundwater user associations in those basins. Even though some 16 basins were declared overexploited, only 2 functioning user associations were active, and 25 years later, most groundwater structures still remain unregistered. There are about 500,000 wells in Spain (Varela, 2006).

Mexico

Mexico fundamentally reformed its water laws in 1992. Water was declared a national property, and each groundwater well required a concession, which, in turn, provided an entitlement to pump a certain quota of water each year. Industrial users and municipalities signed up quickly, while small household wells—too numerous accurately track—were excluded from the beginning. The problem faced by the Mexican authorities was how to track and enforce the groundwater abstractions of the nearly 100,000 agricultural users. An attempt to form Aquifer Management Councils to counsel farmers on alternatives and encourage mutual enforcement did not succeed. In the end, the government used metered electricity consumption as a surrogate for the amount of groundwater pumped, and pursued violators on this basis.

The Aquifer Management Councils failed because they could not attract farmers into an organization set up basically to operate against their interests. A move some years ago to withdraw unused portions of the groundwater quotas led to a pumping frenzy, during which farmers pumped more than they needed in order to not lose their quotas (World Bank, 2004).

to have depleted double its replenishable groundwater by abstracting from the deeper static groundwater table (Nataraj, 2010).

Experience with a regulatory approach is gaining momentum in Maharashtra, where the state's Water Resources Regulatory Authority Act of 2005 provides for volumetric bulk water entitlements which can be transferred, sold, or bartered (Briscoe and Malik, 2006)—though the decisions are unfortunately left to a political body rather than an independent regulator.[36] Tamil

---

36 The experience in Maharashtra, the first state to establish a water regulator, is not promising. Water allocation decisions there were vested in a High-Powered Committee of Ministers, and following a recent amendment of the MWRRA Act 2005, have been shifted to the Cabinet. Thus, a problem at the outset—leaving one of the key decisions to a political body rather than an autonomous regulator—has been perpetuated, if not even aggravated.

Precisely because water allocation decisions cannot be apolitical (any more so than what is happening in the electricity sector), this calls for a neutral forum in which a transparent political debate can take place that leads to undisputed regulatory decisions. States would therefore need to ensure procedural safeguards so that active public participation can take place, for example under the aegis of a truly independent regulator. Maharashtra has not followed this principle despite the high hopes when the MWRRA was introduced, and it is to be feared that other states won't do so either. The result will be growing public dissatisfaction with water allocation decisions, and a renewed retreat into indiscriminate groundwater abstraction because of a sense that there is no place to voice grievances and obtain redress, much less to simply ensure reliable service. This is all the more important as the strong voice of irrigation departments (and well-connected 'front-enders') in the political sphere of most states will drown out attempts by an enfeebled water regulator to achieve changes that, for example, would make AIBP more effective. As one observer has written: "credible regulation would have to truly provide a democratic space for decision-making," and social goals would have to be "an explicit part of regulatory objectives" (Dubash, 2008, p. 53).

Nadu introduced separate legislation just for metropolitan Chennai (Planning Commission, 2007). Municipalities and pollution control agencies could register and license urban tube wells (at least for large users), combining groundwater management with pollution regulation.

Electricity supply as a pivot to control groundwater

So why not price electricity at the marginal cost of supply, as economic reform advocates have been arguing? This would help remedy the fiscal hemorrhaging from electricity subsidies, and it could reduce groundwater abstraction. The answer is straightforward: were this to happen, irrigated area would fall by 15–18 million hectares, making agriculture unviable in most of peninsular India (Shah, 2007a), regardless of the quality of power supply. Chief Ministers that tried to introduce electricity metering for farmers found themselves voted out of office.[37] Other approaches need to be found that meet the same objectives in ways that are more consistent with the institutional constraints, and that will allow a long-term transition to a system of transparent subsidies and lower fiscal burden.[38]

Several states have developed mechanisms to start to deal with this problem. As mentioned earlier, Punjab, which provides free electricity to farmers, is encouraging a shift to high-value orchard crops suitable for water-efficient micro-irrigation, and the state is timing electricity supply to delay paddy transplantation by one month as a way to reduce evapo-transpiration in the summer. Andhra Pradesh (whose Chief Minister was elected on the promise of free power to farmers) provides seven hours of virtually free electricity to its 2.7 million tube-well owners and is promoting intensification of rice to reduce irrigation requirements. In both Andhra Pradesh and Karnataka, sluice gates of irrigation tanks are being sealed to convert the tanks into percolation tanks as a way of replenishing groundwater levels. In both states, however, electricity supply remains highly unreliable and subject to large voltage variations, so that farmers continue to have to spend significant amounts of money to maintain their equipment. Because of the unreliability of the supply, they also tend to pump more than they need when power is available, and leave machines switched on during power interruptions, creating spikes in the system when power is restored.

To get around these problems, Gujarat, in the period 2003–06, implemented what is arguably the most far-sighted and successful experiment with its *Jyotirgram Yojana* scheme. The cost was US$260 million—one-third of the Gujarat Electricity Board's annual loss in 2001–02. The scheme goes "with the grain" of what farmers need but applies some simple yet effective ideas to slowly alleviate the fiscal burden of unsustainable subsidies: a 'managed subsidy versus default subsidy'

---

37 Shah (2009) describes the immediate shift in political alliances as soon as the former Chief Minister of Madhya Pradesh tried to introduce electricity metering for farmers.

38 In the short medium-term, subsidies cannot be decoupled from electricity use (such as in an EU-style subsidy system) because farmer leaders insist on taking credit for ensuring electricity supply to their constituents (Shah, 2009). This will not change until electorates become more urban (at the earliest when electoral districts are redrawn following the 2011 census—typically five or more years later).

(Shah, 2007a). The scheme basically *involves separating feeders/power lines to tube-wells from those to domestic and non-farm users.* As implemented in Gujarat, tube-wells get 8 hours of full-voltage electricity per day on a strict, pre-announced schedule and at a flat tariff, while non-farm connections receive 24/7 assured electricity. At the same time, new connections and pump sizes are being tightly controlled.

As a result of these changes, power supply to agriculture fell from 16 billion units in 2001 to 10 billion units in 2006, groundwater draft wsa reduced by 20–30 percent, and Gujarat government's electricity subsidies have come down from US$786 million in 2001–02 to US$388 million in 2006–07 (Shah and Verma, 2008).[39] The most satisfied group of stakeholders in this scheme are rural housewives, students, teachers, patients, doctors, and all non-farm trades, shops, and cottage industries that get 24 hour uninterrupted electricity supply. Tube-well owners are also pleased with the quality and reliability of electricity supply, though they would of course prefer longer hours. Water markets have shrunk, however, since power rationing is effective and water prices have increased. This has affected, above all, landless laborers and tenants, many of whom have been forced out of farming as a result. The fact that it is the poorest who are most affected (and for whom an income support policy is therefore requested) offers an interesting insight into the benefits of the old policy which, it was usually claimed, benefited mostly the wealthy farmers (Shah and Verma, 2008).

As originally conceived by experts from IWMI, the *Jyotirgram* scheme would have included one further important component which could yet be added. Instead of supplying a straight 8 hours of electricity a day, electricity supply could imitate a high-performing canal irrigation system by providing electricity on a demand-adjusted schedule: up to 12 hours during 30–50 days of peak irrigation demand during the year (depending on the monsoon), and 4–6 hours a day during the rest of the year.

Politically (and technically), this is an extremely successful scheme, as it creates a wide support base among farm and non-farm users, while at the same time controlling groundwater use and limiting subsidies. As agriculture becomes more diversified and incomes rise, the flat tariff (already 70–140 percent higher than before *Jyotirgram* was introduced) will be gradually increased toward the average cost of supply. As described by Shah (2009), "there is nothing that … water laws, tradable ground water rights, ground water cess … can do that *Jyotirgram* cannot do better and quicker" (p. 216).

To be sure, there are skeptics who would like to see a more decisive move toward metering and a per unit tariff, arguing that this has been possible in China and Bangladesh, and that political pressures will shift toward other demands, such as for longer hours of electricity, (Dubash, 2007) and therefore vitiate 'rational flat tariffs.' In the Indian context, it appears, however, that the

---

39 In Gujarat, where 28.5 percent of total electricity generation goes to agriculture, 30 percent of groundwater blocks are critical, semi-critical, over-exploited, or salinity-affected (Planning Commission, 2007; CGWB, 2011a).

Jyotirgram approach is a meaningful first step, since it creates major supportive constituencies who see enough benefits to go along with a gradual increase in tariffs to long-run average levels, to the point in the future where there could be a more or less seamless transition to a per unit tariff.

Community regulation of groundwater abstraction

As Table 3.6 suggests (see last column), different hydro-geological conditions lead to varying approaches to groundwater management. Recent papers and reports (World Bank, 2010; 2009b) speak highly of an effort by over 500 farming communities in seven drought-prone districts of Andhra Pradesh to self-regulate ground water abstraction at low cost, while increasing farm incomes. The approach in this instance is an intensive process of farmer education through 'barefoot hydrologists' who raise awareness about the groundwater situation and about the effects of pumping groundwater at different times. As many as one million farmers participate in a process of crop water budgeting. The key to the program's success is the unique hydro-geological setting for the communities participating: the groundwater in this hard-rock area of Andhra Pradesh is confined to relatively small aquifers, so that the actions of individual farmers or communities have a direct impact on water availability. In such a situation, increased information and awareness can yield good results because each farmer has a stake in the joint resource and her actions cannot escape notice by other affected parties.

Another example often mentioned is that of Hivre Bazaar in Maharashtra (World Bank, 2009a), where, in 1993, a charismatic village figure prohibited the use of bore-wells for irrigation. Since then, a large number of dug-wells have been built, yielding less water for the community but on a more sustainable basis. The actions of the charismatic village leader thus transformed what would have been a competitive behavior for increasing amounts of water at deeper levels, to a cooperative behavior that optimized the benefits for the whole village. There may be many similar examples which have not received wider attention. Each example, though, is unique in that it depends on strong village leadership—a characteristic that is not intrinsically replicable.

Of broader interest are shared well irrigation communities which have evolved over long periods of time and are embedded in the 'texture' of society—such as in large parts of Punjab where the same social structure has survived technological changes from the Persian Wheel to submersible pumps. Water from these shared wells is distributed to each farmer according to land size—including to small and marginal farmers. The key to success here is an elaborate social structure that conveys legitimacy on the actions of the community. As Tiwary (2010) notes, water rights are respected by caste and village Panchayats, and the rules of the traditional shared irrigation system are internalized by all members of the village. "The emergence of new issues and adaptation techniques, frequent conflicts and dispute resolutions, keep reinstating the tenets of social organisation in collective consciousness" (Tiwary, 2010, p. 219).

Re-thinking government investment programs and Centrally Sponsored Schemes

"Perhaps, we can safely say that almost no benefit has come to the people from these projects. For 16 years, we have poured out money. The people have got nothing back, no irrigation, no water, no increase in production, no help in their daily life."

—Prime Minister Rajiv Gandhi, speaking on big irrigation projects to State Irrigation Ministers, August 1986

The analysis above suggests the need for strategies that are specific to the major hydrogeologic regions and that respond to local social and economic conditions.

- *In western and peninsular India,* where the deep aquifers in hard-rock areas require the use of electric pumps: manage groundwater abstraction by rationing electricity along the lines of the *Jyotirgram* program, and engage in a massive effort—at a decentralized level—to recharge groundwater, starting with the most seriously affected groundwater blocks.

- *Throughout the Indo-Gangetic Plain:* emphasize conjunctive management of groundwater, surface water, and rain, with appropriate recharge efforts.

- *In eastern India* and those areas of the Gangetic Plain which now mostly rely on diesel for groundwater pumping: re-introduce electricity supply to agriculture by applying an improved version of *Jyotirgram* that gives priority to marginal farmers for new tubewell connections.[40] Poverty (and/or emigration) is likely to increase substantially in these areas unless electricity is reintroduced for agriculture, because the cost of running diesel pumps is no longer affordable. Rural electrification (with separation of feeders, as in *Jyotirgram*) as a way to better access groundwater is listed as a priority in the Agriculture Chapter of the 11th Five Year Plan (Planning Commission, 2008, vol. III, p. 35).

In the longer term, the shift in the center of gravity of agriculture from the western states toward the northeastern Ganga Basin (eastern Uttar Pradesh, Bihar) may well lead to increased and more sustainable agricultural production. It would also provide substantial regional (interstate) and local equity benefits. Supporting this process will clearly have significant implications for India's agricultural subsidy regime. A quick review suggests that the subsidy burden could well be reduced while (i) providing additional support to agriculture in hitherto neglected areas; and (ii) largely maintaining support (but in a fiscally more responsible manner) in northwestern and peninsular India (Table 3.8). The open political question is to what extent a reduction of food-grain and other procurement in the dominant northwestern states can be offset by alternative and less costly farm support measures. For example, one could explore a neutral subsidy system along the lines of the reformed Common Agricultural Policy of the European Union and/or a Deficiency

---

40 To the extent that *Jyotirgram* gives priority to marginal farmers for new tube-well connections, Shah (2007a) points out that this can imitate the equity outcomes that were intended with land reforms.

Table 3.8	Eastward shift of agriculture—impact on subsidies and capital investments	
subsidy, capital investment	possible fiscal impact	
diesel		↓
electricity	→ (greater coverage but more controlled)	
fertilizer	↓ (decontrol possible over time if other input costs managed; eventually shift to income support subsidy)	
food/MSP	↓ (decontrol food procurement; shift limited government procurement eastward)	
surface irrigation—subsidies		↓
water-saving devices (pipes, etc.)	↑ (will reduce electricity consumption and groundwater depletion; avoid or stop subsidies to manufacturers; avoid confining subsidy for particular crops, such as wheat, as in National Food Security Mission)	
groundwater recharge programs		↑
surface irrigation—capital costs		↓

Source: Author's estimates.

Payment (e.g., as used in the US), which would cover the difference between the market price and the desired level of producer support, with less market distortion.

Groundwater recharge in the Indo-Gangetic Plain and conjunctive management

A shift of agriculture toward the east would reinforce the need to ensure the sustainable use of groundwater in that region. This may require a paradigm shift in how and when irrigation water is used, in order to carefully manage the interplay between surface and groundwater (referred to as conjunctive *management* as opposed to conjunctive *use*). Specifically, the idea involves providing irrigation water through unlined canals during the monsoon season for water-intensive crops like rice and sugarcane. Normally, irrigation departments supply water during the dry season on the assumption that monsoon rain is sufficient for *kharif* crops; excess water during the wet season simply runs off and is mostly not available for crops. The monsoon is often erratic, however, and water-intensive crops can benefit from full and reliable irrigation even in the wet season. By slowing down and diverting water through canals (unused drainage canals with check structures to slow water flow are particularly suitable) that would normally run off in the wet season, water is allowed to seep into the aquifer from where it can be pumped out for a dry season crop. In the process, annual yields increase, as do farmers' incomes, and pumping costs are reduced (IWMI, 2002).

In a similar spirit, the idea was raised several years ago to use Narmada River water to recharge groundwater in North Gujarat. Ranade and Kumar (2004) demonstrate the physical and economic viability to divert unutilized flows from the Narmada main canal into existing canal networks, village ponds, and tanks in North Gujarat to support a decentralized recharge process—at least within the command, if not also outside it (Ranade and Kumar, 2004).

These are just a couple of examples that show how the traditional thinking about the use of canal irrigation systems will need to shift toward a new paradigm in which far more emphasis is placed on aquifer recharge rather than provision of surface irrigation. Stated differently: "Managing the ground water reservoir ought to be the key aim of India's water policy" (Shah, 2008b, p. 45). To be successful, though, conjunctive management requires considerable institutional innovation on the part of irrigation agencies.

Groundwater recharge in western and peninsular India—the dug-well program and redesigning tanks

Nowhere is the urgency of recharging the aquifer more evident than in the hard-rock areas of peninsular India. This is where the government's 'Artificial Recharge of Ground Water through Dug-wells' scheme has been launched, specifically in the 100 districts in seven states accounting for over 60 percent of India's critical and overexploited blocks, and 62 percent of India's electric pump sets.[41] The scheme largely represents a redesign from an earlier 'Master Plan for Ground water Recharge,' along the lines of a detailed proposal put forward by Shah (2008b). In this proposal, Shah spells out all of the elements of a recharge strategy for India, reproduced in Table 3.9. The Master Plan envisaged raising post-monsoon groundwater levels throughout the country to eight meters below ground level (revised from three meters below ground level)—a verifiable objective to which a reasonable time horizon could be attached.

Once modified, old dug-wells (of which there are 9.6 million in India according to the Minor Irrigation Census 2000–01, MWR, 2001) are well-suited to capture monsoon runoff, and necessary O&M is confined to de-silting every 3–4 years. The government program is sensibly aimed at private dug-well owners who would receive a subsidy of 100 percent for holdings up to 2 hectares, and 50 percent for larger holdings. The economics of recharge—at Rs. 1,800–4,500 per ha of irrigated land—are far superior to the Rs. 1,80,000 required to add a hectare of irrigated land in a canal command. Indeed, the savings in electricity from the reduced pumping depth would pay for this several times over (Shah, 2009).

The challenge is how to catalyze a groundswell of self-motivated interest by farmers to engage in the task. Once alerted, farmers can see the benefits of dug-well rehabilitation within the same season and will have a vested interest to support a broad-based effort, as this is similar to accumulating savings in a bank account to protect at least the kharif crop, if not the crops of both

41 The States are Andhra Pradesh, Maharashtra, Karnataka, Rajasthan, Tamil Nadu, Gujarat, and Madhya Pradesh.

Table 3.9 | **Outline of an alternative recharge strategy for India**

key actors	arid alluvial aquifer areas	hard-rock aquifer areas	roles that need to be played by Central Groundwater Board (CGWB), Recharge Special Purpose Vehicle (SPV), other public agencies
farmers		Dug-wells, farm ponds, roof-water harvesting; other private recharge structures.	Vigorous information, education, communication campaign to promote recharge to dynamic waters through dug-wells and farm ponds. Technical support in constructing recharge pits, silt-load reduction, periodic desilting of wells. Financial incentives and support to recharging farmers.
NGOs, local communities		Percolation ponds, check dams, sub-surface dykes; stop dams and delayed-action dams on streams.	Technical and financial support to local communities, NGOs for construction and maintenance. Supportive policy environment and incentives structures. Support for building local institutions for groundwater recharge.
canal system managers		Conjunctive management of surface and groundwater.	Operation of surface systems for extensive recharge. Where appropriate, retrofitting of irrigation systems for piped conveyance and pressurized irrigation; retrofitting of irrigation systems for use of surplus floodwaters to maximize recharge; linking of canals through buried pipelines to dug-wells/ recharge tube-wells for year-round coverage.
groundwater recharge SPV		Recharge canals to capture flood flows or transporting of surplus flood waters for recharge in groundwater-stressed areas. Large recharge structures in recharge zones of confined aquifers.	Creation of SPV to execute, operate, and maintain large-scale recharge structures. Building and operation of large-scale recharge structures in upstream areas of confined aquifers; large earthen recharge canals along the coasts.

Source: Shah (2008b, p. 47).

seasons. In practice, each farmer is aware of the fact that yield and water level drop when his neighbor pumps groundwater. However, unless all farmers in a watershed recharge collectively, it is not in any one farmer's interest to do so. Hence farmers have said that at least ten dug-well owners, and preferably more, must participate before they would be willing to engage in rehabilitating their own dug-well.

The aim is to catalyze a successful mass movement such as the one in inland (hard-rock) Saurashtra. Here, a spontaneous grassroots initiative led to the modification of half a million dug-wells for recharge and the construction of over 100,000 check dams by 2007. A detailed study showed that 10,700 check dams drought-proofed 320,000 ha of farmland, with a payback in 2–3 years and an increase in land values (Shah, 2009). The popular recharge movement was spearheaded by spiritual organizations, and supported by government funds and private investments by diamond merchants. This recharge movement is credited for some of the superior performance of Gujarat's agriculture sector in the past eight years.

To create the seeds of a mass movement for the dug-well recharge program, far more thought will need to be given to the social and community aspects. For example, how can information about success be disseminated quickly? Should the program encourage local businesses (e.g., well drillers)? Should there be a fixed timing for well rehabilitation (e.g., April and May) to encourage joint community action? Who keeps records and tracks progress? Could the Tamil Nadu recharge program be emulated, which has started to maintain electronic records that allow constant interaction with farmers and an assessment of the rate of progress in improving groundwater conditions (Krishnan et al., 2010)?[42]

There also remain technical questions which require further analysis. For example, in water scarce regions, how much water collected in upstream recharge goes lost to downstream users as a result of decreased aquifer flow? Some of the technical issues here are formidable, as it is not just the total amount of water that matters, but when it is harvested and how it reaches the aquifer. This calls for tailor-made technical designs for different geo-hydrological zones. Initial experience with the 100-district government program appears to be mixed.

A key aspect of the Mahatma Gandhi National Rural Employment Guarantee Scheme (MGNREGS) was the mobilization of large numbers of workers for groundwater recharge activities. This scheme has suffered from the usual drawbacks of government-sponsored schemes (CAG, 2008).[43] It will be crucial in the future that work under the program focuses on dug-well and tank revival for groundwater recharge, and also a focus especially on O&M. Advocates remain hopeful that MGNREGS will go at least some way toward supporting the groundwater objective.

---

42 Tamil Nadu is experimenting with percolation ponds and check dams, which in some places have led to a more than 25 percent increase in irrigated area, the return to use of abandoned irrigation wells, improved water quality, and drinking water in the summer (N. Varadaraj, *Experience and importance of Managed Aquifer Recharge in Tamil Nadu, India*, presentation at World Water Week, Stockholm, 8 September 2010). Notwithstanding, Tamil Nadu has been unable to constitute the Ground Water Authority that was called for in the 2003 Ground Water Act. The Central Ground Water Authority (with only a handful of officials state-wide) has been filling the void and approves water-related proposals ("No ground water authority even after 7yrs," *Deccan Chronicle,* March 08, 2010). State-wide, no areas have been notified for regulation of groundwater development according to the Environment Act of 1986.

43 The CAG (2008) findings included: significant leakage of funds, only a small fraction of works completed, use of the program for unintended purposes (above all roads).

Increasing water productivity and enhancing gross irrigated area

While groundwater recharge is essential and unavoidable, the key to mitigating the emerging water crisis and establishing long-term sustainable water use for agriculture is in increasing water productivity, i.e., the crop yield per unit of water consumptively used in evapo-transpiration (Kassam and Smith, 2001). Thus, the IWMI futures model referred to earlier shows that improving water productivity by just one percent a year would assure full self-sufficiency in food-grains in India without the need to increase consumptive water use (CWU); increasing water productivity by 1.4 percent a year would mean that all water demand in 2050 could be met without increasing CWU (Amarasinghe et al., 2007a). Having said that, it is also important to respect limits in consumptive water use in order to ensure that environmental flow requirements are satisfied and the ecological health of river systems is maintained (Smakhtin and Anputha, 2006). In other words, not all river flowing to the sea is 'wasted.'

There is vast scope for improving India's crop water productivity, while also freeing up water to meet urban and industrial demands:

- By shifting the area where crops are grown to areas with higher crop water productivity;
- By diversifying from grain to non-grain food crops, particularly in those areas where water productivity is low;
- By improving cropping patterns and agronomic practices (land preparation, fertilization, timing, poplar agro-forestry[44], etc.);
- By using a range of water saving techniques such as micro-irrigation (efficiency of drip irrigation is 85–90 percent, compared to 50–60 percent with sprinklers, and 30–35 percent with surface canals);[45] or
- By trading between states and/or internationally, depending on comparative advantage in water productivity.

To begin, even though India is already the world's second largest producer of rice (after China), wheat, tea, groundnut, and cotton, yields per hectare of paddy are only 51 percent, and of wheat 56 percent of those in China (FAO, 2009). Much of this shortfall is attributed to poor agricultural practices (e.g., untimely planting) and lack of suitable seeds; some of the shortfall may also arise from the tropical soils in India, which contain less organic matter than soils in the most productive parts of China. Within India, there are considerable differences, with western states like Punjab and Haryana showing much higher yields per hectare than Madhya Pradesh or Bihar (Kumar, 2008).

---

44 Zomer et al. (2007).

45 There is a risk that micro-irrigation favors better educated (and hence richer) farmers who are in a better position to manage the technology and who can afford the necessary water pumps. A study in Maharashtra has shown that the introduction of micro-irrigation actually hastened groundwater depletion because farmers expanded their crop production to include water-intensive high-value crops (Namara et al., 2005).

There are several techniques for improving yields that would need to be widely disseminated. For wheat, there are varieties capable of escaping terminal drought. In zero tillage agriculture, wheat seed is broadcast before the kharif rice is harvested, so organic matter on the ground is maximized. A technique promoted by the National Food Security Mission is to place (rather than to randomly throw) fewer and younger wheat seedlings, with wider spacing, no continuous flooding, and using organic matter. With the same area under cultivation, this practice can triple output using one-third of the amount of seeds.

Similarly, in the System of Rice Intensification (SRI), younger seedlings are transplanted at wider spacing, the soil is kept moist but is not continuously flooded, a rotary weeder is used for mechanical weeding, and more organic compost is used as fertilizer. SRI produces 40–80 percent higher yields with 85 percent less seed and 32 percent less water. Tamil Nadu leads the way—about 20 percent of its rice cultivation area is now under SRI. Expansion to 25 percent of the conventionally-farmed area throughout India would yield an additional 5 million tons of rice (World Bank, 2008a). Just-in-time water availability is critical, again pointing to groundwater as the most reliable source. The Tamil Nadu Agricultural University has designed new tools that reduce labor intensity and that lead to greater acceptance of SRI by farmers. Recent yields from triple harvests in Tamil Nadu have been 10–12 tons per ha per year, using 40 percent less water and fewer pesticides than traditional paddy.

Combining irrigation and water supply—the next frontier

The requirements of micro-irrigation suggest yet new areas in which traditional approaches could be reinvented. By converting canals to pipes and retrofitting reservoirs or building simple overhead tanks, public irrigation systems can be used for delivering pressurized water for on-demand, just-in-time irrigation. This saves scarce farmland, in addition to water. Feeder pipes from single or multi-village schemes (World Bank, 2008c) can deliver water to local NGOs or private entrepreneurs/community groups who can then distribute it to farmers for drip and sprinkler irrigation, and to households. Volumetric pricing of water in this situation permits cost recovery for O&M.

This is an opportunity to push for affordable *private* household water connections, which provide the public health benefits in rural areas that public standpipes do not. The broader benefits in terms of reduced infectious diseases start to kick in when there is community coverage of at least 50 percent for private water, and 40 percent for private toilets (Figure 3.11). The health benefits of private connections arise from the fact that households use (running) water more effectively for hygiene purposes, and there is no contamination of water transported to the home (World Bank, 2000). This would go a long way towards addressing the poor sanitation and health indicators in rural areas: just 12 percent of households have access to piped water; and the infant mortality rate is 62 deaths per 1,000 live births (NFHS-3, 2005–06)—despite Rs. 156,000 crore spent on

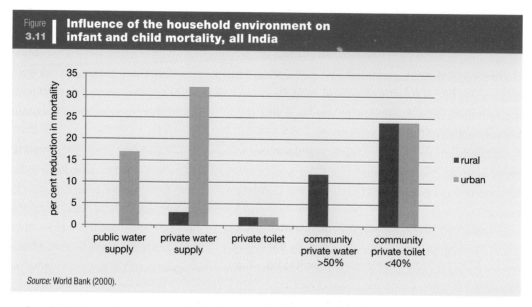

Figure 3.11 | **Influence of the household environment on infant and child mortality, all India**

Source: World Bank (2000).

national drinking water programs in the last 25 years. And it would at last respond to the mandate clearly expressed in the 2009 election, when drinking water was mentioned by the largest number of respondents in the *National Election Survey 2009* as being at the top of the list of pending problems.[46]

Looking at past practice, the irony is that India's traditional approach to planning has actually prevented health improvements in this sector. Although studies show a clear preference and willingness to pay for private connections even among the poor (NSSO, 2006b; World Bank, 2008c), the affordability argument imposed by state and central governments has consistently pushed for public standpipes. Instead, there is enormous scope for locally-driven solutions. In some states, community organizations and NGOs have become active, such as the community-based *Jalswarajya* Water Supply and Sanitation (WSS) program in Maharashtra or the *Byrraju* Foundation's drinking water program in Madhya Pradesh (3i Network, 2008).

Fundamentally, drinking water would be a matter for local government. However, just as in the case of WUAs, where the state does not let go, Panchayat-based Village Water and Sanitation Committees have been unwilling to shoulder the burden of taking over and maintaining facilities

---

46 "How India Voted—Verdict 2009." *The Hindu,* May 26, 2009.

for which they do not feel ownership, and where funds and political control continue to flow through state agencies.[47]

For *urban areas*, the case is often made that additional large water storage infrastructure is required to accommodate the vast needs of growing cities. For example, Delhi benefits from water from the Tehri Dam, though at least 40 percent of Delhi's drinking water is derived from groundwater. Just as with irrigation water, there is a risk that the prevailing supply-oriented paradigm fails to provide the most cost-effective solution to the problem: water supply to greater Delhi is now reaching 150–250 liters per capita per day (l/c/d)—significantly more than, say, the 110 l/c/d for Copenhagen.

Watershed development

Given the strong influence that the monsoon continues to have on GDP, watershed development in areas that have been 'bypassed' by the large investments in the Green Revolution (especially in India's northwestern states) seems to be a vital element for making that "quantum leap" in agricultural growth referred to at the beginning of this chapter.

It is useful to distinguish very broadly between two types of watershed projects (Kerr et al., 2002; Phansalkar and Verma, 2004): (i) projects that aim to increase access to irrigation, particularly in low rainfall areas in western and southern India, where relatively skilled farmers are constrained because of the lack of a reliable water supply; and (ii) projects that are designed to develop rain-fed agriculture in wetter catchment areas, especially in the tribal areas of central India. A major focus of government-supported watershed projects has been on the 64 million ha of degraded land throughout India (Planning Commission, 2008).

Of the two types of watershed projects, those in arid areas have typically received higher government attention and support. They involve various efforts to increase the availability of groundwater through check dams and similar structures, and/or to provide irrigation tanks. An important example is the dug-well and tank redesign program in western and peninsular India, referred to in an earlier section of this chapter. Construction of the required structures lends itself to collective action and the use of labor under MGNREGA, though benefits are usually not equally distributed: landowners benefit more and in proportion to the size of the land holdings.

The development of rainfed agriculture in the more remote, humid (though usually still drought-prone) areas of central India, particularly in tribal areas, has typically received the least attention

---

47 To encourage local initiatives, the central government, in 2002, launched the demand-driven *Swajaldhara* sanitation reform program which involves decentralization and community participation. In Uttar Pradesh, for example, some 90 percent of *Swajaldhara* funds for new capital expenditures reach beneficiaries, compared to 40 percent for supply-driven schemes (Misra, 2005). This new program, however, competes with the older traditional supply-driven programs, including the Accelerated Rural Water Supply Program (ARWSP), which continues to fund most rural water supply activities. The *Bharat Nirman* rural development program followed the ARWSP guidelines as well and reinforced supply-driven programs (Planning Commission, 2008). Aggravating the situation, the central government has also recommended the use of grants to finance O&M, thus removing even the perception that assets are owned by the community and that user charges are necessary to ensure sustainable operation of water supply schemes.

to date. In these areas, there has been a long history of low productivity rainfed farming despite the relatively favorable agro-ecological conditions. The weak agricultural performance has been aggravated by short-term distress migration, specifically by *adivasis,* to find employment to offset the food insecurity arising from erratic monsoon-dependent agriculture, particularly the unreliable *kharif* paddy crop. Populations in upper catchment areas often also suffer the negative consequences of large projects downstream, such as dams and reservoirs, mining activities, and various industries (e.g., Duflo and Pande, 2007).

The key technical constraint in these areas is the lack of suitable irrigation and weak access to agricultural technologies. To improve agricultural performance and reduce the need for migration, the following approaches have been advocated (Phansalkar and Verma, 2004):

- constructing decentralized water harvesting structures—diversion channels and check-dams for protective irrigation and groundwater recharge;
- providing water lifting devices (low-cost high-efficiency diesel pumps or manual treadle pumps) to extract water from local streams as well as from diversion channels and check dams,
- promoting SRI (System of Rice Intensification) to boost paddy cultivation at reduced cost;
- expanding multiple-use upland management—agriculture, horticulture, and forestry;
- supporting vegetable cultivation around the home (possibly with the introduction of inexpensive micro-irrigation systems); and
- especially providing post-harvest support and access to markets for agricultural and non-timber products.

Over the past several decades, the government has kept re-emphasizing the importance of watershed development under a range of central schemes, especially the Desert Development Plan (DDP) and the Drought Prone Areas Programme (DPAP)—both arid-zone programs. Watershed development has been the subject of detailed official analyses, reports, and corresponding guidelines—of which the ninth version (since 1994) was issued in 2008 (GOI, 2008), based on the Report of the Technical Committee on Watershed Programmes in India (GOI, 2006). This latest report pinned considerable hopes on creating a professional, competitively driven institutional framework which would vest strong powers in the Panchayat system and create technical committees answerable to local government, ensuring that disparate watershed programs would be coordinated to achieve a synergistic effect.

In the end, what has emerged, unfortunately, is a variation of the old theme of Centrally Sponsored Schemes and more of what has caused past programs to fail. Additional institutions have been created (an advisory body called the National Rainfed Area Authority (NRAA), State-level Nodal Agencies (SLNA), and District Watershed Development Units (DWDU) for projects covering more than 25,000 ha), but the flow of funds and influence of the state remains unchanged: the state-controlled DWDU or, in smaller watersheds, the existing District Rural Development

Agencies DRDA, has "[t]he full responsibility of overseeing the watershed programme within the district" (GOI, 2008, p. 24) and will be directly responsible for contracting with so-called Project Implementing Agencies. All projects will be sanctioned by the SLNA.

Worse, what little influence the NRAA was going to have to coordinate and infuse new thinking into watershed development has been vitiated by the purely technical composition of that body, when in fact, social and political issues represent by far the most important constraint to improved livelihoods in rain-fed areas.[48] Watershed management is fundamentally a problem of social organization. As a recent article about a watershed program in the Deccan Basalts of central Maharashtra points out (Corbett, 2009),

> [s]ince the late 1990s, both the Indian government and a variety of nongovernmental organizations have funneled some $500 million annually into redeveloping watersheds in drought-prone rural areas. But experts say many such endeavors have fallen short of their goals or proved unsustainable, in large part because they have focused too much on the technical aspects of improving a watershed and too little on navigating the complex social dynamics of farming villages. In other words, no effort gets very far without a lot of hands-on cooperation.

Many observers have argued that at the local level, natural resources are managed by institutions defined by socio-cultural, political, and historical context, which may have no relationship to watershed boundaries. Moreover, primary stakeholders vary depending on the socio-political context and their momentary success in contesting power. As a result, a community-based participatory approach does not necessarily lend itself to overcoming the underlying causes of poverty and non-inclusion, least of all, if that participatory approach is controlled and driven by state authorities. Completed works are therefore not likely to be 'owned' by the community, and sustainability will be in doubt—except in cases where extensive and high-quality NGO 'handholding' has taken place, as was also seen in the case of Water Users Associations.

The efforts to involve Panchayats in the decision-making process and in taking over the completed works is laudable. But no government program can succeed without a recognition that state and community are not separate, and that "village institutions are constituted through processes of state formation" (Mosse, 2003, p. 276). In practice, this means that programs are only likely to be successful if they are clearly designed from the bottom up in the context of meaningful fiscal decentralization.

Floods and dam safety

Flood control embodies some of the same characteristics as canal irrigation development: thousands of crore Rupees have been spent in successive plans on embankments and dams which have achieved little in the way of reducing flood damage. Every year, floods cause hundreds of crore Rupees of property loss, displace thousands of people, and kill hundreds.

---

48 See, for example: Mihir Shah, "Rainfed Authority and Watershed Reforms," *Economic & Political Weekly,* March 22, 2008.

Given the extensive system of embankments that has already been constructed (15,675 km throughout India by 2002), it is difficult to envisage an alternative approach to flood control which would respect ecological constraints and reinforce natural protective systems—wetlands, flood diversion lakes, and adaptation of human settlements and agriculture. However, it has proven difficult, if not impossible, to reliably manage the embankments which are meant to protect flood-prone areas but are regularly breached, usually as a result of poor maintenance—whether accidentally or even deliberately.

The consequences are magnified because the embankments simultaneously obstruct natural drainage and destroy the natural process of building river deltas and flood plains. Soils around villages trapped between embankments become water-logged and natural soil fertility is reduced. The problem is aggravated by a vicious cycle of silt buildup from Himalayan rivers, causing embankments to have to be successively heightened (up to 30 feet above ground level) and making them often fail to withstand the heavy pressure from further silt accumulation during monsoon rains. North Bihar is one of the worst affected areas, where the continued construction of embankments has increased the flood-prone area almost three-fold over the past 50 years (to 6.9 million hectares), placing three-fourths of the population at risk every year (Roy, 2008).[49]

Nor have upstream dams solved the problems of floods because they, too, have been poorly managed. Sudden releases of water to reduce water levels prior to the monsoon have breached downstream embankments and caused flooding in numerous places throughout India, including some western states. There is also an inherent conflict between different objectives: for power generation and irrigation, dams should be kept full; for flood control, they should be relatively empty. It is no surprise that power and irrigation tend to assume priority—all the more so, as most dams are silting up quite rapidly and reducing the storage available for hydro-power and irrigation.

Dam safety is also a problem that merits considerable attention. A procedure to regularly monitor the safety of large dams is being put in place as a result of legislation being passed by the government[50], hopefully averting possible future disasters such as might be caused by a breach in the large Bhakra Dam (which recently displayed a dangerous deflection), or an earthquake triggered by the filling of a large reservoir (as happened in 1967 during the impoundment of the reservoir behind the 103 meter Koyna Dam in Maharashtra, killing at least 200 people).[51]

The World Bank's recent Project Performance Assessment Report on a Dam Safety Project highlights the pervasive problem with poor dam maintenance: subsequent to the project's closing, "basic dam safety facilities are not being maintained or replaced to the extent needed" (World

---

49 A devastating flood there in 2008, caused by a preventable burst in an upstream embankment for the Kosi tributary to the Ganga, displaced as many as 3 million people and killed some 1,000 people. ("Bihar's Tragedy." *India Today,* September 15, 2008; "That sinking feeling." *Down To Earth,* September 20, 2008.)

50 "Bill on dam safety gets Cabinet nod." *The Hindu,* May 13, 2010.

51 A recent paper in the Chinese journal *Geology and Seismology* makes the case that the Great Quake of Sichuan (magnitude 7.9), which killed some 80,000 people, was caused by the impoundment of the Zipingpu Reservoir near the epicenter of the quake (Lei et al., 2008).

Bank, 2009c, p. ix). This is particularly a problem where dams affecting India are located across the border in Nepal (such as is the case with the Kosi Dam which causes continual problems in Bihar).

The collusion of vested interests of contractors, engineers, and politicians makes it difficult to break the cycle of building more embankments and dams that fail to serve the intended purpose and aggravate the problem (yet pay off handsomely as a 'third crop'). Moreover, flood control is the responsibility of the Irrigation Department, which of course is apt to seek technical solutions rather than to adopt a management approach. To paraphrase the title of a famous book describing the corruption in addressing drought problems in India: "Everybody Loves a Good Flood" (Sainath, 1996). It is no surprise that most floods can be traced to human failure to manage water systems that have been allowed to become technically too complex for a centralized and weak government apparatus.

Indeed, most recent major floods in India appear to have their origin in inappropriately operated dams (e.g., multiple floods in Orissa and Punjab from the Hirakud, and Bhakra and Pong Dams) or breached embankments.[52] A review of the July 2010 (and 1993) floods in the Ghaggar River Basin, affecting Punjab and Haryana, showed dozens of breaches of embankments and canals. A lot of damage was caused by water being channeled through poorly designed, unused, and unmaintained canals, which have not been commissioned because of inter-state water disputes (having cost up to Rs 1,000 crore to be built). In addition, a lot of damage to property is caused because settlements have been encouraged in flood plains and flow paths.[53]

Globally, new thinking is emerging in the engineering community on how to effectively address recurring floods, moving much closer to an *adaptive approach* long advocated by ecologists. In the United States, for example, the Army Corps of Engineers (which has traditionally been in charge of building flood control structures) is officially changing its approach. Now, the Corps is "allowing more flooding to occur, while working with local and state governments to manage development on surrounding land to reduce economic damage when flood happens" (Barrett and Ball, 2011). Most importantly, floods are being effectively managed with natural flood plains—tracts of land purchased by NGOs and the government and turned back into natural wetlands that absorb and filter water from rainfall and floods. Retention ponds are being added in

---

52 Does India have better flood protection than Pakistan, given the extreme severity of the recent floods there? In fact, problems with regard to the quality and upkeep of embankments in the two countries are remarkably similar. What appears to have aggravated the Pakistan floods (despite the fact that the weather event causing them was not as severe as originally believed), is the extreme—politically-driven— deforestation in the last few years by Pakistan's timber mafia in sensitive areas along the rivers. Large logs that had been stacked in ravines in preparation for being smuggled out were swept into rivers, destroyed bridges, pierced poorly maintained embankments, and clogged up major reservoirs. In addition, as in India but possibly to a greater extent, growing numbers of poor people had settled in flood plains. Medha Bisht, "Pakistan Floods: Causes and Consequences." *Institute for Defence Studies and Analyses,* August 19, 2010. Kamila Shamsie, "Pakistan's floods are not just a natural disaster." *The Guardian,* Thursday 5 August 2010.

53 Swarup Bhattacharya and Vineet Kumar, "Unprecedented floods in Ghaggar Basin." *Dams, Rivers & People,* June–July 2010, p. 2.

II/3

agricultural areas, and more channels are being built to imitate natural river branches to help drain rivers to the sea (Barrett and Ball, 2011).

In India, this kind of approach would apply as much in cyclone-affected areas (such as in Orissa) as in areas prone to flooding from rivers. The following measures represent essential ingredients of a successful flood control policy:

- *Detecting and forecasting* to prepare the population and reduce damage. This is both the most important and the most cost-effective approach, but it assumes that decision-makers at all levels take note and act accordingly (the Central Water Commission has a system of 175 stations which in the 2008–09 flood season, provided 6,675 flood forecasts).[54]
- Developing *land use policies and zoning plans* in close collaboration with the local population.
- Building *water escapes* that allow the inundation of specific areas on a planned basis and compensate farmers for agricultural losses.
- Providing *flood insurance*.
- *Modifying building design and infrastructure* to withstand flooding and prevent harm to people and property.
- *Adapting agricultural practices*, such as expanding the use of tube-wells for the *rabi* season, to reduce water-logging and the propensity for flooding; crop diversification, including to aquaculture as is increasingly practiced in Orissa.
- *Restoring the ecological system* of lakes, ponds, marshes, and flood plains to act as a sponge for excess water; protecting forests and/or promoting mixed agro-forestry (e.g., by *adivasis*) in catchment areas to prevent erosion.

The bottom line

Unless you change direction, you are apt to end up where you are headed.

—Old Chinese Proverb

The art of reform is not to say how things should be…but to define a sequenced set of actions which lead in the right direction and are politically and administratively feasible

—Don Blackmore, former Chief Executive, Murray Darling Basin Commission

At the risk of felling all the trees in order to see the forest, the following six broad ideas emerge from the foregoing analysis. They are advanced as a summary of the chapter and a basis for discussion.

---

54 Appropriate planning could yield enormous benefits: for example, relief and damage repairs from a single drought (2003) and flood (2005) in Maharashtra absorbed more of the state's budget (US$4 billion) than the entire planned expenditure (US$3 billion) on irrigation, agriculture, and rural development from 2002–07 (World Bank, 2008b).

Move beyond supply-side solutions

Focus on water resource management, not water resource development.[55]

- Move from a 'product' mentality (new infrastructure) to a 'service' mentality (predictable provision of water). Use water efficiently, harvest it locally wherever it makes sense economically, and reduce transport.[56] Most farmers have long ago moved to a new paradigm and will only support programs and projects which can benefit them in their current situation.

- Allow private initiative to fill the void that government cannot fill, such as in the case of water service providers, water pipelining, or spontaneous initiatives to recharge aquifers. Farmers should think of themselves as clients demanding a reliable product. Embrace the changes that emerge from spontaneous initiatives (swayambhu) and guide them with appropriate incentives.

- Stop mourning (or glorifying) projects that no longer have the appeal they might have had in colonial times.

Break the institutional incentive of irrigation departments to obsess about irrigation potential created, when only 20 percent ends up being utilized due to social and political constraints to expanding irrigation beyond outlets of branch canals.

- Consider significantly reducing the mandate of irrigation departments to the provision of bulk water—a step that would be consistent with the vanishing number of irrigation engineers remaining in irrigation departments, despite their enormous size).[57]

- Encourage independent irrigation service providers with clear authority and accountability to convey water in the most efficient manner appropriate (not to be confused with the practice of converting irrigation departments into *Nigams* which tend to retain the same old culture).

Think long-term.

- Storing and replenishing groundwater (i.e., 'banking' groundwater) is often far more cost-effective than building and (not) maintaining surface water structures, quite aside from the fact that it offers water where and when farmers want it. By using electricity supply (with separate feeders for agricultural use) to control groundwater draft, promoting sound recharge to reduce the need to pump from deep wells, and refining conjunctive uses, the

---

55 China is slowly embarking on this shift from water resource *development* to *management,* albeit building on a tradition of strong local government. Ultimately, sound water resource management in India will also require strong local government with its own capacity to make decisions—not driven by state functionaries responsible for managing central or state-designed programs.
56 This happens to be the guiding principle of Australia's new water policy—the country that has one of the highest water storage capacities in the world in relation to mean annual river flows, and that is now starting to question the merits of this paradigm.
57 Historical evidence suggests that it would be a mistake to fund programs to strengthen the capacity of irrigation departments to implement Command Area Development Programmes with the help of Water Users' Associations. Going into the 12th Five Year Plan, the large amounts of central funds available to irrigation departments for major and medium irrigation projects will ultimately perpetuate the prevailing institutional mentality, regardless of any attempt to design other incentives.

cost of using electricity can be much lower than the implied energy cost of constructing and maintaining new surface irrigation structures while continuing unsustainable ground-water use. At the same time, the economic return is greater as a result of the higher agricultural productivity. Many countries are adopting groundwater banking today.

Don't pass regulations that cannot be enforced.

- To provide incentives for the right balance between extraction and conservation, apply indirect mechanisms (above all, managing electricity supply along the lines of the *Jyotirgram* scheme in Gujarat) that are consistent with the enforcement capacity of government. Even the wealthiest countries with a well-developed enforcement system have shied away from unenforceable groundwater regulation or economic charges.

Look for the point of greatest resistance, then think innovatively.

- The long-established paradigm within which official agencies operate often makes them blind to alternative approaches. For example, if land acquisition is the greatest impediment to building canals because farmers don't want to part with valuable farm land, then allowing piped water transfer would appear to be the solution—all the more so, as entrepreneurs are emerging all over India to provide services in this regard. Piped water also allows volumetric pricing and can be used for micro-irrigation and household water.

Don't throw the baby out with the (ground)water.

- Groundwater depletion is serious in parts of India (especially Haryana, Punjab, parts of Rajasthan, around Delhi, and in several locations in peninsular India) and requires urgent remedial measures. But there is also considerable scope for additional groundwater development the country, particularly in areas where there is good annual replenishment or where it can even help relieve potential water-logging.

Top water priority: recharge aquifers in peninsular and western India

- *Start from the problem, not the solution.* The need for aquifer recharge is non-negotiable because groundwater requirements will not decline. This is no different than in many other countries facing the same conditions.

- *Do the economics (without sunk costs).* Groundwater irrigation is more efficient than surface water irrigation in terms of capital and O&M costs, water per unit of output, and it can produce higher-value crops. Even so, it could be uneconomical if pumping costs are too high. Groundwater recharge can pay for itself in terms of saved electricity if it can help raise the water table significantly. This will call for a new way of looking at surface water systems.

- *Reorient surface water systems* (tanks, canals) where appropriate to strategically replenish aquifers. The challenge is to optimize conjunctive *management* of surface and groundwater.

To compete with groundwater, surface irrigation must serve many functions and offer water on demand.

There are many reasons why canal construction programs in India have been woefully behind schedule and over cost. Beside the usual constraints affecting all major infrastructure projects, the key reason is that what is being offered does not respond to what is demanded. Groundwater pumping by individual or groups of farmers has filled the void. However, as the example of irrigation service providers in Maharashtra shows, surface water schemes can also meet farmer demands if the state can give the right signals and allow water entrepreneurs with a clear mandate and narrow responsibility to meet the needs that irrigation departments cannot meet.

To remain relevant, surface irrigation systems will increasingly need to:
- support conjunctive management within existing command areas;
- contribute towards aquifer replenishment; and
- provide reliable water supply for extensive irrigation as well as pressurized water for micro-irrigation and even householdwater purposes.

To achieve this, a wide range of techniques can be employed:
- a judicious mix of lined and unlined canals to replenish aquifers (e.g., at the time of rehabilitation, seepage 'joints' could be added to lined canals where appropriate);
- sealed or perforated rubber pipes suitable to the local hydro-geology (water percolating from perforated rubber pipes may still represent a savings compared to the amount of water evaporating from shallow surface canals, and it recharges the aquifer);
- low or high pressure rubber pipes depending on crop (micro-irrigation) and household water requirements;
- partial or complete sealing of the sluices of tanks and small reservoirs to turn them into percolation ponds (practiced in Tamil Nadu and other states in the south since the mid-1980s); or
- construction of a series of farm ponds as intermediate storage facilities—a practice successfully applied in Rajasthan and elsewhere.[58]

Extend electricity supply to regions now using diesel for groundwater pumping
- As recommended in the 11th Plan, work with states to replicate a program like *Jyotirgram* in the Eastern Ganga Basin (where virtually all groundwater irrigation is based on diesel or kerosene pumps), as part of a broader package of infrastructure, especially to stem the increasing marginalization of poor farmers who cannot afford the rising cost of diesel fuel. Using groundwater in this area is often desirable to prevent water-logging and reduce flooding.

---

58 Similar to the 'melon-on-the-vine' configuration prevalent in mid- and southern China.

- Experiment with incentives (procurement, offsetting subsidies) to shift water-intensive crops from water-scarce northwestern states toward north-central India. Consider new kinds of market-neutral subsidies to compensate for the loss of grain subsidies in the traditional agricultural states of Haryana and Punjab.

Increase water productivity

Improving water productivity by 1.4 percent a year would mean that all water demand in 2050 could be met without increasing consumptive water use. Steps in this direction could involve:

- shifting where crops are grown to areas with higher crop water productivity;
- diversifying from grain to non-grain crops, particularly in areas where water productivity is low;
- improving cropping patterns and agronomic practices (land preparation, fertilization, timing, etc.). For example, the System of Rice Intensification (SRI) can produce 40–80 percent higher yields with 85 percent less seed and 32 percent less water.
- using water saving techniques such as micro-irrigation.

Change the culture and provide critical data

Take a fresh look at the many uncoordinated, silo-type government agencies dealing with water.

- Merge surface and groundwater objectives with the goal of managing water *services,* not designing new projects (taking care not to allow groundwater agencies to be overpowered by surface water agencies).
- Implement a detailed, real-time performance management system for all public water infrastructure initiatives.
- Narrow the mandate of irrigation departments, e.g., to the provision of bulk water.
- Change the culture through a comprehensive education and training program based on curricula developed *within* India that emphasize social, environmental, economic, and legal analyses to promote a culture of sound project appraisal and an open-minded evaluation of alternatives.
- Improve the capacity to draw up contracts that provide incentives for demand-driven solutions while safeguarding the interests of users and the public.

Beware of so-called "sector reform."

- By virtue of the funding they provide, 'Water Sector Reform Projects' and centrally sponsored schemes tend to reinforce the prevailing corporate mentality, even if they contain budget-matching incentives. Cost recovery and financial independence will not 'depoliticize' irrigation departments, as finance is not the constraint to better performance. The culture will only change if the objectives and deliverables of the institutions change (e.g., to deliver bulk water on a strict and reliable schedule from main and branch canals to private irrigation service providers). Meeting these new objectives does not depend on additional funding, but on dramatically changed objectives and incentives.

Urgent real-time data requirements.

- Last but definitely not least, far more information on water issues is required on a real-time basis to make informed decisions—from preventing floods to advising planting times, from river flow to groundwater status at the micro level. Among the most important data are those on groundwater depletion and the potential for replenishment, if only to understand the complementarities and trade-offs between surface and groundwater use. Information from the Central Ground Water Board remains contradictory despite refined estimation methodologies discussed in 2009 (CGWB, 2009a; CGWB, 2009b).[59]

India's long-term outlook for the water sector may be less dire than some would argue. There will be significant stresses in particular locations. With a paradigm shift in how the problems are viewed, however, new and sometimes unexpected solutions will come into focus. In that sense, the 11th Plan report is correct when it states that "[t]he problems that seem to loom large over the [water management and irrigation] sector are manageable and the challenges facing it are not insurmountable" (Planning Commission, 2008, vol. III, p. 44).

---

59 Information on groundwater remains difficult to interpret. For example, CGWB refers to sub surface storage potential (591 bcm, assuming saturation of the vadose zone to 3 meters depth below ground level), retrievable storage potential (436 bcm), annual replenishable groundwater resources (433 bcm), net annual groundwater availability (399 bcm=433 bcm minus natural discharge during non-monsoon season), feasible groundwater storage capacity (214 bcm), and retrievable groundwater storage capacity (160 bcm)—based on 2004 data. Moreover, there is an even weaker understanding of the large deep (non-replenishable) aquifers and what might be the longer-term impact of tapping that resource.

Review of Agricultural Extension in India

Chapter 4

Marco Ferroni and Yuan Zhou

Introduction[1]

Extension, like agriculture, is at a crossroads in India. Agriculture (the predominant sector in terms of employment and livelihoods) exhibits sluggish growth, food price inflation, widening socioeconomic disparities between irrigated and rain-fed areas, and slow development and uptake of new technology. Relatively low rates of total factor productivity (TFP) growth in recent decades are among the sources of concern related to food security, sustainability, farm incomes, and the scope for poverty reduction. Policies and programs for agriculture and rural development are attempting to address these challenges. The task is formidable and problems of implementation abound. To be successful, the process of structural transformation in agriculture, which is well underway, requires accelerated and inclusive productivity growth, income convergence between those employed in agriculture and the rest of the economy, and farming methods that produce sufficient food and fiber without further undermining the natural resource base. Many inputs and enablers are needed to achieve this, extension not the least of them.

The purpose of extension is to disseminate advice to farmers. Gaps in knowledge contribute to the yield gap in biophysical and economic settings. Services and purchased inputs such as seeds and synthetic complements are essential productivity-enhancing tools. However, their effective use requires knowledge, which advisors need to articulate and communicate to farmers. The knowledge farmers need goes well beyond production. It includes price and market information, post-harvest management techniques, and an understanding of product quality determinants and safety standards. Some farmers marshal and command the needed knowledge on their own. The 'resource-poor' majority of farmers (growers of a large share of the nation's food) depend on science-based extension from outside to complement their local knowledge for improved farming and prospects for sales. How, therefore, can one best get meaningful advice to farmers and create learning environments that help achieve the desired outcomes and results?

Extension in India has a mixed record of achievement. The literature is clear in recognizing agricultural extension as a factor in promoting productivity increases, sustainable resource use and, more broadly, agricultural development (Singh, 1999). But the public provision of extension

1 Comments on an earlier draft by Hans Binswanger-Mkhize, Partha R. Das Gupta, K. D. Kokate, P. N. Mathur and others are acknowledged. Section 5 is based on unpublished research carried out for the Syngenta Foundation by Fritz Brugger.

has on balance fallen short of expectations. Research-extension-farmer linkages are seen to be absent or weak in many instances. At the same time there are duplications of efforts, with a multiplicity of agents attending to extension work without adequate coordination (Planning Commission, 2008).

India is not alone in this predicament. Delivering meaningful extension is not easy. Farmers living in widely dispersed communities can be difficult to reach. Farmers' information needs vary across locations, making extension challenging. Supply side rationing may be a problem in the sense that there are likely to be too few extension agents relative to the number of farmers. On the demand side, self-selection on the part of larger, more commercial farmers may bias outcomes. Extension service budgets may be inadequate. Issues of motivation, competence, performance, and accountability of extension institutions and their agents may affect results (Anderson, 2007).

Many countries have neglected extension in the past. Discomfort with the difficulties and some of the results may be among the reasons why. More broadly, governments around the world have neglected agriculture as a whole (World Bank, 2008), and extension has suffered as a result. But agriculture, and by implication, extension, appear to be 'coming back' as governments and other organizations recreate awareness of the sector's role in providing jobs and livelihoods, food security, and other benefits. In India, agriculture is back 'on the map' because of much discussed performance shortfalls that need to be addressed. Extension follows suit. A National Seminar on Agriculture Extension took place in New Delhi in February 2009 to discuss the state of extension and five specific topics: knowledge management for agriculture extension, public extension with a focus on convergence of extension systems, the role of information and communication technology (ICT) and mass media in agriculture extension, private sector initiatives including public-private partnerships in extension, and farmer-led and market-led extension systems. The well-attended national seminar was the first of its kind in many years (Ministry of Agriculture, 2010). More recently, preparations for India's 12th Plan (2012–2017) have focused extensively on extension.

This chapter builds on that discussion. The objective is to review the issues and performance of extension, attempting—to the extent possible—to derive policy and action implications by looking at extension from the dual point of view of primary production and linking farmers to agricultural and food value chains. The chapter recognizes that extension has become pluralistic in India in the sense that a large number of private, 'third sector' (i.e., NGOs, foundations), and informal service providers now coexist with the public system. Market-based extension offered by agrodealers, input suppliers, and buyers of products is massively on the rise—a phenomenon that is promising, given the emergence of supply and value chains as drivers of an increasingly complex sector.

The chapter is organized as follows: Section 1 discusses the role of extension in achieving productivity growth in agriculture. Calculations for India are reported that have implications for the state-wise and crop-wise allocation of public expenditure on research and agricultural extension. Section 2 maps extension models and evolving needs in India. Sections 3 and 4 discuss modes of extension delivery by the public sector, the commercial sector, and NGOs. Section 5 discusses mobile applications in agriculture, the 'up and coming' tool for scaling up and linking farmers to supply chains. The scale up of innovations and links to markets for the majority of small and marginal farmers remain a serious problem. In the concluding section, the chapter develops recommendations to resolve this challenge.

Knowledge, information, and agricultural productivity growth

Four types of technologies have raised yields in agriculture and animal husbandry in the past: better crop varieties and livestock genetics, fertilizer and feed, mechanization, and chemicals underpinning crop protection and animal health. Knowledge and information constitute the necessary fifth ingredient for technical progress to take hold. Knowledge and information include agronomic and animal husbandry know-how and data on aspects such as soil characteristics, the weather, and markets and prices.

How important is knowledge as a determinant of productivity growth? The studies that attempted to answer this question found the contribution to be vital, as common sense would suggest. For example, Evenson and Fuglie (2009) determine that country-level TFP growth rates in agriculture are significantly influenced by 'technology capital', an index that measures both the capacity to develop or adapt new technology and 'the capacity of users (farmers) to master the new techniques'. The authors' 'technology mastery' sub-index, which includes the number of extension workers per 1,000 hectares as a component, clearly contributes to TFP growth in their model, beyond certain thresholds of public investment in R&D. This suggests that both aspects of technology capital are needed to drive TFP growth in farming: the capacity to develop technology and farmers' ability to use it. The latter can be enhanced by extension advisory services, in addition to schooling more broadly.

The literature on yield gaps is similarly clear about the role of knowledge and information in reducing gaps in the real world of farmers' fields.[2] For Ladha et al. (2003), crop management (which reflects the state of farmers' agronomic know-how, at least to a degree) is an important category of causes of the gap between potential and farmer-achieved yields in rice wheat growing systems of the Indo-Gangetic Plains. The yield depressing factors named in this study that are at least in part amenable to treatment by improved organization and know-how include low water use efficiency, water logging, nutrient mining, imbalanced fertilizer use, and pests that are not adequately addressed.

---

2 Yield gap can be defined as the difference between realized productivity and the best that can be achieved with current genetic material and available technologies and management.

A team at Wageningen University recently studied yield gaps for major crops and world regions, defining five production constraints and inviting a group of experts to assign weights to them to reflect their relative importance. The experts were experienced crop specialists from national and international research institutions (see Hengsdijk and Langeveld (2009) for methods and results).

Figure 4.1 shows the Hengsdijk and Langeveld estimates of the contribution of their five production constraints to the yield gap for maize in different parts of the world, including South Asia. The constraints are (i) limited water availability, (ii) limited nutrient availability, (iii) inadequate crop protection, (iv) insufficient or inadequate use of labor or mechanization, and (v) deficiencies in knowledge that result in inadequate crop management. It is instructive according to this study to see that in South Asia the knowledge constraint (which agricultural extension, presumably, could ease) accounts for about one-fourth or 2 tons/hectare of the estimated yield gap of close to 8 tons/hectare for maize.[3] The authors acknowledge the difficulty of measuring and comparing yield potentials and actual yield across a range of conditions. Their results are indicative in character. But the relative contribution of the different factors accounted for in Figure 4.1 seems plausible, and the point about knowledge as a constraint on yield is clear.

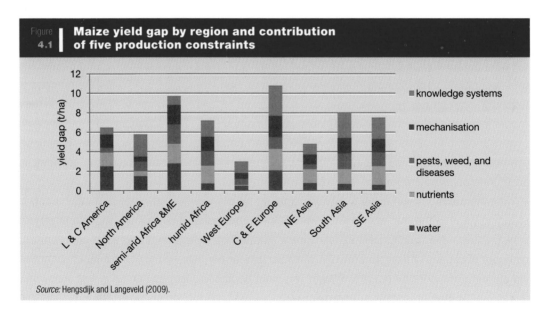

Figure 4.1 | **Maize yield gap by region and contribution of five production constraints**

Source: Hengsdijk and Langeveld (2009).

In their analysis of yield gap among rice growers in the Northeastern zone of Tamil Nadu, Lekshmi et al. (2006) find that gaps are likely due to degraded, less fertile soils, pockets of

3 The potential yield against which region-specific average actual yields were compared was derived from simulations by the IMAGE model in the Global Agro-Ecological Zoning Project (see Hengsdijk and Langevelde, 2009).

endemic cropping systems, and a low adoption rate of high-yielding technologies by farmers. On the latter point, the study notes that 'the intervention of technically sound, well trained and equipped extension personnel at the grass root level is lacking'. The study then states that the cost of agricultural inputs is high and positively correlated with yield gaps of paddy.

Studying cereal yield gaps globally, Neumann et al. (2010) distinguish between growth-defining, growth-limiting, and growth-reducing factors while stressing the importance of management and, by implication, extension, to contain the latter two. Labor is a determinant of agricultural production, as noted by the authors. Its quality, as shaped by education and agricultural support services—including extension—is critical to the success of the farm enterprise. Other factors tested by Neumann et al. for their effects on yield include irrigation, land and slope management, and access to markets. The authors contend that crop production is only profitable if it is not too distant from markets, implying that agriculture tends to be less productive in more remote regions.

If knowledge and other factors such as agricultural research investment are important to reduce yield gaps and promote productivity growth, and if extension helps disseminate technology and knowledge, then is there enough public investment in research and extension to bring out the full potential effect of these forces on productivity growth? A study by Chand et al. (2011) concludes that the answer is negative for key crops and regions across India, but that there are also important instances of crops and states where productivity growth is high. The econometric analysis of TFP growth in this study includes 'research stock' and 'extension stock' as arguments, defined as the sum of weighted (public) research investment of five years and extension investment of three years over the periods reviewed. Chand and co-authors use this analysis to map TFP growth by crops and states, providing a basis to prioritize public research and extension resource allocations (Table 4.1).

The table distinguishes between five productivity growth categories for combinations of crops and states: negative, stagnant, low, moderate, and high. If the first three are regarded as priorities for increased investment in agricultural research and extension, then the action implications for public resource allocation by crop and geography are clear: rice productivity, for example, is lagging in nine states that are identified in the table; pulses are lagging in six states; oilseeds in ten; fiber crops in six; and maize, wheat, and other cereals in eight. Resource allocation decisions for public research and extension must be based on multiple criteria to be sure, but the information in Table 4.1 is a good place to start.

Like TFP growth, crop yields vary greatly between states as shown in Table 4.2. The table (which implies that the scope to raise yields is in general rather large) offers data on the partial productivity measure of production per unit of land at a particular point in time. A simple way to spot the lagging states is to identify those reporting below-average yields. Since, intuitively, knowledge gaps are among the factors contributing to yield variations in similar agro-climatic zones, there is scope for states to learn from each other and share know-how on narrowing the

Table 4.1	Trends in TFP growth in various crops in selected states of India, 1975–2005

crops	total factor productivity growth category				
	positive				negative
	<0.5%	0.5–1.0% (low growth)	>1.0–2.0% (moderate growth)	2.0% (high growth)	
cereals					
rice	KN, MP, HY, BH, OR, WB	AS, KR, UP	AP, TN	PB	
wheat	BH,WB	MP, RJ	HY, PB, GJ, UP		HP
maize	MP	UP	BH	AP	HP, RJ
jowar		TN	MH, AP		MP, RJ, KN
barja		UP	HY	RJ, TN, GJ, MH	
pulses					
gram	MH, MP, UP		HY	BH	RJ
moong			AP	RJ	MP, MH, OR
arhar		GJ, KN	MH, MP	AP	TN, UP, OR
urad	MH	UP	AP	RJ	MP, OR, TN
oilseeds					
rapeseed & mustard	UP	AS	RJ	MP	WB, PB, HY
groundnut			MH, GJ, AP	OR	TN, KN
soybean		MP, RJ	UP		MH
cash crops					
sugarcane					BH, KN, HY, AP, MH, TN, UP
fiber crops					
cotton	PB	HY	GJ, MH	AP	
jute	AS	WB, OR, BH			

Note: AP: Andhra Pradesh, AS: Assam, BH: Bihar, GJ: Gujarat, HP: Himachal Pradesh, HY: Haryana, KN: Karnataka, KR: Kerala, MP: Madhya Pradesh, MH: Maharashtra, OR: Orissa, PB: Punjab, RJ: Rajasthan, TN: Tamil Nadu, UP: Uttar Pradesh, WB: West Bengal

Source: Chand et al. (2011).

differences in yield and the factors underpinning it, such as extension. Extension clearly does not work equally well across states.

In some states such as Punjab where landholdings are large and irrigation is widely practiced, yields are high, and raising them further can be challenging. Technological breakthroughs are needed to increase (if not even just to sustain) the current level of TFP. In other areas, where holdings are small and irrigation restrictive (e.g., West Bengal, Orissa), the pressing need is to make existing knowledge and know-how reach large numbers of farmers. Each state's challenges are different, and there are large disparities within states, often linked to irrigation capability and to the proximity of production catchment areas to markets. The incentives to adopt high-yielding technology improve with irrigation and the proximity to cities and markets, raising the returns to extension, particularly in high-value crops. In remote areas where market access is limited, the choice of crops is constrained and extension needs to focus more on staples such as pulses and grains.

Table 4.2 | Yields of key crops in major producing states, 2008–09

state	rice	wheat	maize	jowar	bajra	gram	arhar	rapeseed& mustard	groundnut	sugarcane	cotton
	ton/ha	ton/ha	ton/ha	ton/ha	ton/ha	ton/ha	ton/ha	ton/ha	ton/ha	ton/ha	ton/ha
Andhra Pradesh	3.25		4.87	1.56	1.02	1.41	0.46		0.88	78	0.43
Assam	1.61	1.09						0.54		38	
Bihar	1.60	2.04	2.68			0.93	1.18	0.96		44	
Chattisgarh	1.18					0.83					
Gujarat	1.74	2.38	1.48	1.20	1.37	1.01	0.99	1.14	1.40	70	0.51
Haryana	2.73	4.39		0.51	1.77	1.04		1.74		57	0.69
Jammu & Kashmir		1.74	2.01		0.59						
Jharkhand	2.03	1.54	1.41				0.62				
Karnataka	2.51	0.92	2.83	1.18	0.70	0.55	0.53		0.59	83	0.36
Kerala	2.52										
Madhya Pradesh	0.93	1.72	1.36	1.19	1.37	0.98	0.80	1.03	1.14	42	0.23
Maharashtra	1.50	1.48	2.38	0.88	0.77	0.68	0.60		1.12	79	0.26
Orissa	1.53			0.63		0.66	0.86		1.16	60	
Punjab	4.02	4.46	3.40					1.22		58	0.74
Rajasthan		3.18	1.74	0.58	0.83	0.78		1.23	1.67		0.41
Tamil Nadu	2.68		4.39	0.83	1.48		0.61		1.99	106	0.28
Uttar Pradesh	2.17	3.00	1.50	1.01	1.61	1.01	0.91	1.12	0.71	52	
West Bengal	2.53	2.49	3.78			1.04		0.76		93	
All India	2.18	2.91	2.41	0.96	1.02	0.90	0.67	1.14	1.16	65	0.40

Source: NSSO (2005) and Agricultural Statistics at a Glance (2010).

Rainfall is often referred to as a factor limiting production, but according to the rainfall map of India (Figure 4.2), nearly half the country receives rather copious precipitation in excess of 1,000 mm per year, although the distribution of that rain may be erratic. The high rainfall regions include states with less than average yields of rice and wheat and irregular production between years. If rainfall is not fully determining, what other factors contribute to the observed low levels of yield? Farmers' practices and states of knowledge are part of the answer. Consider rice as an example, where farmers are frequently following paddy rice practices (e.g., flooding, puddling, and transplanting), even though the conditions under which they work may not be suited to this. Under rain-fed conditions (especially in upland and medium areas), drought spells can be fatal for rice. Farmers may not be aware that there are productive methods, such as direct seeded rice on unpuddled soil and aerobic rice. China is very active in developing and spreading these practices. But in India, improved methods are stuck in research institutions and do not spread. Extension is needed to disseminate the right practices among farmers.

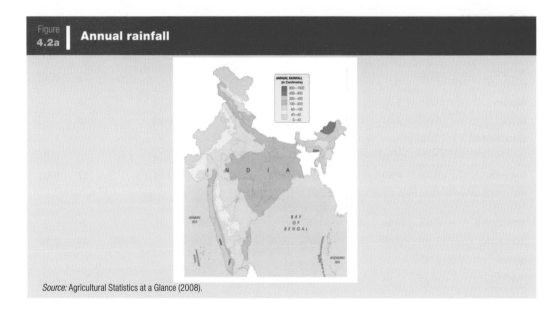

Figure 4.2a | **Annual rainfall**

Source: Agricultural Statistics at a Glance (2008).

Extension is a conveyor belt that brings knowledge and information to bear on farming. The effectiveness of extension varies across states and is influenced by the presence or absence of irrigation and the location of areas of production in relation to the market, among other factors. Given the known contribution of knowledge and information to agricultural productivity growth, stepped up public investment in research and extension for the benefit of productivity-lagging states and underperforming agricultural activities may well be called for, as suggested by Chand and co-authors. However, additional funding alone is unlikely to be enough. As demonstrated

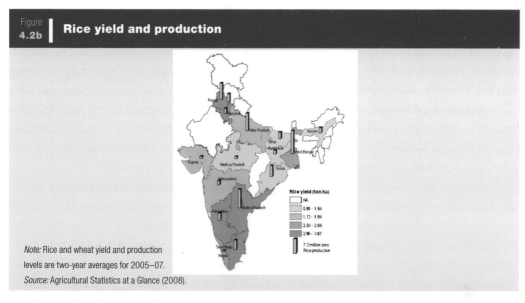

Figure
4.2b **Rice yield and production**

Note: Rice and wheat yield and production levels are two-year averages for 2005–07.
Source: Agricultural Statistics at a Glance (2008).

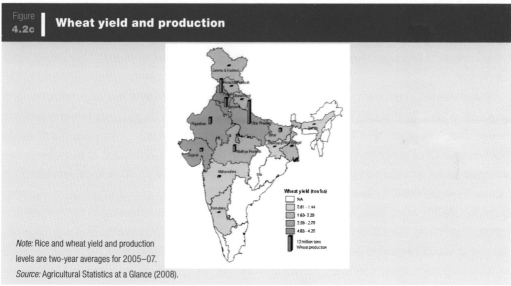

Figure
4.2c **Wheat yield and production**

Note: Rice and wheat yield and production levels are two-year averages for 2005–07.
Source: Agricultural Statistics at a Glance (2008).

later in this chapter, it is also necessary to reform the methods and ways by which research and extension are planned and delivered.

Extension models and evolving needs

Agricultural extension practice has evolved over time, following similar patterns and trends across the globe. The Training and Visit System (T&V) was an early anchor in the past 40 to 50 years. Promoted by the World Bank from the 1970s, T&V reflected a belief in the role of the state as the main actor in development. Under the unified, top-down approach of T&V, existing efforts and organizations were merged into a single national service to promote the adoption of high-yielding (Green Revolution) technologies. The system experienced success in a number of countries, including India, at least for a period of time. It took a 'campaign approach' to raising food production that resonated in settings where farmers' needs and the promoted technologies matched. But there are indications that, T&V failed to generate impact on the promised and required scale. In good measure, this may be so because it was a supply-driven system that promoted messages developed by research scientists with little input from farmers, the users of technology. T&V was abandoned, or at any rate became toothless, in the 1990s.

Since then, agricultural extension has evolved towards pluralistic models and modes (Birner and Anderson, 2007; Neuchâtel Group, 2000). New thinking includes the delivery of extension services in the context of decentralization, and aspects such as outsourcing, cost-recovery, and the involvement of the private sector and NGOs. This is mapped in Table 4.3, which shows the possible combinations of provision of extension services and financing. Extension can be offered by public sector bodies such as Ministries of Agriculture, the private sector (for example, consulting firms, seed and input companies, and buyers of products), and non-profit entities such as NGOs, commodity boards, or farmer-based organizations. Financing can come from the public purse, donors, user charges paid by farmers, or private firms. The latter may provide extension in the context of product sales to farmers or stewardship schemes to reduce agricultural input supply risk.

The challenge in the case of pluralistic approaches is to identify the mix of possibilities and business models best suited to supporting agricultural and rural development cost-effectively in ways that take local conditions into account and recognize the role of farmers in innovation (Anderson, 2007). Farmer participation in the development and dissemination of technology has emerged as an important theme in extension practice over the years. This finds expression in Farmer Field Schools, for example, and the Agricultural Knowledge and Information Systems approach (AKIS), which stresses the merits of direct links between farmers and agricultural scientists. The Farmer Field School model revolves around group-based learning and was originally devised to teach integrated pest management to rice farmers in Asia. Versions of Farmer Field Schools operate in many countries, including India, but not usually as an organized nationwide system of extension (Davis, 2006). Participatory methods seek to convey knowledge to enable farmers to become self-teaching experimenters and effective trainers of other farmers (Anderson, 2007). Farmer Field Schools differ from earlier T&V technology transfer-based extension because they are participatory as opposed to operating from premises that expect farmers to adopt generalized recommendations formulated outside the community.

Other thinking in extension, compatible with the AKIS and Farmer Field School approach, stresses innovation systems and market-based, demand-driven extension. The innovation

Table 4.3 | Options for providing and financing agricultural advisory services

provision of service	financing of service				
	public sector (various levels of decentralization possible)	private sector: farmers (individuals)	private sector: companies	third sector: nongovernmental organizations (NGOs)	third sector: farmer-based organizations (FBOs)
public sector (various levels of decentralization possible)	(1) public sector extension (various degrees of decentralization)	(5) fee-for-service extension, provided by public sector	(9) private companies contracting public sector extension agents	(11) NGOs contracting public sector extension agents	(15) FBOs contracting public sector extension agents
private sector: companies	(2) publicly financed contracts or subsidies to private sector extension providers	(6) private extension agents, farmers pay fees	(10) information provided with sale of inputs or purchases of outputs	(12) extensions agents from private company hired by NGOs	(16) FBOs contracting extension agent from company
third sector: NGOs	(3) publicly financed contracts or financial support to NGOs providing extension	(7) extension agents hired by NGO, farmers pay fees		(13) extension agents hired by NGO, service provided free of charge	
third sector: FBOs	(4) public financial support to supplied to extension provision by FBOs	(8) extension agents hired by FBO, farmers pay fees		(14) NGO financing extension agents who are employed by FBO	(17) extension agents hired by FBO, service free to members

Source: Birner and Anderson (2007), adapted from Anderson and Feder (2004), Birner et al. (2006), and Rivera (1996).

systems concept proposes inclusive ways of thinking about the participants and the institutional context in which the generation, diffusion, and use of new knowledge take place (Rajalahti, 2008). Demand-driven systems (which may be managed and financed by farmers themselves) seek to make sure that innovation follows the market's lead. Swanson and Rajalahti (2010) use the term 'farmer-based extension organizations' to refer to demand-driven systems which, they note, may as a downside come to be dominated by large-scale, commercial farmers who do not necessarily represent the priorities and needs of their smaller peers. Bringing the rural poor into these schemes is likely to require special efforts and skills.

Market-oriented extension for specific crops (sometimes referred to as 'commodity-based advisory systems') may be provided by contractors, parastatals, farm cooperatives, and others (in particular, agribusinesses) with a stake in the value chain. Participating farmers may pay for the advisory services and underlying research, with fees based on the quantity and value of products sold. The cotton-based advisory system in Mali (West Africa) is an example. The Gujarat Cooperative Milk Marketing Federation, a state-level association of milk cooperatives in Gujarat, provides extension services and training to 2.8 million members who pay for the services through the price they receive for their milk.

Market-oriented extension is relevant in economies that are experiencing growth and changes in consumer preferences that create markets for high-value products. It is the growing market (not new technology) that stimulates the uptake of innovation in this case. China and, to a lesser extent, India have been effective in making some of their extension market-driven (Swanson, 2009). Rapid economic growth in their non-agricultural sectors has boosted demand for high-value products that create new opportunities for farmers. Extension workers may find themselves challenged under these circumstances if they lack training in marketing, methods of farm and post-harvest management, and financial services. Success under the market-driven approach manifests itself when farmers can organize themselves as producer groups or sales cooperatives, access knowledge and needed resources, and sell profitably into predictable supply chains.

Different extension models and approaches exist around the world. Birner et al. (2006) argue that there is no single best method for providing need-specific, purpose-specific, and target-specific extension advice. The right approach depends on the policy and infrastructural environment, the capacity of potential service providers, the farming systems and potential for market access, and the characteristics of local communities, including their willingness and ability to cooperate with agents of agricultural extension. Different approaches can work for different sets of conditions. To fit a particular situation, agricultural extension needs to be flexible and able to accommodate local needs (Raabe, 2008).

In India today, these local needs have everything to do with the rapid transformation of agriculture that is visible almost everywhere one looks. Market liberalization and globalization are driving Indian agriculture out of the staple-based subsistence system of the past towards a high-value, information-intensive commercial enterprise (Adhiguru et al., 2009). In this new world of agriculture, farmers are interacting with different information sources to help them produce and sell products and deliver safe commodities of good quality to consumers. As noted by Adhiguru and co-authors, the information requirement that ensues is demand-driven, and as such, different from the supply-led public information system that was appropriate during the Green Revolution. The grand challenge now is to improve farmers' access to the right kind of timely knowledge and information, and to reach all farmers. There is a role for both public and private information systems in this situation, as illustrated in Figure 4.3 where public providers in India appear in the boxes on the left-hand side of the diagram and private sources of extension in those on the right. Public and private information systems should complement each other and operate in partnership rather than at cross-purposes or duplicatively at the expense of underserved areas. To the extent that private extension by for-profit and non-profit actors is on the rise, the public sector's role should become subsidiary in nature, focus on lagging areas and types of farming, create conditions to attract the private sector there, and formulate and deliver rules and quality control. 'Cyber extension' and cell phone-based applications are there to support the process.

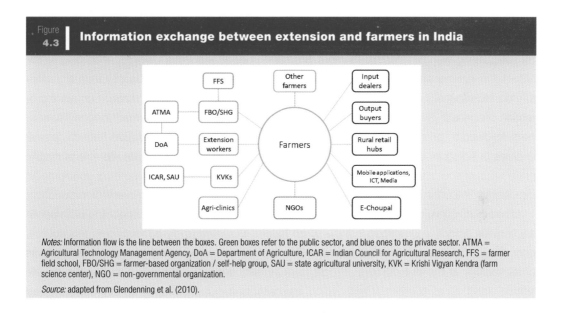

Figure 4.3 | **Information exchange between extension and farmers in India**

Notes: Information flow is the line between the boxes. Green boxes refer to the public sector, and blue ones to the private sector. ATMA = Agricultural Technology Management Agency, DoA = Department of Agriculture, ICAR = Indian Council for Agricultural Research, FFS = farmer field school, FBO/SHG = farmer-based organization / self-help group, SAU = state agricultural university, KVK = Krishi Vigyan Kendra (farm science center), NGO = non-governmental organization.

Source: adapted from Glendenning et al. (2010).

The sources of information and extension advice accessed by farmers in India are varied and suggestive of some interesting patterns.[4] Some 40 percent of farmers of all sizes access information on modern agricultural technology from one source or another, according to all-India data collected by the NSSO (Table 4.4). As reported by Adhiguru et al., and with reference to the table, access to information from any source increases with farm size. Progressive farmers, input dealers and mass media (radio, TV, newspapers) are the most important sources of information. Sources such as (public) extension workers, primary cooperative societies, and output buyers/processors are much less important on average and, according to this source, are biased towards larger farmers in the case of extension workers and cooperative societies. Other public programs, including government demonstrations, village fairs, farmers' study tours, and Krishi Vigyan Kendra farm science centers (KVKs) are of minor importance as sources of extension and are clearly biased against small farmers in this all-India assessment. The private sector in the form of progressive farmers and input dealers is more important than the public sector as a source

---

4 This discussion is based on Adhiguru et al. (2009) and analysis of farm level data collected by the National Sample Survey Organization (NSSO) in its 59th round in 2003.

of extension information for all farmers, including small farmers. The NGOs' reach of farmers is modest according to this source and displaying somewhat of a bias against small farmers, too.[5]

The role and importance of the different surveyed sources of extension information varies in relation to the type of information sought. The main aspects of cultivation on which farmers seek information refer to seed, fertilizer application, crop protection, and harvesting/marketing, according to the NSSO. In animal husbandry, healthcare and feeding practices top the list. At the national level, extension workers stand out as a relatively important source of information on seed, along with progressive farmers, the mass media, and input dealers, according to the NSSO. On fertilizer and animal feed, input dealers are consulted more frequently than any other source. Newspapers and radio are the important sources for obtaining information on plant protection chemicals. The main reported source of information on harvesting/marketing is newspapers, followed by progressive farmers. The role of extension workers is negligible here (cf. tabular analysis in Adhiguru et al., not shown). The NSSO survey did not investigate the role of mobile phone-based sources of information, which (as demonstrated in a previous section) constitute an increasingly important guide to harvesting and marketing in agriculture and livestock production.

The NSSO survey suggests that the paradigm of pluralism in extension (involving both public and private actors) is practiced in India to an extent. But, worryingly according to this source, only about 40 percent of farmers access off-farm information regarding improved components of technology at the all-India level. Progressive farmers and input dealers, and thus the private sector, stand out as sources of information, as mentioned, but questions may at times be raised about the quality of the information they supply. The public sector is present, as discussed in the next section, but farmers' access to its mechanisms and resources, including extension workers and KVKs, seems to be low. This is a matter deserving attention as we proceed.

Public extension in India

Public extension has a long and distinguished history in India going back to the pre-Independence and the pre-Green Revolution eras. Extension went through distinctive stages over time, evolving with national priorities (Singh and Swanson, 2006). Thus, the food crises starting in the late 1950s prompted a refocusing of extension from 'rural development' to agricultural production intensification and food security. The combination of Green Revolution technologies in the late 1960s and the 'single line of command' T&V system from the mid-1970s helped bring about food self-sufficiency during the 1980s and beyond. Analyzing some of the effects of T&V in advanced agricultural regions, Feder and Slade (1986) found that the method greatly increased the number and frequency of contacts between farmers and extension workers, who were an important source of knowledge about farming practices. T&V helped make yield increases in wheat and rice possible. After allowing for other factors affecting farmers' performance and

---

5 In the NSSO survey, small farmers were defined as operators farming up to 2 hectares of land.

Table 4.4 | **Access to information from different sources across all farm sizes in India (%)**

sources	farm size			
	small	medium	large	all India
any source	38.2	51.0	53.6	40.5
other progressive farmers	16.0	20.0	20.8	16.8
input dealers	12.6	14.8	18.3	13.2
radio	12.4	16.4	16.8	13.1
TV	7.7	15.3	22.4	9.4
newspaper	6.0	10.3	15.9	7.0
extension workers	4.8	9.8	12.4	5.8
primary cooperative societies	3.0	6.2	8.0	3.6
output buyers/food processors	2.1	3.6	3.4	2.3
government demonstrations	1.7	3.4	4.6	2.1
village fairs	2.0	2.4	2.38	2.0
credit agencies	1.6	2.8	3.4	1.9
others	1.6	2.1	2.0	1.7
participation in training programs	0.7	1.9	2.3	0.9
Krishi Vigyan Kendras	0.6	1.0	1.7	0.7
para-technicians/private agencies/NGOs	0.5	1.0	0.8	0.6
farmers' study tours	0.2	0.3	0.6	0.2

Source: Adhiguru et al. (2009).

solving the attribution problem with a research design that included control groups, Feder and Slade found yield differences of about 7 percent over three years that were attributable to T&V. T&V strengthened the state-level extension machinery and energized a young and growing extension staff. It was a movement that, for a time, revitalized the system of agricultural research and extension in the face of significant challenges—just what is needed again today.

As mentioned in the previous section, however, doubts about the methods and extension value of T&V began to arise for a number of reasons, including the apparent limitations of the approach in less well-endowed agricultural settings. Poverty and malnutrition remained widespread in lagging rural areas and indeed grew, prompting a search for new solutions in the 1990s. Many state-specific and centrally driven innovations were introduced (Sulaiman, 2003, provides an overview). Subsequent Plan documents explored the role of extension under a liberalized regime. Extension implications for agribusiness sub-systems were among the concerns, as was the role of extension in addressing crop-wise and region-wise disparities in growth, natural resource degradation, and vulnerable areas and people (Academic Foundation, 2004).

A breakthrough of sorts emerged in the form of the Agricultural Technology Management Agency (ATMA), as the 21st century dawned. ATMA was piloted through the Innovations for Technology Dissemination component of the World Bank supported National Agricultural Technology Project (NATP) that became effective in 1998 and concluded in mid-2005. The new thrust represents a decentralized approach that emphasizes local solutions, diversification, market-orientation, farm income, and employment growth, operating through state-level and local institutions. This is very different from T&V, except in one respect: like T&V, ATMA is intended as an organizing framework, a unifying thrust that would encourage coherence and convergence among extension actors and create incentives not only for institutional reform, but for improved performance of processes and institutions. The approach would integrate extension activity across the line departments and decentralize decision-making through 'bottom-up' procedures that would link research and extension and involve farmers, NGOs, and the private sector in planning and implemention at the block and district levels (Singh and Swanson, 2006). This section explores ATMA and other drivers of public extension and assesses system performance, including the aspect of public-private interaction in extension.

ATMA

ATMA is an autonomous organization registered under the Societies Registration Act of 1860, able to receive and dispense government funds, enter into contracts, maintain revolving funds, collect fees, and charge for services. A Governing Board determines program priorities and assesses impact. The heads of individual ATMA jurisdictions (Project Directors) report to the Board. The project directors chair the respective ATMA Management Committees, which include the heads of all line departments and the heads of research organizations within the district, including the Krishi Vigyan Kendra (KVK) farm science centers and Zonal Research Station (ZRS). The original organizational structure of ATMA is shown in Figure 4.4.

Under the NATP project, the ATMA program was implemented as a pilot in 28 districts in seven states. By 2006, ATMA had been adopted in some 60 districts (about 10 percent of the total) and was programmed to be expanded to all rural districts within five years (Singh and Swanson, 2006).

Perhaps not surprisingly, however, implementation bottlenecks began to emerge. According to Kapoor (2010), these include qualified manpower constraints at block and village levels, lack of formal mechanisms to support delivery below the block level, insufficient technical and financial support (the support provided during the pilot stage having weakened over time), and lack of a clear operational framework for implementation of public-private partnerships. Additionally, according to this source, the links between ATMA bodies, ICAR, the SAUs, and the KVKs are weak. ATMA, therefore, is not the desired 'magic bullet' some believed it might become. As a framework, ATMA is arguably on the right track, but it has to cope with problems of alignment of stakeholders and partners. A question one may ask is whether the incentives and capabilities built into the thrust are compatible with the need for flexibility and responsiveness on the ground.

In view of the system's implementation constraints, the government issued new guidelines on ATMA in June 2010. The *Guidelines for Modified Centrally Sponsored Scheme* 'Support to State Extension Programmes for Extension Reforms' note that the system does "not provide the dedicated manpower support at State, District and Block levels" that is required (Government of India, 2010). The new guidelines, therefore, provide for modifications to strengthen specialist and 'functionary' support at different levels; making sure that the 'farmer friend' model (linking farmers and extension agents) works in practice, in particular by filling block-village gaps; revising the 'ATMA Cafeteria' (or list of extension activities to choose from); better enabling Farmers' Advisory Committees to advise administrative bodies at the different jurisdictional levels about extension needs; and delegating powers to State Level Sanctioning Committees for them to approve the state extension work plans (SEWPs). (This is required for the release of ATMA funds). The guidelines include a new organizational chart that articulates sets of activity and fund flow at state, district and block levels (Figure 4.5).

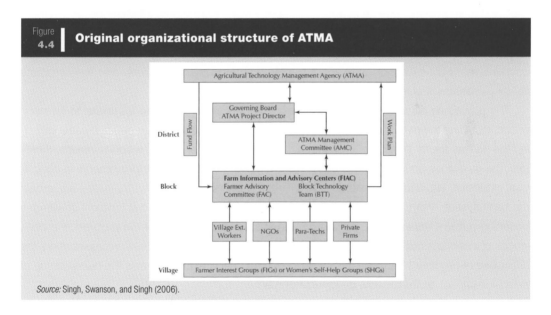

Figure 4.4 | Original organizational structure of ATMA

Source: Singh, Swanson, and Singh (2006).

The guidelines provide for convergence in four areas: manpower and extension-related work under different programs and schemes; public agricultural research and extension at different levels of implementation; convergence with development departments to ensure that the extension

II/4

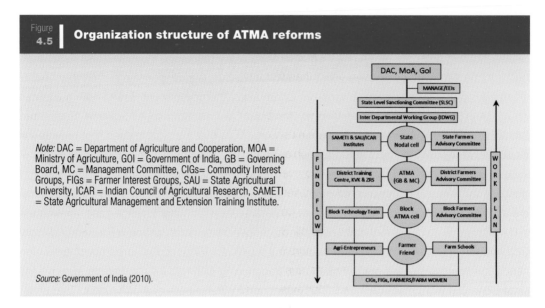

Figure 4.5   Organization structure of ATMA reforms

Note: DAC = Department of Agriculture and Cooperation, MOA = Ministry of Agriculture, GOI = Government of India, GB = Governing Board, MC = Management Committee, CIGs= Commodity Interest Groups, FIGs = Farmer Interest Groups, SAU = State Agricultural University, ICAR = Indian Council of Agricultural Research, SAMETI = State Agricultural Management and Extension Training Institute.

Source: Government of India (2010).

activity forms a coherent whole; and convergence with and involvement of the non-governmental sector. The intention in this latter area is to 'ensure promotion of multi-agency extension strategies, and to implement scheme activities in Public-Private-Partnership (PPP) mode'. A minimum of 10 percent of 'scheme allocation on recurring activities at District level' is meant to be incurred through the non-governmental sector—NGOs, farmer organizations, Panchayati Raj institutions, cooperatives, para-extension workers, agri-entrepreneurs, input suppliers, and the corporate sector (Government of India, 2010).

The guidelines also clearly attempt to increase the system's responsiveness to farmers' needs, including the needs of small and marginal farmers, among other aspects, by strengthening the 'farmer friend' provision. With respect to both convergence and responsiveness, therefore, the guidelines convey a sense of the government's dedication to improvements in agricultural extension. Nevertheless, implementation—the quality of which varies by state—will remain the central issue. Ways and means will need to be found to overcome the inherent challenges besetting public extension that derive from the scale and complexity of the problem, the challenges of instilling a culture of accountability to farmers in a multi-tier extension organization, the difficulties of alignment between knowledge generation and extension, and the dependence of extension impact on the broader policy environment.

Comparison with China

To gain perspective, a comparison with China may assist. With ATMA, India is decentralizing public extension and adapting it to local needs, partnering with non-governmental providers and other organizations, setting the stage for improved quality control, and (as far as the central government is concerned) operating as a residual force. This appears to be the path China is following, although there are many differences in agriculture and agricultural extension between the two countries, not to mention the levels and nature of government spending in the sector. China has many more public extension agents than India and a much lower village-per-extension staff ratio of 1.1 to India's 6.4 (Annex 4.1).

One of the differences between the two countries is that in China, technology, agriculture and extension have evolved to the point where extension no longer needs to focus primarily on staple grains and basic aspects of technology such as improved (often hybrid) seeds, fertilizer, and crop protection. Farmers know about these sources of support; average grain yields in China being twice what they are in India. Extension focuses on new frontiers in the production and marketing of high-value products. Delivery appears in general to have been fairly effective, but (as argued in Annex 4.1) this is not to say that all is well. Soil and water stewardship are two very important issues that extension needs to address more effectively in both countries, and new approaches must be found to foster the environmental sustainability of farming.

In India, there is a continued need for extension to focus on grains, pulses, and oilseeds in lagging areas, while at the same time cover high-value products in the supply chains that already exist or are being formed. Infrastructure, such as electricity and roads, is less well-developed than in China, reducing the relative effectiveness of extension, however well-conceived. But the two countries display similarities in the sense that decentralization as well as reliance on multiple extension participants are on the agenda. Studying the two approaches comparatively may be instructive for planners of extension as ATMA runs its course under new guidelines.

KVKs and State Agricultural Universities

Farmers face a range of extension providers in the public and the cooperative sector, including those under ATMA, the state Agriculture Department village extension officers, public radio broadcasts, crop fairs, IFFCO extension by the Indian Farmers Fertilizer Cooperative, and KVKs and SAUs. The mission of the KVK farm science centers of ICAR is to test and transfer technology to farmers. KVKs, of which India currently has close to 600 (on average about one per district), serve the purpose of linking research and extension. They are ICAR's leading vehicle for extension, but most KVKs are small institutions with some 20 scientific and administrative staff operating under a program coordinator. The effective reach of KVKs is therefore very limited, as seen in Table 4.4. Their method of operation—focusing on adaptive research, field trials, testing, and field demonstrations—seems well-suited to the task. But districts are large and heterogeneous as far as agricultural conditions are concerned, and KVKs would have to be more numerous and better endowed to make their mark on the required scale.

The State Agricultural Universities (SAUs) are much larger entities than individual KVKs, with fully-fledged agricultural research capability. But they, too, are a limited resource compared with the size of the farm population that should be reached. The extension activity of the SAUs operates through state-level agricultural entities, but sometimes reaches out to farmers directly. The organizational structure varies by state. Like KVKs, the SAUs are important, but under-resourced elements in public extension.[6] According to the National Academy of Agricultural Sciences (2005, quoted in Glendenning et al., 2010), aspects of concern related to SAU extension include 'centralized agendas' and information that does not adequately reflect local needs. Similarly, as with KVKs, the extension focus of SAUs tends to be limited to aspects of primary production, at the expense of the post-harvest and marketing dimension.

Evidence from the field

Two recent studies in parts of Uttar Pradesh (UP) and Madhya Pradesh (MP) shed light on agriservices (including extension) in those areas, highlighting issues for analysis and policy design (Reardon et al., 2011a; 2011b). The purpose of the studies, among other aspects, was to analyze from what types of suppliers farmers obtain their inputs and services, including extension. The role of rural business hubs (such as those described in the next section) was a subject of the investigation. Samples of 810 farm households were drawn from three study zones in UP and MP, respectively, in or near the catchment areas of business hubs. The samples were not intended to be representative at the state level. Field work was carried out in 2009. Key findings related to extension are as follows (quoted from source):

The UP survey shows that only 18 percent of households from all farm sizes had access to extension from any source, public or private, during the period surveyed. This differed between regions, with the more commercial areas in the west and center showing higher levels of access to extension than those in the east. Many respondents indicated that they were unable to find extension advice at the right time. Those who did get extension generally reported a high satisfaction rate, so the main issue seems to be access. Of total uses of extension, only 7 percent were from state extension officers. Other public extension, taken together, (i.e., KVK, all-India radio, university extension, plant protection unit) amounted to 18 percent, meaning that of the meager quantity of extension accessed by farmers, only 25 percent came from the public sector. The UP study states that 'on paper' public extension is in place. The recommendation is to make it more effective and accessible.

The MP survey yielded more favorable conclusions, perhaps because the sample included the Malwa plateau, which is dominated by commercial agriculture with volumes of high-value vegetable production. Eighty percent of households reported using extension from some source (public or private), with little variation over sampled regions. Non-use, the study states, seemed to

---

6 The 11th Plan document states that the SAUs are important loci of regionally relevant research, but are so poorly funded by their own state governments that many of them are in chronic overdraft and almost all rely mainly on ICAR funding for research (Planning Commission, 2008).

be driven by low farmer demand in sampled west and central zones, whereas in the east, it was more a result of delivery and quality problems in extension (the share of state extension workers in all extension was less in the east [29 percent] than in west and central zones [41 percent]). Smaller farmers used extension slightly more than larger farmers, but farmers not using extension were more likely to be small. Reported satisfaction with extension was very high, with timeliness identified as the main 'major bottleneck' in all zones. Public extension from extension workers and KVKs emerged as relatively significant at 49 percent of all direct extension uses, the remainder being covered (i) by public indirect provision (such as radio), and the private sector, which amounted to 25 percent of uses.

Many questions arise from these results, some of which (specifically, the high extension coverage and satisfaction) are somewhat surprising for a state such as MP with low reported productivity growth on average in key crops. It should be noted, however, that the sample covered some of MP's more developed parts. Farming conditions and the farmer clientele vary widely within and across states—compare Bundelkhand and Malwa for evidence on this. Could there be bias in how public extension responds? Is public extension more dynamic in commercially vibrant agricultural settings than in lagging ones? (It probably is.) Does it compete with the private sector in commercial areas? Does it modernize itself in the context of such competition and/or cause the private offer to be sharpened? Or is it complementary, filling information needs that the private sector fails to cover or is not trusted to supply? Are there thresholds of public extension that need to be attained before farmers switch to private sources of extension? The above example from underserved UP, where (like in MP) the sample was drawn from relatively more developed areas, does not seemingly confirm this: what little extension there was came disproportionately from the private sector, specifically input companies, rural business hubs, and sugarcane processors in this case. A question begging to be asked is what extension looks like in the poorer parts of UP if it is as limited as reported in the more developed parts.[7]

Public extension may be lagging because of leadership shortfalls at the local level even as funds are available—just as it may be excelling with the right coincidence of drive and motivation. Unpublished evaluations of extension under ATMA by government entities find fault with many aspects in some states, including (according to one such evaluation):

- insufficient percolation of the planning process down to village level
- insufficient focus and attention to extension in districts, haphazard and inadequate mobilization of farmer and community interest groups
- failure to link ATMA structure at the district level to the corresponding KVK
- failure to create synergy between line departments
- tardy allocation and release of funds impacting extension at district and block levels
- overburdening of project directors of ATMA with 'multifarious' activities

---

7 An online conversation with Thomas Reardon in June 2011 helped spawn some of this discussion.

II/4

- neglect of opportunities to create synergy with the private sector.

This is a long list and one hastens to add that there are instances of favorable evaluative assessments too.

One such instance has been documented with reference to agricultural reform in Bihar (Singh et al., 2009). Based on data generated from 540 farmers over a period of three years (2005–2007), this study judges the extension reforms introduced during the NATP period of ATMA (the pilot phase) to have been quite effective (note that the sample of 540 farmers is not representative for the state of Bihar as a whole). Interaction with farmers and need-based training of scientists and extension workers sharpened the focus of research to meet location-specific requirements of growers, according to the study. Adoption of improved technology and practices progressed across all categories of farmers, leading to diversification of farm enterprises and added yield and incomes. The study documents 'reduced adoption lag' and growth in incomes, although (as might be expected) increases in income were higher in more advanced districts where base income was relatively high (cited from source).

Perspective

Students of agricultural extension in India state almost unanimously that the pilot phase of ATMA was initially a success that was later diluted. Many factors account for this, including the 'lab-to-land' relationship. The once strong link between research and extension is generally weak today, but the example from Bihar just cited shows that it should be possible to make it strong and functional again at the district and local level where it counts. It is at these levels, too, that qualified and adequately led and empowered male and female staff is needed—people who understand agriculture and farming, accept the principles of devolution, and are trained in notions of farmer-led and market-led extension. Qualified agronomists willing to work in the field are in demand; there are not enough of them. Another aspect that is needed refers to workable solutions to the challenges of aligning the multiplicity of actors and schemes by different entities at varying jurisdictional levels, not to achieve uniformity, but to arrive at a measure of coordination for best performance overall.

The UP and MP surveys referred to above illustrate the by now well-known fact that public extension has evolved from a monopolistic stance to a situation characterized by the presence of many non-governmental actors. This should be seen as a sign of success: it is because of past public research and extension efforts, among other aspects, that agriculture has developed to a point where choices are available to many farmers. The farmers are keen on innovations, many of which come from the private sector nowadays. The private sector, in turn, is keen on business opportunities, which are not necessarily confined to the dynamic sites but may extend to poorer regions as well. Growing herbicide sales by private companies in traditional rice growing areas in a context of rising rural wages serve as an example. Mechanization is another example as labor becomes scarce, and so is the demand for good direct-sowing techniques instead of transplanted rice. Micro-irrigation and mechanization are spawning whole new service industries in rural areas, bringing innovation to farmers, and illustrating the growth of demand and opportunities for agribusinesses of many kinds. How the public sector reacts and adjusts to this is a major issue in the quest for productivity growth to ensure food security, environmental sustainability, and greater equity and poverty reduction.

The private sector: commercial providers and NGOs

Agricultural extension by commercial companies is advancing rapidly in India. Segments of the private sector offering extension as part of their value proposition include the crop science industry, seed and input companies, distributors and agrodealers, service providers of various kinds, food processors and retailers, and mobile operators and the content providers with whom they partner. Contract farming is an increasingly important vehicle for agricultural extension. The term used in the literature for extension in this context is 'embedded services', where companies deliver information with the sale of inputs or the marketing of products (Feder et al., 2011).

Input providers and product aggregators present information services to farmers to differentiate their offer, foster safe and effective use of products and technologies, expand market share, and ensure the supply of commodities on a timely basis in the quantities and qualities they seek. Companies may work independently or in partnership to develop integrated offers on the input side (covering, for example, seed, fertilizer, crop protection, and irrigation products), or they may foster growth in value chains through forward linkages with buyers of produce. Such private-private partnerships may be complemented by public-private or for-profit/non-profit cooperation, in which companies link up with public providers of extension and/or NGOs. The government and NGOs can help kick-start markets for extension linked to input and/or output markets by delivering 'patient services' outside commercial channels, in addition to their work in settings that offer no incentive to the commercial sector at this time. Extension by commercial actors on the input and the output side of farming and NGOs are the topics of this section.

Both the commercial and partnerships arenas follow a variety of models for delivering and financing extension. Commercial actors may supply extension to farmers or farmer-based organizations by offering information services and inputs in contract farming or 'outgrower' schemes. This may include sending agronomists into farmers' fields. Alternatively, commercial organizations may hire the services of, or partner with, third parties. The possible partners include NGOs, consultants, research institutes or universities, as well as public providers of extension. As for financing, farmers and their organizations, input suppliers, and product buyers may pay for tailored services from a range of possible providers. They may also benefit from public or donor-funded schemes that hire for-profit or NGO providers to offer services for free. A further model is one in which farmers obtain commodity-specific extension advice related to production contracts.

Payment for extension under contracts may be through the prices paid to farmers that would reflect the cost of the extension service.

Extension by input and technology providers

As previously discussed, input and technology providers are a frequently consulted source of extension advice through their commercial links with input dealers. Agrodealers and the farm input suppliers selling through them have an interest in pre-sale and sometimes continuing after-sale advice to growers. This is because, for best results, farmers need not only inputs, but also knowledge about their proper use. More than products, input suppliers really sell effects that are expected to materialize as a consequence of the combined application of the synthetic good and knowledge. Brand reputation and market share are codetermined by the quality and relevance of the advisory services offered to farmers. The problem from the industry's point of view is cost: how to extend product-related knowledge cost-efficiently to large numbers of farmers who each buy only small amounts? The challenge for regulators and the public, on the other hand, is reliability—the veracity of the information and the integrity of products in markets such as agrochemicals, where counterfeits abound and can be useless or even dangerous.

There are an estimated 282,000 input dealers in India. They are pillars of their communities in rural or semi-rural areas and have every interest to offer quality services to their farmer clients. But this requires training, perhaps together with lead farmers, who as a category, are known to serve as multipliers of agricultural know-how and good practice. The National Institute of Agricultural Extension Management in Hyderabad (MANAGE) offers a training program for input dealers leading to the Diploma in Agricultural Extension Services for Input Dealers (DAESI). So far, however, only a minute fraction of all input dealers have been trained (about 3,000). The DAESI diploma covers four modules: agronomy; extension and communication methods; individual and business development; and laws relating to seeds, fertilizers, agrochemicals, and consumer protection. A list of trained input dealers by district and other information are available on the MANAGE website.[8] MANAGE also offers training to other stakeholders and providers of extension.

Another way to train input dealers is by association with large organized input sales and extension schemes, of which there are a number of private ones in India today. The *Mahindra Krishi Vihar* (MKV) 'one-stop farm solution center' by the Mahindra & Mahindra Ltd. tractor and utility vehicle company is one such scheme. Early centers were started in 2000 with the establishment of the Mahindra ShubhLabh Services subsidiary. The subsidiary's stated mission is to 'tackle deficiencies in the farm sector, including low consumption of quality inputs, lack of mechanization, scarcity of farm finances and low awareness of scientific farm practices'.[9] The farm solution centers are arranged in 'hub and spokes' fashion; farmers access services through spokes

8 http://www.manage.gov.in/daesi/daesi.htm.
9 The quotes and information in this discussion of Mahindra Krishi Vihar are taken from Sulaiman et al., 2005.

at the village level. The centers operate on a franchise basis. They provide farmers with services that include quality inputs, the possibility to rent farm equipment, credit in partnership with banks, farm advice by trained field supervisors who visit fields, and off-take of crops through contracts with processors. Dovetailed with the extension advice in this model are the distributorships and retailing of fertilizer and agrochemicals in partnership with the respective manufacturers.

A study of MKV that deserves mention as one of the few that assess the results of extension based on primary data from the field—even if the sample size was small—offers the following insights (Sulaiman et al., 2005): (i) farmers are willing to pay for an integrated set of services that gives them access to quality inputs; (ii) farmers working with a private extension service provider (in this case, MKV) can substantially increase their yields and farm income; (iii) the increases are attributable to field-specific technical advice on application of the right inputs at the right stage of crop growth; (iv) as a private organization, MKV has been able to develop a sustainable and profitable business selling extension services related to both production technology and linkages to markets; (v) the apparent success of this model is in some measure due to MKV's flexible 'learning by doing' approach; and (vi) a private extension approach of this type focuses mainly on medium- and larger-scale farmers.

Hariyali Kisaan Bazaar (HKB), run by DCM Shriram Consolidated Ltd. (the fertilizer, seed, and sugar conglomerate), is an example of a business that seeks to provide 'end-to-end agri-solutions' to farmers. The offer is built around a package of agri-inputs, extension, credit, and produce marketing services. HKB operates a chain of more than 300 rural retail stores across eight states following the model depicted in Figure 4.6. Rural stores cater to 15,000 or more farmers each. HKB has evolved over the years into a 'rural super bazaar' which provides fuel, credit, insurance, and mobile phones, in addition to agricultural inputs, all under one roof. Factors that affect the volume of business include the quality of the monsoon, as seen in 2009, when business slowed down as rainfall dropped far below average. HKB have since their inception displayed a strong ability to lead and react to opportunities in the market, balancing efforts at consolidation and expansion. Understanding the needs of the farmer and an ability to build trust are among the hallmarks of HKB.

Other examples of privately driven extension by input suppliers include:

- *Tata Kisan Kendra*—now called *Tata Kisan Sansar*—(TKS) by Tata Chemicals Ltd. This is a 'one-stop farmer solution shop' providing operational and advisory support to farmers, initially in the states of Uttar Pradesh, Haryana, and Punjab. TKS is a franchise-based 'hub and spokes' model of outlets; extension includes soil testing, remote diagnostics and house brands for seeds, cattle feed, pesticides, and sprayers. There are currently 32 hubs catering to 681 Tata Kisan Sansars and covering approximately 2.7 million farmers in some 22,000 villages across 88 districts in different parts of the country.[10]

---

10 Information taken from http://www.tatakisansansar.com/

Figure
4.6

Hariyali business model

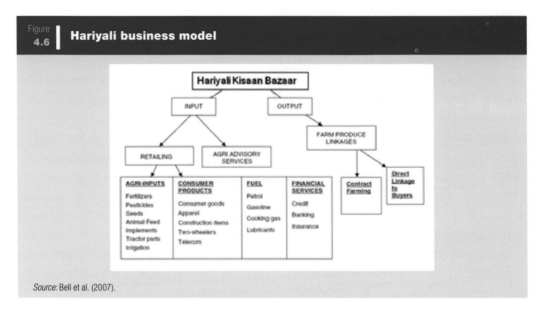

Source: Bell et al. (2007).

- *Godrej Agrovet* is a chain of rural outlets, each serving some 20,000 farmers. Godrej Agrovet partners with other companies to extend its product range. Its 'one-stop solutions' model offers agricultural equipment; consumer goods; technical services; soil and water testing; veterinary, financial, and post office services; and pharmaceuticals. The Godrej agri-services and retailing business was started in December 2003 in Manchar, Pune District. As of 2010, over 60 centers had been set up across the country. Godrej has announced its intention to have at least 1,000 stores across India that would offer a broad range of farming and consumer services and goods.[11]

- *Jain Irrigation* builds awareness regarding micro-irrigation at the Jain High-Tech Agriculture Training Institute. Farmers, students, government department officers and NGOs with an interest in agriculture receive training on topics that include watershed management, water resources and irrigation management, fertigation, and modern methods of crop cultivation. Jain Irrigation also has a team of experts in agronomy and engineering who mentor client farmers.[12]

Input and technology providers seem to have converged on the one-stop solutions model (or business hubs) for rural communities illustrated by examples such as MKV, TKS, Hariyali, and Godrej Agrovet. As shown below, versions of these models are also being applied by aggregators and processors offering extension services. Glendenning et al. (2010) note that 'the impact of these approaches on smallholder farmers has not been evaluated, but these services could

---

11 See http://www.afaqs.com/news/company_briefs/index.html?id=8986_Godrej+Aadhaar+launches+agri-services+cum+rural+retail+s tores+in+Gujarat.
12 See FICCI, Corporate Interventions in Indian Agriculture, New Delhi, October 2010.

possibly provide better-quality inputs and technical services than those offered by the local input suppliers upon which most farmers currently rely'. The private efforts 'provide products for purchase and offer information to farmers on the products they sell, along with agronomic advice'. This would clearly appear to be relevant and helpful for farmers, even if the impact in terms of farm-level and aggregate outcomes remains largely unassessed.

Extension by aggregators and processors

The main setting in which aggregators and processors of products impart extension advice is contract farming, the role of which is growing in Indian agriculture. Gulati et al. (2008) make the point that while 'front end' activities in the agricultural and food system (such as wholesaling, processing, logistics, and retailing) are rapidly expanding and consolidating, the 'back end' activities of primary production have been continuously fragmenting. Contract farming, the authors note, holds the potential to link both ends and create viable business opportunities for farmers and agribusinesses alike.

Contract farming is sometimes faulted for being an exclusive arrangement bypassing farmers not in the contract. However, public or NGO-sponsored support systems are also 'exclusive' to the extent that they do not reach all interested farmers (a form of rationing). Contract farming is not without risk, for example, when parties fail to honor the contract, side selling occurs, prices paid to farmers are low and quantities purchased below expectations. But contract farming does offer significant potential to improve production and farmers' lives through more predictable links with the market.

The literature on contract farming is large, and varies in its opinions. A study by the International Food Research Institute (IFPRI) on contract farming for poultry production in Andhra Pradesh is representative of the supportive view, finding that '... contract production is more efficient than non-contract production. The efficiency surplus is largely appropriated by the processor. Despite this, contract growers still gain appreciably from contracting in terms of lower risk and higher expected returns. Improved technology and production practices, as well as the way in which the processor selects growers, make these outcomes possible. In terms of observed and unobserved characteristics, contract growers have relatively poor prospects as independent growers. With contract production, these growers achieve incomes comparable to that of independent growers' (Ramaswami et al., 2006).

Among those who have studied contract farming for organic crop production, Gahukar (2007) identifies a need for it not only because of the advantage of organized sales, but also as a vehicle to train farmers in the guidelines and protocols they need to follow. The author calls for stepped-up efforts to convince farmers about the economic benefits of the approach, but cautions that to have a future, contract farming must be profitable for all parties.

In a Punjab study, Singh identifies different models (among them corporate-led, state-led, consortium-led, and franchise), finds merit in contract farming, and stresses the need for extension related to both production and marketing of crops (Singh, 2005). On the matter of agreements, the author states that it is often not the contract per se which makes or breaks episodes of contract farming, but how the arrangement is practiced in a given context. Monitoring mechanisms are needed, as is a voice for all parties.

Some examples of contract farming and 'value chain integration' by companies include:

- Contract farming in wheat is practiced in Madhya Pradesh by Hindustan Lever Ltd. (HLL), Rallis, and ICICI (MANAGE, 2003). Under the system, Rallis supplies agri-inputs and know-how, ICICI provides credit to farmers, and HLL (the processing company) offers a buyback arrangement for wheat. Farmers benefit by having an assured market and floor price for their wheat, in addition to a timely supply of quality inputs and technical advice at no extra charge. HLL enjoys a more efficient supply chain, while both Rallis and ICICI benefit in the form of an assured clientele for their products and services.

- PepsiCo practices contract farming in tomato, Basmati rice, chilies and groundnuts in Punjab, and potato in a number of states including Punjab. In West Bengal, where PepsiCo has initiated a small project in coordination with the Syngenta Foundation for Sustainable Agriculture in Bankura, a modified version called *contact farming* is being pursued—a model that is also catching on in other states. PepsiCo ensures technology transfer through trained extension personnel, and supplies agricultural implements free of charge and quality farm inputs on credit. In return, it obtains agreed quantities of quality produce from farmers at a pre-defined price. An 'aggregator' or intermediary hired by PepsiCo organizes the participating small farmers and consolidates their output in bulk batches. Contracted farmers also have the opportunity to manage risk associated with growing potatoes with a weather index based insurance product that is sold through ICICI Lombard and managed by Weather Risk Management Services.

- Adani Agrifresh produces apples in Himachal Pradesh for the New Delhi market, sourcing its entire requirement from about 4,000 farmers at the present time. The extension training focuses essentially on post-harvest practices, because apples must be in the cold chain within 24 hours of harvest. Assured prices, said to be generally 5 percent above the market, are announced on a weekly basis for different grades of apples (FICCI, 2010).

- FieldFresh Foods Private Ltd. practices contract farming with over 3,500 smallholders in Maharashtra and Punjab, where it provides guarantees to purchase produce grown within a specified quality range. The prices to be paid for given quality specifications are announced in advance. Baby corn is a key product collected for export and domestic sales. The company offers (and monitors compliance with) detailed production protocols, and sensitizes farmers to adequate input use and minimum residue limits. Lead farmers are designated as mentors and to manage demonstration plots, recruit farmers, and provide advisory services and post-harvest and logistics support (FICCI, 2010). A 2010 case study by the Yale School of Management describes FieldFresh's difficulties as it tested different sourcing models, finally choosing contract farming as the best approach.[13]

The above indicates that private initiatives span a variety of regions and crops and are being implemented by both large- and medium-size organizations (FICCI, 2010). A large number of such private initiatives are at work. The message that emerges is that this is an active area that deserves to be studied both for the methods of extension and their impact on productivity and incomes by farm type and size. The report by FICCI notes that companies face significant challenges on account of the smallness of farm holdings and the vagaries of nature that plague production. But the benefits of the projects for farmers are frequently deemed to be notable, too. They include productivity gains, price advantages and learning effects such as how to comply with (international) standards and norms.

Extension through mixed partnerships

While contract farming 'carries the essence of the farm-firm linkage' (Gulati, 2010), the incentives for it to arise and be practiced in ways that are attractive for all parties do not come about naturally, but need to be built. This is particularly true for contract farming with smaller, resource-poor growers. Mixed partnerships including one or more non-profit actors may be necessary to bring contract farming to this group. The non-profits might organize farmers in groups and initially provide extension services for free, nurturing the process and helping to build trust with the farmers (that their product will be sold) and the buyer (that there will be agreed qualities and quantities of product to be bought).

An example of a mixed partnership in agricultural extension is the arrangement (dating back to 2001) between the Dhanuka Group (an agrochemicals company) and the government of Madhya Pradesh, with MANAGE, the National Institute of Agricultural Extension Management as an advisor. Under this partnership, the aim was to work together in one district (Hoshangabad) on a wide range of topics. These included: soil testing; training of farmers in soil fertility issues and fertilization; seed treatment and the sourcing of quality seed; diagnosis of pests and diseases; safe and effective use of crop protection products; organizing farmers into groups; conducting group meetings and demonstrations of various kinds; and researching markets to identify potential wholesalers, processors, and retailers to whom direct sales would be possible, bypassing middlemen. Extension in the district was to a large extent privatized under the partnership. Agricultural production responded well, to the point where the National Productivity Council honored the district with its best productivity award in 2004 (Singh, 2007).

---

13 Yale School of Management, FieldFresh Foods, Yale Case 10-036, December 2010.

Mixed partnerships of this kind (public-private and/or for-profit/non-profit) abound, but rarely appear in the literature. Basix, the 'livelihood promotion institution' established in 1996, links extension services paid for by farmers with microfinance products offered by its own for-profit financial arm. Poultry Coop is a for-profit venture that pays for live birds procured from small farmers after deducting the cost of feed and services, including advisory services. It continues to be supported by its founder, the NGO, PRADAN, which is further discussed below. The Agriclinics and Agribusiness Centers (ACABC) provide agricultural advisory services to farmers through agricultural graduates known as 'agripreneurs'. While returning mixed reviews on different aspects of the scheme, studies indicate that the agripreneurs can be a solution with greater ability to meet farmers' needs than the public extension system (Glendenning et al., 2010).

Perspective on extension by commercial providers and mixed partnerships

An important question that arises in the context of extension advice from agrodealers and input suppliers relates to the quality of the information they provide. Input companies are often said to be promoting their brand whereas agrodealers are thought to push sales with an eye on the margins as to whether this is in the interest of farmers. It would be good if this were tested neutrally with the right study design, sampling, and survey-based tools. There is a more favorable hypothesis that would then be assessed—that private solutions are responsive to farmers' needs by their nature. Input dealers have an incentive to offer good services and advice, because doing so is the foundation for their reputation and business. Seed and technology providers have sales forces that pay repeat visits to farmers to cultivate business. They understand that honest advice linked to the products they sell creates a competitive edge. Buyers of products provide advisory services to farmers as part of their procurement drive. And since it is crucial that extension and R&D are closely linked in both directions, commercial extension is likely to be an effective means to deliver private sector R&D results that meet farmers' needs.

Extension by commercial actors may not reach resource-poor marginal farmers, but neither does public extension in large measure. Partnerships with non-profit entities can create conditions where smaller and poorer farmers are reached. The Poultry Coop—PRADAN—is an example. This is a frontier with much untapped potential. It can be pursued in the context of government efforts under ATMA, where the need for public-private partnerships is recognized, or independently of it. But lining up incentives and clarifying accountabilities and roles in partnerships are hard work. NGOs and more generally, non-profits, the subject discussed next, can play a catalytic role.

Extension by NGOs

NGOs are very important sources of support for small farmers in India. As with government organizations, however, their numbers are insufficient as service providers in community-based extension to cover all those seeking advice. NGOs range considerably in size, from small, local entities to large organizations with multi-state reach. Their level of professionalism and knowledge of agriculture vary but their social commitment is typically high. Many dedicate themselves to forming self-help groups or farmer-based organizations that may become focal points for demand-driven agricultural extension. These self-help groups and farmer-based organizations, and the NGOs that help bring them to life, are often supported by outside sponsors and donors. Box 4.1 explains how one such external entity, the Syngenta Foundation for Sustainable Agriculture, partners with small, local NGOs to deliver extension services for productivity growth and improved links to markets. In the projects in question, the NGOs' presence preceded that of the Foundation. Progress in community organization and social programs was already notable by the time they teamed up with the Foundation to address agriculture.

NGOs, such as Basix, PRADAN, and BAIF, are at the larger end of the scale in India, and are perhaps better referred to as social entrepreneurs. They operate in numerous states (Figure 4.7), have been active for many years, and work according to established approaches and methods. Basix (a microfinance institution) works with more than 3.5 million customers, of whom over 90 percent are rural poor households and about 10 percent are urban slum dwellers. Basix operates in 17 states, 223 districts, and over 39,000 villages. It has a staff of more than 10,000; 80 percent of the employees work in small towns and villages.[14]

Basix intermediates extension services for farmers across eight crops (cotton, groundnut, soybean, pulses, paddy, chili, mushrooms, and vegetables) as well as dairy operations and rearing of goats and sheep. The purpose is to improve farming and find market outlets and value-adding activities together with the farmers who pay for the services. The agricultural, livestock, and enterprise development services are made available by 1,000 livelihood service providers, who work like extension agents for 200–400 customers each (Glendenning et al., 2010). According to Basix, its services reach around 800,000 farmers and involve productivity enhancement, risk mitigation, local value addition, and alternative market linkages for synthetic inputs, bio-inputs, and outputs.

The NGO, PRADAN, is a leader in the promotion of self-help groups in India. In crop and livestock development, the organization focuses on productivity enhancement, diversification, and links to markets as core strategies. PRADAN was established in 1983 according to the belief that the way to conquer poverty is by enhancing poor people's livelihood capabilities and giving them access to sustainable income-earning opportunities.[15] Specific areas of engagement include increasing the productivity of the main cereal crops to improve food security, and diversification into cash crops such as pulses, oilseeds, and vegetables. Horticulture is becoming increasingly important in the livelihood programs of PRADAN. All projects share the goal of enhancing the

---

14 Taken from http://www.basixindia.com/.
15 Taken from http://www.pradan.net/.

| Box 4.1 | Extension with multiple partners: The Syngenta Foundation in India |

In 2004, the Syngenta Foundation for Sustainable Agriculture (SFSA), together with Syngenta India Limited (SIL), initiated work in India to address problems facing smallholder farmers. Since neither had prior experience in extension *per se,* they started small to improve their understanding of the subject. A pilot project began in Chandrapur in central India, in partnership with Maharogi Sewa Samiti (MSS), the Leprosy Service Society founded by the social visionary Baba Amte. Farming experiments in the MSS community demonstrated very quickly the possibility of significant increases in crop productivity from improved agronomic practices. The trials also indicated the cash-generating potential of vegetables in just one year. The Somnath campus of MSS was only growing rice when the project started. Rice production was intensified, but soon truckloads of vegetables also went to the Chandrapur market. It was then that the SFSA-SIL team felt confident about reaching out to farmers on a larger scale. A fully-fledged agricultural extension program emerged, which now operates in some 14 locations across four states.

Thus, for example, in early 2006, three projects started in disadvantaged areas of Bankura, Kalahandi, and Jawhar. Each project runs in partnership with a local NGO that had been working with rural people but not in agriculture. The first task was to reorient their approach to include farming. In each project a small extension team was put in place, led by a qualified agriculturalist and assisted by field workers consisting of local youth. Then the process of capacity building of the targeted farmers as well as the extension teams began. Advanced crop technologies were passed on through farmers' workshops, trials in farmers' fields, and demonstrations. When it was realized that knowledge alone would not suffice, steps were taken to make available the recommended inputs and tools (paid for by farmers). Seed multiplication by farmers was introduced to improve availability and bring down prices. Encouraging results began to emerge. By following improved methods, including the System of Rice Intensification (SRI), farmers achieved significant yield gains. Vegetable cultivation turned out to be a remunerative option for many of them. A striking feat was achieved by a Bankura tribal couple who earned a net income of INR 10,000 (approximately US$200) by growing tomato on just 337m^2 in 2008. Good seed and the right choice of variety, coupled with agronomic support, made the difference.

As SFSA decided to scale up activities and reach thousands of smallholders, it took feedback from farmers and a commissioned external evaluation into account. Intervention steps were broadened, for example, to include watershed management and rainwater harvesting, working with self-help groups in clusters of villages, holding farmers' fairs, and building learning communities. Intense discussion with partner NGOs about crop technology, agronomy, work methods, and principles of learning with and from farmers became a hallmark of the approach.

The program has come a long way since it began with a few hundred farmers in 2004. The outreach now covers about 45,000 farmers, not including those who have graduated from the program. About one-third of the smallholders reached have become successful vegetable growers. The projects are also helping farmers tap into government schemes. Linking groups of them to markets is an explicit objective that needs to receive more attention in the future. Activities to help make projects self-sustainable are being strengthened. For example, farmers' groups are pursuing additional income-generating enterprises such as the production of hybrid seed for sale.

management of natural resources, and hence greater environmental sustainability, particularly in hilly regions.

PRADAN, like Basix, works with many local partners. It has, for example, an association with the government of Madhya Pradesh in rural development, women and child development, forestry, sericulture, and agriculture, with a focus on land and water development and agricultural productivity. For the latter purpose, PRADAN works through the Rashtriya Krishi Vikas Yojana scheme and, at times, in association with ATMA. The association with Madhya Pradesh is of note in light of the state's public-private partnership with for-profit Dhanuka, referred to above. Partnerships can take many forms, but the general purpose is always the same—to pool assets and capabilities for given purposes, in this case the delivery of agricultural extension and the creation or provision of links to markets in ways that bring value to farmers.

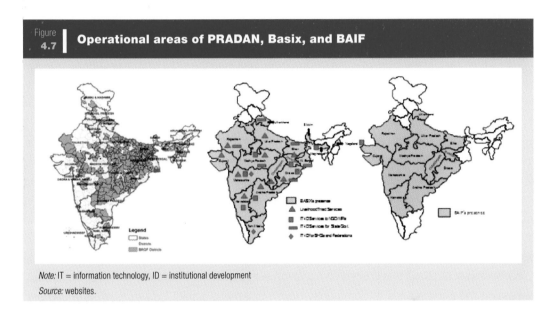

Figure 4.7 | **Operational areas of PRADAN, Basix, and BAIF**

Note: IT = information technology, ID = institutional development
Source: websites.

According to its 2007–08 Annual Report (the most recent one available on the internet), PRADAN's 41 field teams worked with 142,000 families during the year (up from 113,000 the year before) in more than 3,400 villages. Like Basix, PRADAN receives funding from a number of sources, which include the Indian government, Indian philanthropic and corporate bodies, as well as international donors and philanthropic organizations.

The BAIF Development Research Foundation (also Bharatia Agro-Industries Foundation) is another large NGO operating in 12 states and working in agriculture and livestock development, in addition to other sectors. BAIF's areas of work also include water resources development, sericulture, agroforestry, post-harvest product management and marketing, cattle feed and forage production, microcredit, and applied research (for example in cattle reproduction). As in the case of PRADAN, the fostering of rural self-help groups is important.[16] BAIF was established in 1967, has more than 3,000 employees, including a strong contingent of scientists, and today operates

16 See http://www.baif.org.in/aspx_pages/index.asp.

from some 750 BAIF centers across India. BAIF works through associate organizations and is said to reach out to 2.5 million farmers in more than 45,000 villages, many in tribal, mountainous, and dry-land areas. BAIF is recognized by the government of India through numerous sector ministries which have recommended that states learn from the BAIF experience and develop programs in association with this NGO.[17]

One of BAIF's activities is the 'wadi' program to establish orchards, supported by soil and water conservation work on degraded land in tribal communities. This program currently covers over 5,000 villages, benefitting more than 150,000 families in six states. BAIF facilitated the formation of farmers' cooperatives and federations of self-help groups. 37 of these organizations now form the national Vasundhara Agri-horti Producers Company Ltd. (VAPCOL). VAPCOL supports its members in the development of products, processing, and the supply chain. 2008–09 sales were worth some US$17 million. The wadi program is being replicated with technical assistance from BAIF under a special Tribal Development Fund established by the National Bank for Agriculture and Rural Development, NABARD.[18]

Interventions of this kind are clearly relevant as vehicles to bring 'extension' to resource-poor, marginal farmers. Our observations in the field suggest that Basix, PRADAN, BAIF, and others like them, large and small, are spearheading needs- and demand-driven extension, going for what works based on systematic assessments of opportunities and constraints, organizing women's groups and farmers, and fostering innovation in participatory ways, with an eye on the market for activities reared in the primary sector and value addition. It is an approach that would seem to ensure that farmers' information needs are met, to paraphrase the title of the paper by Glendenning et al. (2010) that is cited on occasion in this chapter. Of course, the literature formally assessing the performance of community-based extension is thin. Problems of 'elite capture' and deficits in the performance of service providers may exist. But the observed approach and dedication of many NGOs is notable and the number of farmers reached is large, as the figures referred to suggest. The number of farmers in need is of course still much larger, illustrating the problem of limited availability of competent service providers noted in Section 2 and discussed by Feder et al. (2010). Consequently, much more is needed along the lines of the activities of contributors such as Basix, PRADAN, and BAIF. It is hoped that the public sector, the donor community, and domestic and international philanthropists with the means to support providers such as these take note and consider offering their support.

Mobile applications in agriculture

Mobile applications have the potential to revolutionize the linkages and transactions between farmers and service providers of many kinds. They can be a resource for agricultural extension,

17 Based on http://sapplpp.org/links/baif.
18 Source: http://dev.ikf.in/baif/our_programmes_land_based_livelihood.asp.

but are not yet widely discussed even in the more recent literature on extension that is cited in this chapter. This section attempts to fill this gap by exploring the current and potential roles of mobile applications in both extension and the task of bringing farm produce to the market. This chapter charts the evolving 'ecosystem' of mobile communications in agriculture and assesses the experience in India so far. It focuses on the scope and risks, recognizing that any overview of this highly dynamic field is quickly out of date.

Mobile applications in agriculture (sometimes referred to as 'mAgriculture') are about the delivery of agriculture-related information and services via mobile communications technology, in particular mobile phones, smartphones, PDAs, or tablet devices such as the iPad. 'mAgriculture' is thus different from, or a special subset of, 'eAgriculture'. This broader field involves the delivery of agriculture-related services using information and communication technologies that require access to personal computers and the internet. It may also involve wireless devices as well as techniques such as remote sensing and geospatial information systems that capture and present data linked to locations.[19]

For 'mAgriculture' to materialize, farmers must have access to cell phones. Mobile applications bypass those without cell phones, except where community level solutions are available. Mobile phone access constitutes a lower hurdle than the prerequisites for 'eAgriculture', namely owning or having access to an internet-connected personal computer, predictable supplies of electricity, and understanding how to use the computer. Teleconnectivity is growing rapidly[20] and, assuming continued fast expansion in rural areas, could empower large numbers of farmers by providing access to information on farming and supply chains.

Nevertheless, in the short term, 'mAgriculture' remains constrained on both the demand and the supply side. Demand-side factors include deficits in connectivity, illiteracy (a problem in text-based communication such as SMS, not voice-based interaction), low average levels of education, and poverty. Supply-side limitations are related to product timeliness and relevance, marketing and pricing, and the suppliers' business models, as discussed below. Products must be helpful to the farmer and available in the right language. The scope and need for innovation in the realm of content remain huge.

The benefits of 'mAgriculture' extend potentially to all aspects of extension, service delivery and linking farmers to specific markets. The benefits include access to information at lower cost, reduced asymmetry of information, increased transactional efficiency, improved agribusiness process management, and higher producer productivity and incomes.

---

19 The Agropedia system of digital content organization is an example of 'eAgriculture' (www.agropedia.iitk.ac.in). Agropedia was launched in January 2009 as a one-stop shop for information on Indian agriculture. A 'knowledge organizing platform … to leverage the existing agricultural extension system', Agropedia offers, *inter alia,* knowledge modules of chickpea, sorghum, pigeon pea, and groundnuts developed by ICRISAT. Other partners include SAUs, ICAR, some NGOs, some KVKs, NRSA, TATA Chemicals, FAO, technology partners, and others.
20 By January 2011, India's total wireless subscriber base was 771.18 million, of which 33.6 percent was rural (TRAI, 2011).

The importance of lowering the cost of information is difficult to overstate: knowledge and information deficits are key constraints in agriculture. Research in Sri Lanka found that the cost of information from planting decision to product sale in wholesale markets can be as high as 11 percent of the overall cost of production (de Silva and Ratnadiwakara, 2008). Much of the discussion of mobile technologies in agriculture is optimistic for this reason, even though it acknowledges accessibility as a challenge in harnessing the full potential (Bhavnani et al., 2008). Some of the literature takes issue with the assumption that technology is an autonomous force for good, arguing that it can reinforce existing dependencies and forms of control (Leye, 2009). A third approach occupies the pragmatic middle ground, viewing technology not as an end in itself, but as an enabler of positive developmental outcomes in the context of the right policies and mentoring (Fourati, 2009), an idea with which the authors of this chapter agree. Meaningful products, the right delivery arrangements, and client mentoring make all the difference.

Mobile applications can serve a wide variety of needs. These include extension in the narrow sense of advice on farm production, transactions in markets for inputs and farm outputs, the sale and administration of financial and other services, and the collection of data for research. Figure 4.8 identifies business processes that offer opportunities for mobile applications along the value chain. This chapter focuses on 'extension proper' and market transactions for inputs, services and outputs.

Figure 4.8 | Farming activities from a business perspective

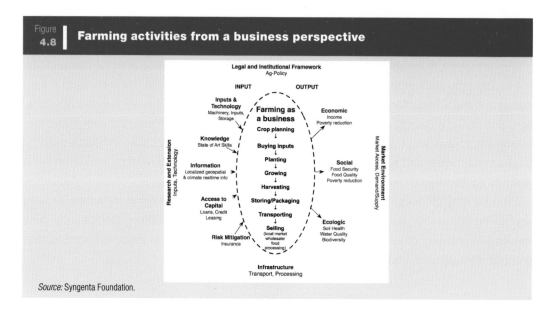

Source: Syngenta Foundation.

The complexity of mobile applications depends on their particular goals (Figures 4.9 and 4.10). Low-complexity applications enable the one-way provision (by voice or text message) of information such as weather forecasts or price data that are generated automatically or stored in databases. Medium-complexity applications, in turn, involve services for decision support that work with location-specific information. Providers can, for example, use these to develop soil fertility-related recommendations or crop-specific disease warnings based on local climate data. Information in these applications essentially flows one-way, but is focused on specific clients.

Figure 4.9 | **Business process offering opportunities for mobile applications**

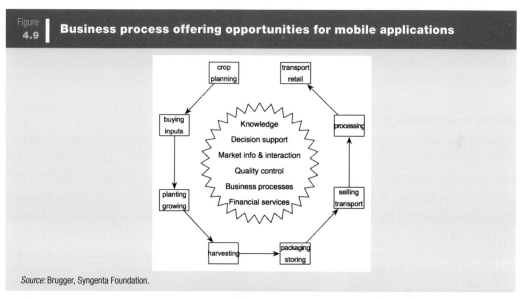

Source: Brugger, Syngenta Foundation.

Figure 4.10 | **Complex applications**

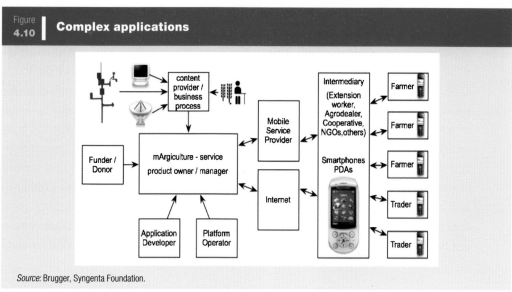

Source: Brugger, Syngenta Foundation.

Systems of high complexity involve information flows in more than one direction. They are transactional, permitting the administration of user-generated information, customized feedback and advice, remote diagnostics, and the management of individual accounts for farmers by service providers, input dealers, product aggregators, and traders. Mobile banking, the transfer of money by cell phone, and crop insurance that runs on a mobile platform (as in Kenya[21]) are examples of complex mobile solutions already available, albeit in some instances, still on an experimental basis. Below are some examples of different application complexity levels covering extension, market links and services. Table 4.5 lists the cases discussed and provides web addresses and qualifiers that characterize the ventures and their respective business models, many of which are at the pilot stage.[22]

Mobile applications in extension

Extension calls for applications that disseminate knowledge to address skills gaps in agriculture and promote learning. Such applications are on a continuum from 'mLearning' to 'mFarming', where 'm' refers to mobile communications. Under 'mLearning', knowledge of farming and agricultural techniques is disseminated to subscribers with the possibility for interaction and group learning among farmers. Digital Green, for example, can be said to facilitate 'mLearning' with its videos 'of farmers, by farmers, and for farmers' and its hundreds of mediated screenings in villages and rural settings (see below). 'mFarming', on the other hand, is about services and individual decision support with the help of local, contextually relevant information. mKrishi and e-Sagu (see below) are applications more in the realm of 'mFarming'.

- **aAqua ('Almost All Questions Answered')** is an internet-based discussion portal initiated in 2003 by the Developmental Informatics Lab of the Indian Institute of Technology in kiosks and cybercafés in Pune. aAqua is more an example of 'eAgriculture' than 'mAgriculture', except that it offers access to its platform via SMS as an additional service. It is an open forum where users have created more than 90 percent of the content themselves, uploading text, photographs, and videos to the site. A farmer can ask a question on aAqua from a kiosk or cybercafé; other farmers or experts view the question and reply (in English, Hindi, or Marathi). Different discussion groups cover aspects of crop cultivation, animal husbandry and dairy, market prices, and other topics. There is rapid retrieval of documents and images using keyword-based searches assisted by query expansion and indexing techniques. aAqua, a non-profit venture, operates on the basis of freely accessible software and only a small initial investment. It can be replicated quite easily, but has not gone to scale: the number of registered users was about 17,000 by early

---

21 See www.kilimosalama.org.
22 The description of the cases in the discussion below is based on information gleaned from these web addresses.

Table 4.5	**Sample of India's mobile applications in agriculture (2011)**								
	business model	medium	complexity	info flow		business model	medium	complexity	info flow
extension aAqua	non-profit	internet/ text	low	interactive	Kisan Call Center (KCC)	government	voice	low	interactive (call center)
Avaaj Otalo	non-profit	voice	medium	interactive	mKrishi	commercial		high	interactive
	non-profit	video	medium	one-way	eSagu	non-profit		high	interactive
Nokia Life Tools (NLT)	commercial	text	low	one-way	Nano Ganesh	commercial		low	one-way
IFFCO Kisan Sanchar Ltd (IKSL)	commercial		medium	one-way and helpline	**market facilitation** Reuters Market Light (RML)	commercial	text	low	one-way
					e-Choupal	commercial		low	one-way

Note: Entries in this table are a sample of mobile applications only.

Source: Authors.

2011. Poor Internet connectivity in villages and illiteracy appear to be among the conditions working against scale up.

- **Avaaj Otalo** is a voice-based system for farmers to access and discuss relevant and timely agricultural information by phone. The system was designed in 2008 as a partnership between the IBM India Research Laboratory and the Development Support Center (DSC), an NGO in Gujarat supported by different donors. Avaaj Otalo is an important and promising experiment in voice information services for small farmers. A must-read description by Patel et al. (2010) arrives at optimistic conclusions on the suitability of voice as a medium for online communities in the rural developing world.

  By dialing a phone number and navigating through simple audio prompts, farmers can record and respond to questions, and they can access content assembled by experts. In addition to the question-and-answer forum, Avaaj Otalo offers both an announcements board of regularly updated topics and access to past programs of DSC's popular weekly radio show where listeners can call in to discuss their experiences related to the advice heard on the air. Farmers, Avaaj Otalo learned, are extremely interested in listening to other farmers' questions and the corresponding discussion in interactive fora. Avaaj Otalo was initiated as a pilot with 63 farmers in Gujarat in 2009 and received 3,500 responses in the first month. The application was launched across the state in 2010 with a publicly accessible number. The number is toll-free at the time of publishing (airtime cost being borne by DSC). This raises issues of financial sustainability that are discussed by Patel and co-authors, along with possible solutions.

- **Digital Green** is a non-profit organization with funding from the Bill and Melinda Gates Foundation and the Deshpande Foundation. It disseminates agricultural information to small and marginal farmers through digital video (see Ghandi et al., 2009). The approach offers significant potential to improve the effectiveness and reach of extension programs by delivering targeted content that is scalable to large numbers of farmers. The application is 'mobile' in the sense that the product is portable, but it is not a cell phone driven solution. Some 1,200 videos on agricultural techniques have been produced since operations started in 2008. Farmer groups and extension providers can access the library and use films sequentially to build farming capacity over time and as a learning resource in community interactive settings. An innovative IT solution supports up to 100,000 concurrent users anywhere in the world, enabling offline operation in low and limited bandwidth locations (internet connectivity is needed to synchronize user data with the global repository). Statistics on the number of screenings and farmers involved, and case-by-case stories of impact, are updated frequently on Digital Green's website. More comprehensive evaluative assessments are not available as of today. The number of farmers involved has increased rapidly, reaching 42,000 in early 2011. But maximizing impact will depend, to a large extent, on the nature and effectiveness of Digital Green's partnerships with extension providers, be they NGOs, governmental agencies or the private sector. In the non-profit domain, Digital Green is partnering with large, well-established organizations such as BAIF and PRADAN. It will be interesting to see how video as a medium is incorporated into, and is allowed to shape, the methods of extension of these and other organizations.

- **Nokia Life Tools (NLT)** was launched in India in 2008 and in Indonesia and China in 2009 as a commercial application aiming to supply a range of agriculturally relevant resources on low-cost Nokia phones. Information is pushed to subscribers via daily text messages in up to 10 languages in two categories of service—'basic': available across India for Rs. 30/month, and 'premium': available in 10 states for Rs. 60/month. The agriculture segment of NLT covers commodity prices in a large number of mandis for crops chosen by the subscriber, data on seed and other input prices in locally relevant markets, weather forecasts by postal code, and agricultural and animal husbandry tips and techniques. Data on the number of subscribers could not be verified for this chapter. The application, which works as long as there is GSM coverage, is very promising, but depends on successful solutions to the problem of collecting accurate data with sufficient 'granularity' to be helpful to users. NLT is partnering with private and public institutions in the quest for user-relevant information. Partners have included Reuters Market

Light and e-Choupal (see below), some NGOs, input suppliers, microfinance institutions, and some state marketing boards. Impact assessments are not available at this time.

- **IFFCO Kisan Sanchar Limited (IKSL)** emerged as a partnership between mobile operator Bharti Airtel and the Indian Farmers Fertilizer Cooperative Ltd (IFFCO) in 2007. The remit of IKSL is to improve farmers' decision-making capability by providing information on market prices, farming and animal husbandry techniques, fertilizer, weather forecasts, and rural health initiatives. Five free voice messages in local languages and customized for different jurisdictions are sent to subscribers every day, except Sunday. A 24-hour farmer helpline completes the service. IKSL markets this as part of a special mobile package on Airtel's network with an IFFCO Kisan branded SIM card for which farmers pay a one-time activation fee. The voicemail service is free, but helpline queries are charged at the rate of 1 Rs./minute. IKSL targets the millions of farmers that populate IFFCO's 40,000 member societies. On a cumulative basis, close to 3 million SIM cards have reportedly been activated; some 0.7 million farmers were active customers in late 2010. There is potential to go to scale in this partnership, which in its early days received a launch grant from the GSMA Foundation. A market research firm interviewed some 8,000 respondents in 2009 to assess their satisfaction level. The service received good ratings on parameters such as clarity and relevance of messages in comparison with other sources of information. Individual descriptions of impact on crop yield and farmer income are available on IKSL's website, but formal assessments of impact remain to be published. As in the case of NLT and other applications, it is probably still too early for this. The IKSL model is promising, yet also raises questions on many demand- and supply-side aspects that determine the size of this market for farmers of different kinds and economic means.[23]

- **Kisan Call Centers (KCC)** were launched in 2004 by the Department of Agriculture and Cooperation of the Ministry of Agriculture to deliver extension services to the farming community across the country. The purpose is to respond instantly to issues raised by farmers in 22 local languages in all states. Calls are toll-free and handled in two categories. Level 1 answers most calls. On Level 2, subject matter experts answer the more difficult items within a prescribed number of hours. Figure 4.11 shows the generic workflow of call centers such as KCC.

KCCs report that farmers' demands for information relate to the suitability of weather conditions to farm operations, fertilizer application and pest management, the sourcing of quality inputs and credit, and crop insurance and market support systems. KCCs have good call-related statistics, possess nodal agencies that monitor their activities and

---

23 Following IKSL's success, two similar ventures between a phone operator and a fertilizer company were recently launched: Reliance Communications and Krishak Bharati; and Bharat Sanchar Nigam Ltd and National Fertilizers Ltd.

Figure
4.11   **Call center workflow**

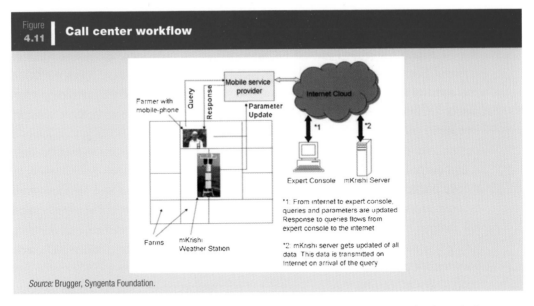

Source: Brugger, Syngenta Foundation.

conduct agent training, have state level monitoring committees and a knowledge management system for their agents, and conducted an evaluation study in 2006–07. This study found good levels of client satisfaction, but provides no quantitative information on impact.[24] KCCs have cumulatively answered more than six million calls so far. According to senior officers in early 2011, KCCs plan various forms of expansion, both geographical and social. Future additional customers will include farmers in the North-East, and farm women and illiterate individuals. KCCs can be financially viable as long as the Ministry of Agriculture provides support.

- **mKrishi** is a personalized, integrated rural services platform launched by Tata Consultancy Services (TCS) in late 2007. The goal is to raise on-farm yield, reduce input cost, provide better market linkages, and foster rural entrepreneurship. The platform is complex. It combines multiple technologies to bring together information regarding local weather, fertilizer requirements based on soil conditions, pest control, and current food grain prices in local markets in a rich content format to the farmer's low-end mobile handset. It allows farmers to send queries, images, and voice-activated SMS, and it provides customized responses in the relevant language.[25] Customization is in part made possible by automated weather stations and sensors that are deployed in villages and linked to a central server. A Frequently Asked Questions database handles many of the queries. More sophisticated questions are forwarded to experts who work with a system that resembles email and enables them to see photos and other local information. Farmers

---

24 See http://www.docstoc.com/docs/36523062/Impact-Evaluation-Study-of-Kisan-Call-Centres.
25 Quoted from http://www.csr360gpn.org/magazine/feature/mkrishi-connecting-indias-rural-farmers/.

receive responses within 24 hours (Pande et al., 2009). mKrishi is adaptable to illiterate farmers who can make queries from a cell phone using voice-specific functions. Figure 4.12 provides an action chart.

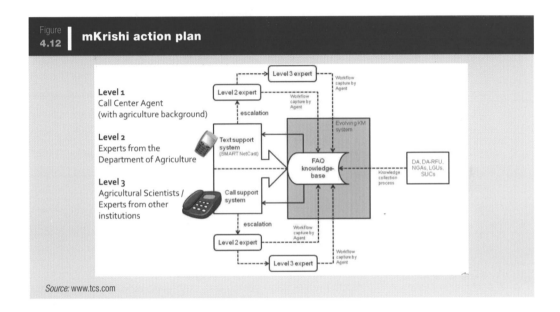

Figure 4.12 | mKrishi action plan

Source: www.tcs.com

mKrishi is a fee-for-service application that was used by some 5,000 farmers in early 2011. Willingness and ability to pay have emerged as issues discouraging a greater number of users, prompting TCS to reduce subscription fees. TCS works with partners and stakeholders, including local entrepreneurs, input companies, and NGOs, to commercialize the service. The application's impact on farm productivity and income will be a function of the relevance and affordability of the advice and products and marketing support intermediated through these partnerships.

- **Nano Ganesh** involves irrigation control through mobile phones. Mobile operator Tata Teleservices and the agro-automation company Ossian are enabling farmers to activate and monitor irrigation pumps remotely. The system uses a low-end Nokia phone and a mobile modem connected to the pump's electrical starter. While not an 'extension' application according to strict standards, this is an interesting technology that could serve as a platform for additional services. The technology was developed to deal with conditions of erratic power supply. Farmers routinely have to walk or drive several kilometers to water their crops, only to find that in situ there is no electricity to power their pumps. Nano Ganesh allows farmers to dial a code from any phone to a mobile modem attached to the pump. This informs them whether electricity is available and allows them to switch the pump on or off remotely. The system should help save time, water, electricity, and fuel.

- **e-Sagu** was started in 2004 as an initiative of the International Institute of Information Technology (IIIT), Hyderabad, and Media Lab Asia. The purpose is to deliver timely, personalized advice to farmers for a nominal subscription fee. A team of agricultural experts at IIIT and an 'agricultural information system' constitute the 'brain' of e-Sagu. Local centers equipped with a weather station serve as intermediate assembly points and each cover about ten villages. Lead farmers work as coordinators, collecting farm registration and farm management and agronomic data. They visit participating farms weekly to observe and photograph crop status. This information goes to the main center, which prepares farm-specific advice. Transmission to and from the main center is by e-mail from connected local centers or on a compact disc dispatched by courier. Lead farmers deliver the advice to the farmers in their charge who are in this way mentored on a 'query-free' basis at regular intervals from pre-sowing operations to post-harvest management and precautions. The system is efficient in that agricultural scientists can now give advice without visiting crops, which enables them to advise more farmers.

  e-Sagu has served several thousand farmers so far. Assessments noted a positive correlation between the adoption of e-Sagu advice and both crop yield and savings due to the more judicious application of fertilizer and crop protection products (Ratnam et al., 2006). e-Sagu reports that its operation can be financially sustainable, but that will depend on the future business model, including e-Sagu's ability to combine forces with a strategic partner. e-Sagu—more an 'eAgriculture' than a 'mAgriculture' application—has missed out on advances in mobile technology in recent years that could enable it to gain efficiency and relevance as a solution in 'mAgriculture'.

Mobile applications for better market access and services

- **Reuters Market Light (RML)** is a leading commercial information service for farmers delivered via SMS. The information includes market prices, weather updates, news on agricultural policies, and advice to match each stage in the farming cycle. Farmers can personalize the information with reference to types of crops, region, and their local language. RML sells its service through mobile operators, agri-retailers, credit societies and rural banks, input companies, and others with a business or non-profit stake in agriculture. By the end of 2010, RML offered information throughout approximately 1,400 different markets on 440 different crops and varieties, and specific weather forecasts for 2,800 locations. RML has hundreds of thousands of farmer subscribers in 13 states. A 2009 study by the Indian Council for Research on International Economic Relations (ICRIER) indicated that all RML customers interviewed had benefitted to the tune of 5 percent to 25 percent of their annual income.[26] Mittal et al. (2010) found that the price information given by RML is accurate and of good consistency. According to them, this explains the high degree of confidence in RML expressed by farmers in their survey.

- **e-Choupal,** an 'eAgriculture' application, is reported here because of its importance as a successful platform to create a virtual market and address infrastructural and other bottlenecks. These problems affect the transparency and functioning of many markets in India. e-Choupal was launched in 2000 by the agri-division of ITC Ltd, the Indian Tobacco Company. A network of rural commerce hubs equipped with a computer connected to the internet serves some 600 farmers in surrounding villages. A local person acting as a sanchalak (coordinator) runs the village e-Choupal. Farmers go there to obtain daily updates on crop prices in local mandis, procure seed, fertilizer, and other products (including consumer goods), and sell their crops for prices offered by ITC. Through its bulk operation, ITC typically pays more than farmers would receive from traditional traders. Thanks to its system, ITC operates at a cost advantage, controls the quality of what it buys, and obtains direct access to farmers and information about conditions on the ground. ITC reports that it recovers its equipment costs from an e-Choupal in the first year of operation and that the venture as a whole is profitable. In an attempt to leverage its brand, e-Choupal partners with, and opens up rural markets for, third parties in sectors ranging from seed, implements, and other inputs to consumer goods, finance, insurance, and other services. Charging for access to its platform helps ITC recover spending on infrastructure and operations. In 2010, e-Choupal's 6,500 village kiosks served some four million farmers who grow a range of crops in 40,000 villages across ten states. Expansion to 100,000 villages in 15 states is planned.

Perspective

Mobile phones and 'mAgriculture'/'eAgriculture' can raise productivity and farm incomes when the information is timely and of good quality, and when farmers believe they can trust the advisory relationship. This is the conclusion of the empirical study by Mittal et al. (2010) referred to previously—the first investigation of the impact of mobile phones on Indian agriculture. The study also states, however, that the full potential of mobile telephony will only be realized with improvements in content, supporting infrastructure, access to financial services and markets, and farmer education. Resource-poor small farmers face greater barriers than larger farmers in deriving benefits from mobile applications because of their more limited ability to use and leverage the information that can be accessed.

Some additional insights documented in the study are as follows (see Mittal et al., 2010):

---

26 Quoted from http://en.wikipedia.org/wiki/Reuters_Market_Light#ICRIER_study_in_2009.

- *Mobile devices as an instrument of information dissemination:* Despite continuing connectivity gaps, farmers view mobile devices as the instrument of choice to gain access to agriculture-related information. Interviewees felt that, because of its more personalized nature, mobile telephony had the potential to be a more reliable source of information than other available sources.

- *Type of information sought by farmers:* Interview data show that farmers access information on their mobile devices in the following order of topics: seed, mandi (output) price, fertilizer application, crop protection, harvesting and marketing, and implements and tools.

- *Impact of mobile devices on agriculture and small farmers:* Almost all interviewed farmers reported increases in convenience and cost savings from using their mobile devices. But reported usage and benefits varied by states. Farmers in Maharashtra reported far higher use of their phones and mobile-enabled information services than those in Uttar Pradesh and Rajasthan. The authors ascribe differences to variations in infrastructure, financial services, and the mobile-enabled information services available in the three states.

- *Impact of mobile devices on traders/brokers:* Mobile phones are a critical resource for traders and brokers. They enable them to shift tonnage across markets in response to price differentials—in the process, optimizing their daily earnings and also smoothing out supply. Mobile devices also facilitate the numerous services as advisors and intermediaries that traders provide for farmers.

- *Mobile devices and market transparency from the farmer's perspective:* Market information accessed by mobile phone influences farmers' selling decisions. Market information improves farmers' ability to negotiate better pricing terms from local traders.

Mobile applications and examples of 'eAgriculture' such as the ones listed and discussed above are important to agriculture and farming in India. But some of the available services are more successful and have a greater impact than others. Why is this so and what is needed to overcome the demand and supply constraints on the dissemination of relevant, mobile-enabled information services for farmers, including small farmers? This is the concluding topic of the chapter.

The applications listed and discussed above vary considerably in their ownership and business model (government, non-profit, commercial), technical complexity, direction of information flow (one-way versus interactive), the medium employed (voice, text, video), and the type of information pushed (general as opposed to customized). There are also differences in the availability of farmer mentoring to help interpret and clarify the action implications of mobile information. It can be assumed that mentoring raises the value of the information, particularly for resource-poor smallholders. It may also play a role in building trust.

Mentoring, which is part of some of the mobile offers, can take many forms and vary in intensity. IKSL's helpline compensates, to a degree, for the limitations inherent in its one-way communications model. However, Mittal and co-authors also found that awareness of customer support options tends to be low, and that farmers therefore do not often contact the information provider with further queries. Avaaj Otalo offers mentoring via interactive voice communications and its radio call-in feature (radio remains relevant in the 'mAgriculture' world!). Digital Green's videos provide raw material and vehicles for live group mentoring. E-Choupal's hubs and *sanchalaks* are focal points for mentoring. The more complex mKrishi and e-Sagu applications offer mentoring by definition because of the two-way communications characteristic they share. Much more could be achieved in these cases, though; mKrishi is at work with its business partners to build up the mentoring part of its approach. The need for mentoring is naturally greater where the mobile information is weaker or more ambiguous. Mittal and co-authors identified differences in the subscribers' perception of RML and IKSL in this respect. The RML service was perceived as providing information that was well-tailored to subscribers' needs and easy to access, whereas users generally saw IKSL as a bit more 'hit and miss' and sometimes lacking in relevance to farmers' needs. In some sense, this is not surprising: agronomic information is more difficult to convey through mobile means than market price data, driving home the need for mentoring and the required technical and organizational arrangements to make it possible in the first place.

Mentoring and the feeling that they are being taken care of are important for clients. However, other factors are important too. They include richness and clarity of content, accessibility, and value for money in the case of commercial applications. Accessibility comprises aspects over which the suppliers of applications have no control (literacy, initial skills, and educational level of prospective users) and others they can shape (language offerings, accessibility across a range of handsets, timeliness of service, training, and support functionality). The cases listed and discussed above cover a range of conditions and approaches in this realm.

From the supplier's point of view, the viability of mobile offerings depends on a range of factors, including (i) the business model and how the partners in the venture work together; (ii) cost considerations and how to finance content development and the maintenance of the technical platform; and (iii) revenue generation and the required size and scale. Systems that run like premium services on a subscription basis (e.g., RML) need high volumes to generate enough revenue for the operator. On the other hand, services that are more content focused and offer higher levels of individualization may be difficult to scale up. The providers of these services may have to tap other funding sources in addition to the user fees they can charge. Business and stakeholder partnerships may be the solution here, while philanthropic donors may be the necessary support for non-profit ventures. The exit strategy for donors and eventual cost recovery should be considered from the outset; it is not clear how this is handled in the non-profit ventures identified above.

The next few years will be a fast-moving and defining period for mobile applications in agricultural extension and, more generally, the integration of mobile-based information services into processes of development and economic growth. The cases discussed above show a trend towards content-rich location-based information and technological integration. The functionalities offered by 'mAgriculture' ventures have evolved from market information to weather forecasts and related news, and further to targeting agricultural extension, spreading know-how about crop cultivation techniques and livestock production. Providers such as mKrishi are now tackling the challenge of customizing information. Among other aspects, this takes the form of 'tele-agriculture' where data go for analysis and remote diagnosis, with experts replying to farmers, offering them personalized solutions as in e-Sagu, or more automated information and advice as pioneered by mKrishi. In these set-ups, some types of intermediaries—extension agents, namely—continue to play important roles. In e-Sagu's case, the agent collects localized information and translates the experts' advice for the benefit of the farmer. With mKrishi, the agent would be an entrepreneur who invests in an agriculture-related business besides supporting the farmer. Novel and exciting business prospects are emerging, boding well for the future of agricultural extension if and when farmers can combine (or are assisted to combine) information with other resources on which they can act.

Conclusion and recommendations

The beginning of this chapter confirmed the vital role of knowledge and information as co-determinants (with other factors) of productivity in agriculture. One-fourth of the yield gap for maize in South Asia is due to knowledge deficits according to an estimate cited. Extension is key in this situation and a much debated topic in India today, as the country seeks to modernize its farms and achieve the higher and sustained levels of TFP growth that are the hallmark of success in agriculture over time.

The preceding pages discussed encouraging developments and continued shortfalls in extension. The scope for mobile applications to make information available to farmers and communicate with them is vast and chased by innovative actors, a major bright spot in extension. Interactive next generation platforms now emerging are expected to redefine the environment of service provision to farmers and facilitate links to markets. Community-based knowledge and information services for farmers, fostered by professional NGOs, are another bright spot, where farmers at the lower end of the endowment spectrum are empowered through their organizations and community structures to innovate, diversify, and assume greater control over their lives. Extension by commercial providers is growing rapidly, to the point where they are already the main source of agronomic information in the segment of farmers that accesses such information. Input dealers and 'progressive farmers' are the first 'ports of call' for many, far ahead of governmental extension workers and public institutions such as the KVKs on an all-India basis.

Questions are sometimes raised about the quality of the information given to farmers by the private sector. Expressions of doubt with respect to quality can be countered with reference to the long-term business interest of private operators, which would seem to demand high quality services and information. This issue should be studied openly and impartially with the aid of appropriate surveys so that the debate can move on from conjecture to objectivity and facts.

Two large and interrelated issues remain unsolved: coverage of small farmers, and the public sector's role and effectiveness in extension. Small farmers represent an untapped opportunity for food security and agricultural growth. Their productive potential could be multiplied, and sustainably so, with the right kinds of technology, services, mentoring, and access to markets.[27] But this is not happening on the required scale. Data cited in this chapter suggest that public providers of extension reach, at most, 6 percent of farmers operating up to 2 hectares of land. All providers, taken together, reach some 40 percent of farmers of all sizes, typically with at least somewhat of a bias in favor of larger and relatively better-endowed growers. The task is to expand coverage to all farmers that operate under conditions where there is potential for growth in cropping and livestock production. For the NGOs, rising to the challenge requires doing more of what they already do, as described in this chapter, with adequate resources in the form of trained and motivated manpower and operating funds. For the private sector, expanding the offer will depend on infrastructure, availability, and quality of other public goods. For the public sector, the need lies in experimentation, documentation, replication, and scaling up of what appears to work, including the instances of success recorded during the pilot phase of ATMA. For all participants, expanding coverage means leveraging each other's skills and contributions through judicious partnerships.

The entity that is currently challenged in extension is clearly the government—at all levels. The private sector is taking off in the context of dynamic opportunities in agricultural supply chains shaped by economic growth and the expansion of demand. Its presence will become increasingly pervasive, requiring the government to reassess its role. The public sector's ways in extension have evolved in the context of institutions that emerged from the experience of the Green Revolution, and food security considerations that centered on the supply of basic grains for consumption at farm level and public procurement. This mindset is no longer relevant. Agriculture has moved on, subsistence farming is at least aspirationally a memory of the past, and high-value products and processed foods are increasingly displacing the staple commodities of old (although the staples still need to be supplied). Rising rural wages (and rising food prices) are raising farmers' demand for technology and services, and the private sector readily responds. How should the government adjust?

---

27 The situation is more complicated for the segment of marginal farmers who are so asset-poor that their prospects for economic advancement may lie more in the labor market than in the intensification of production.

The answer lies in filling the many remaining gaps, in partnership with private for-profit and non-profit actors, and with a view to paving the way for eventual agriservice delivery on a commercial basis to currently underserved areas that can be linked to the market.

A new look at the reasons for laggardness by crop categories and states might help to get organized for this task (see Section 1). Funding may emerge as a constraint from this analysis, but other aspects, including political commitment to agriculture at state and local levels, institutional issues, management, organization, and implementation, are likely to show up as more immediately binding and more intractable constraints. Indeed, the problem of implementation is widely recognized and cited as a bottleneck in Indian agriculture and rural development, leading to the subtle question of how skills and motivation—not to mention a sense of mission and renewal—can be injected into the relevant administrative levels where this is needed, such as to produce the desired outcomes in farmers' fields at scale. ATMA appears to have assembled the right conditions for this during its pilot phase, as argued in this chapter. Under new Guidelines issued in 2010, it must now combine framing the task from above with the needed resources, guidance, and empowerment below. Block-level ATMA organizations have a key role to play. They include farmers' representatives and community and local structures that originate proposals that would then go to the district level for approval. But there are questions as to how well this process works. Rendering it functional, representative, transparent, and effective is a top priority today.

The action implications that can be formulated first involve the institutional dimension: ATMA must become more functional in the settings and states where extension is not performing well. The public sector's role is to develop innovative ways to foster the delivery of services in the relatively lagging states and areas within states with the aim to reduce the agricultural performance gap between them and the rest of the country.

Public-private partnerships should be seen as a 'tool of choice' to this effect, and the opportunities of cooperation with and through professional NGOs and non-profit service providers should be fully utilized. Already, there are instances where professional NGOs provide content to the government's extension agenda, sometimes against payment from the government or as conduits for subsidies channeled to farmers. This is a positive development that could be expanded systematically, together with the provision of government funds for needs such as soil testing labs, the establishment of field trials and demonstration plots, farm machinery, and cold storage facilities that arise in the context of NGOs' work with farmers. In other words, the outsourcing of extension to adequately resourced professional NGOs under monitored contractual arrangements that specify the targets and standards to be adhered to and the outcomes to be produced.

Other action implications that arise from the discussion in this chapter include the need to strengthen and adequately fund the SAUs, KVKs, and other points of delivery of extension, including agriclinics and agripreneurs. Extension and prioritized agricultural research that addresses what farmers need should be functionally linked. Needs-oriented thinking and action on the ground should be rewarded and made part of the culture of agricultural R&D programs and institutions at all jurisdictional levels, but particularly at the district and local level where farming actually occurs.

The centrality of agronomic support to farmers should be recognized and properly delivered, allowing adequate time for mentoring in the field, working through community structures and local personnel trained for the purpose. The centrality of markets should be recognized, as well as the need to incorporate sourcing of inputs and prospects for product sales into the extension agenda in cooperation with the private sector—such as rural business hubs where they exist. Priority should be given to the training of input dealers, due to their documented importance as sources of extension advice for farmers. The work of MANAGE, in this respect, offers a model that could be applied at scale.

China's Extension System

When China started its rural economic reforms and the 'Household Responsibility System' in 1978, it needed to reorganize and strengthen its system of agricultural extension. By the mid-1980s, China had established a comprehensive nationwide extension network, with a five-level hierarchy: national, provincial, city, county, to township. The national agro-technical extension and service center (ATEC) is responsible for long-term strategy, with implementation taking place at the other hierarchy levels. The extension system is organized according to agricultural sub-sectors. Most counties have set up extension stations for crop, livestock, aquaculture, agricultural machinery and economic management. Depending on local conditions, more specific crop stations, such as for cotton or tea, may be established as well. By the mid-1990s, the total number of trained extension staff exceeded one million, with more than 90 percent working at county and township levels.

At the end of the 1980s, the system became overstaffed and inefficient, due in part to the proliferation of specialized stations or extension hubs (Huang et al., 2000). This created a financial burden for local governments. To resolve this, the central government began to implement a series of reforms. So-called 'commercial' reforms to introduce a new funding model began in the early 1990s. Extension workers were classified into three categories: fully-funded agents (government payroll), partially-funded agents (government pays part of base salary), and self-funded agents (base salary comes from commercial activities and grants). Counties had flexibility in implementing these reforms. In most cases, crop stations are categorized as fully-funded agents while input supply stations, for example, for seeds and pesticides, are classified as self-funded agents. Livestock and aquaculture related stations are often classified as partially-funded agents. As a result, the funding for extension activities from local, provincial, and the central government has been reduced.

In order to improve efficiency and coordination among extension stations, another reform was initiated with a view to merging various specialized hubs into one-stop shops. In the process, for example, many counties merged their crop management technology, plant protection, and soil and fertilizer technology stations into single crop extension service centers (Hu et al., 2009). Figure 4.A1 shows the structure of the extension system after this reform. The integration of crop-related stations was relatively straightforward since these are directly under the administration of agricultural bureaus. But the next step, with mergers across different bureaus (e.g., livestock, aquaculture), proved more difficult because of administrative barriers. The process towards complete mergers is still underway.

In an attempt to reinvigorate the extension system, the central government enacted another reform in the late 1990s (Hu et al., 2009). This shifted administrative rights (the so-called 'three rights' of personnel, finance, and materials) from county agricultural bureaus to township governments, with a matching shift in budgetary burden. Unfortunately, this reform broke former productive links between the county and township agricultural extension stations. Township agents are frequently called upon to respond to administrative duties that have nothing to do with agricultural extension, including family planning, budget management, elections, and legal matters. The current extension system also faces problems in other respects, for example, the lack of competency and updated knowledge of agro-tech extension staff who struggle to keep up with farmers' rapidly evolving needs.

Public agricultural extension in China has been reformed repeatedly to improve efficiency and better serve the country's huge number of small farms. The system has been quite successful in promoting the adoption of new agricultural technologies including improved crop varieties. Since the economic reforms began in 1978, there has been a steady increase in the production of cereals and high-value horticultural and livestock products. For example, cereal yield has risen from about 3 to 5.5 ton/ha on average, resulting in an increase in cereals production from 290 to 480 million tons. The volume of fruits and vegetables increased 8.6 times between 1979 and 2008, that of animal protein (meat products) 5.6 times. Extension has played a significant role in these achievements by disseminating modern technologies. China continues to be concerned about national food security, but farmers no longer have to sell specific quantities of cereals to the government at reduced prices. Farmers generally grow the crops that reflect their aspirations, given their land, labor and other resources and their access to markets.

The key factors that facilitated the rapid transformation of the agricultural sector in China include rapid economic growth, changing consumer food demand, investment in rural infrastructure, vocational agricultural training for young people in rural areas, and the transformation of the agricultural extension system to better serve the needs of farmers as they diversify and intensify their farming activities (Swanson and Rajalahti, 2010). These factors are visible in other Asian countries as economic growth unfolds. However, the participation of small farmers in the agricultural transformation depends on specific policy and operational issues—for example, whether the public agricultural extension system is prepared to organize farmers into producer groups to enable them to pursue high-value horticulture, livestock, fisheries, and other enterprises that will help increase their household income.

In China, small farmers are increasingly organized into producer groups for different high-value crops and products. There are mainly two types of producer organizations: associations

Figure 4.A1 | Structure of the agro-technical extension system in China

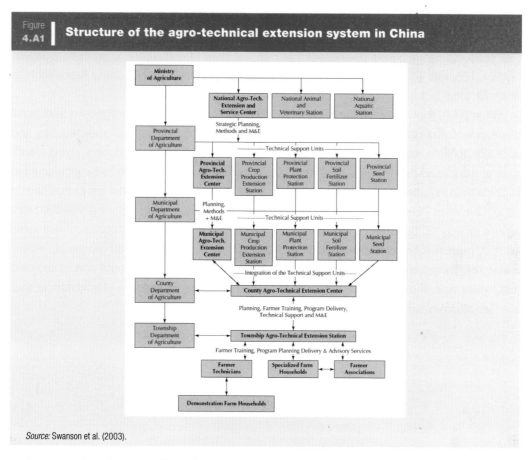

Source: Swanson et al. (2003).

and cooperatives (co-ops). New Cooperative Laws enacted a few years ago allow farmers to organize themselves and to link up with agribusinesses. There are many types of co-ops, such as vegetable marketing or fish producer co-ops, etc. Most pursue mixed activities that include farmer training, and processing and marketing of produce. By 2005, there were over 150,000 specialized farmers' co-ops in China. About half of them focus on crop cultivation, one-third on animal husbandry and aquaculture, and the rest on machinery and other sectors. Many villages have one or more co-ops. Geographically, the co-ops are relatively more developed in the middle and eastern parts of the country where commercialization is more intense. The producer organizations usually understand their members' real needs better, and when necessary, invite technicians from extension stations or research institutes to provide technical and managerial training. The co-ops and associations have played an important role in promoting agricultural development and increasing farmers' income.

Apart from public and cooperatively led extension, agribusiness enterprises (seed companies, pesticide and plastic film manufacturers, for example) also participate in agricultural extension in

China. Multinationals, such as Monsanto and Syngenta, have R&D and extension departments in the country. Their extension workers visit farmers in the fields and guide them on how to apply their products. Small private businesses that may not have the resources to proceed independently collaborate with existing extension networks (including the CropLife China Association) to reach farmers.

Since 2006, "New Countryside Construction", a new national strategy for agriculture and rural development, has been put forward to address the widening rural-urban income inequality and stimulate domestic consumer demand. It seeks to improve agricultural production, living standards, and public administration in rural areas. Agricultural extension seems to be better resourced in China than in India, at least as far as the numbers of extension personnel are concerned: In 2006, China had 787,000 extension staff in the public extension system, including 560,000 technicians serving 637,000 villages for a ratio of 1.1 villages per technician. This contrasts with some 100,000 public extension workers in India for 638,000 villages, a ratio of 6.38 villages per extension agent. There are documented performance shortfalls in both systems, but it should be noted that, unlike India, the Chinese system is present in all counties, irrespective of how remote they are. See http://www.syngentafoundation.org/index.cfm?pageID=734.

Agricultural Research for Sustainable Productivity Growth in India

Chapter 5

Partha R. Das Gupta and Marco Ferroni

Introduction[1]

There is vast untapped production potential in Indian agriculture. Yield gaps could be progressively and sustainably closed, production increased and natural resources used judiciously, with the right technologies and adequate incentives and support for farmers. Technology is an enabler and the product of research, which is the topic of this chapter. Agricultural research has a great record of achievement in India, having spawned the Green Revolution over fifty years ago. It has gone through different phases since then, and is now in the public eye. This new attention stems on the one hand from the country's ambitious agricultural growth target of 4 percent per annum. It also reflects the need and opportunity to respond to a range of current developments. These include the leveling-off of productivity growth of key food crops (Singh and Pal, 2010), new options in science, technology and the organization of research, changes in the agricultural economy and the demand for food, as well as challenges linked to natural resource stress and climate change, and redefined public and private roles.

This chapter examines the performance of India's public and private agricultural research system and offers recommendations in light of the evolving agricultural and scientific context and need. Following this introduction, Section 2 offers a brief historical overview. Section 3 analyzes the organization and achievements of India's National Agricultural Research System (NARS) and the challenges before it in the coming years. Section 4 covers private agricultural research. Section 5 discusses molecular plant science and biotechnology, which will play key roles in the transformation of Indian agriculture and should give rise to novel forms of public-private cooperation. Section 6 concludes with recommendations.

Historical overview

In the face of looming food shortages soon after Independence in 1947, Jawaharlal Nehru remarked: 'everything else can wait, but not agriculture' (Swaminathan, 2006). The First Five Year Plan, launched in 1951, assigned top priority to irrigation, power, agriculture and community

1 The authors acknowledge valuable inputs received from Ramesh Chand, Malavika Dadlani, M.J. Chandre Gowda, C. Ramasamy, P.K. Joshi, Pravesh Sharma, Krishna M. Singh and others in India and from Viv Anthony, Paul Castle, Mike Robinson, Tanja Wenger and Yuan Zhou at the Syngenta Foundation for Sustainable Agriculture.

development. Land reform to improve property rights for farmers was initiated and investment in agricultural infrastructure, including fertilizer manufacturing, stepped up. The land reforms of the mid-fifties remain one of the most important milestones in the social history of India (Bhalla, 1989).

Backed by these measures and others to come, the average annual growth rate of agricultural production rose from about 1 percent in the first half of the twentieth century to about 2.6 percent in the post-Independence period (Tripathi and Prasad, 2009). While production increments in the first two decades after Independence were achieved largely by bringing more land under cultivation, the mounting pressure of a growing population made intensification and efforts to increase productivity essential. The First Plan instigated irrigation infrastructure development and fertilizer production. However, the existing crop varieties, in particular of wheat and rice, did not respond favorably to increased fertilizer dosages under irrigation. New types of seed were needed.

In their search for fertilizer-responsive varieties, Indian agricultural scientists found a first solution in the form of semi-dwarf, high-yielding varieties brought in from Mexico in the early 1960s. These were the result of Norman Borlaug's breeding program initiated in 1943 at what would later become CIMMYT, the International Maize and Wheat Improvement Center. First tested under a national demonstration program in 1964, the vastly superior performance of the varieties from Mexico prompted the government to import a large quantity of seed for planting in 1966. Farmers adopted the new varieties so fast that within two years national wheat production rose from 12 to 17 million metric tons—a feat referred to as the 'Wheat Revolution'. High-yielding, semi-dwarf varieties of rice such as IR-8 arrived soon after, the result of collaboration between ICAR (the Indian Council of Agricultural Research) and IRRI (the International Rice Research Institute).

Success in wheat and rice paved the way for innovation in other crops (NAAS, 2009). The NARS created improved hybrids of sorghum, pearl millet and maize, later replaced by superior versions developed jointly with CIMMYT and ICRISAT (the International Crops Research Institute for the Semi-Arid Tropics). Significant achievements emerged in horticulture, sugarcane, cotton and some other crops. India also became the world's largest producer of milk. The breakthroughs of this era can in good measure be attributed to agricultural research and extension, but were also supported by farm credit and smallholder development programs in addition to public investment in irrigation and rural infrastructure. Agricultural growth occurred despite macroeconomic policies and market regulations deemed 'distortionary' and detrimental to the sector (Evenson et al., 1999). Critically, there was political will to 'win' in agriculture, and a vision that inspired scientists and the farming community alike.

The 'golden period' gave way to a phase of deceleration as policy reforms called for rationing of public investment in agricultural research from about 1990 (the post-WTO period). There was concern that the returns on investment in research might decline in a situation in which the 'easy' gains from the Green Revolution had already been reaped through the wide adoption of high-yielding varieties (HYV) and high input use. There was a perception that the newer versions of crop varieties developed by public institutes were unable to surpass the maximum yields already achieved (Evenson et al., 1999). While public investment in agriculture fell from the mid-1980s to the early 2000s, subsidies rose (Figure 5.1). Private investment in agriculture rose during the second half of the 1980s, fell during the early part of the 1990s and then accelerated to multiples of public sector values as a share of agricultural GDP (Figure 5.2). Annual growth in output of crops and livestock decelerated during the 1990s in the context of reforms that included currency adjustments and the liberalization of commodity imports and domestic trade. Total factor productivity (TFP) growth declined.

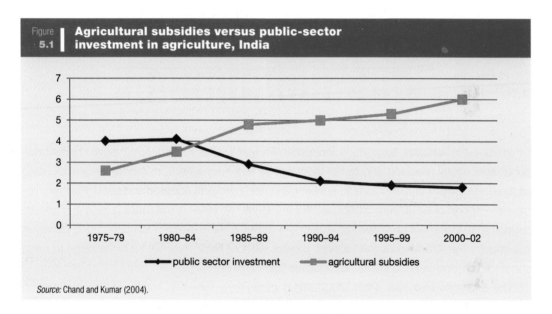

Figure 5.1 | **Agricultural subsidies versus public-sector investment in agriculture, India**

—◆— public sector investment   —■— agricultural subsidies

Source: Chand and Kumar (2004).

Reacting to this downturn, the Government launched a New Agricultural Policy (NPA) in July 2000, introducing the 4 percent per annum agricultural output growth target. According to Chand (2000, 2001), the studies underlying the NPA attributed declining productivity growth to three factors: the decline in public investment in agriculture (including agricultural research), slowdowns in yield gains from the earlier HYVs, and progressive degradation of natural resources. The NPA recognized the private sector's role in agricultural research, the need for training and capacity building, the effect of post-harvest management as a factor affecting the nation's food supply, and the relevance of modern supply chains to which farmers can be linked. The NPA called for a renewal of the forces of innovation and a knowledge-based vision of agriculture. It did not spell out how and when to achieve the declared policy goals. The Tenth Five Year Plan (2002–2007) endorsed the NPA's annual agricultural growth target and formulated a public investment program to support it.

Figure 5.2 | Gross fixed capital formation in agriculture as a share of GDP-agriculture at current prices (%)

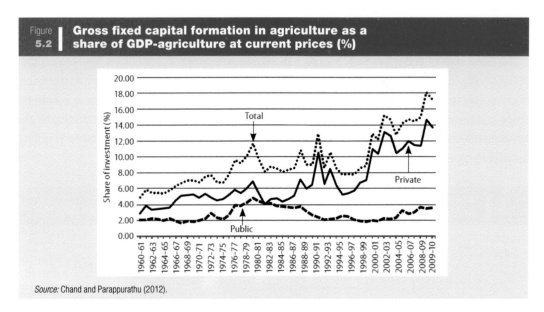

Source: Chand and Parappurathu (2012).

In 2006 the National Agricultural Innovation Project was launched. NAIP aimed to revitalize public agricultural research and revise the strategy and approach from fundamentally focusing on the development of technology to ensuring that technology is adopted and used. Partnership-based methods of work were introduced for this purpose. NAIP is still running, and it would be premature to comment on its overall impact. As noted in the Eleventh Plan document, however, in a country the size of India even a large project such as NAIP may not be big enough to bring about the far-reaching change its protagonists had in mind.

The Eleventh Five Year Plan (2007–2012) pledged a renewed thrust of policy attention and investment to support the 4 percent annual growth target of agricultural GDP (Planning Commission, 2008). In the short run, the declared focus was on improving yields by disseminating and efficiently using existing technology. With reference to the longer term, the Plan stressed the need to adopt 'a comprehensive view of the functioning of the agricultural research system' and making 'systemic changes' during the Plan period. The Plan stated that 'agricultural research is underfunded but lack of resources is not the only problem [...] available resources also have not been optimally utilized because of lack of a clearly stated strategy that assigns definite responsibilities [and] prioritizes the research agenda rationally' (Planning Commission, 2008). The Plan projected public spending on research to grow but also stressed the need to increase the accountability of State Agricultural Universities and the relevance of their research output. It stipulated the need for synergy between NARS and other key institutions of the national science and technology establishment, including CSIR (the Council of Scientific and Industrial Research), DBT (the Department of Biotechnology of the Ministry of Science and Technology), and others.

Average agricultural growth during the Eleventh Plan is estimated at 3.3 percent per year (Planning Commission, 2011)—short of target, but approaching it and higher than the figure of 2.5 percent during the Tenth Plan (Sharma, 2011). Food grain production in 2010–2011 reached the record level of 241 million metric tons (23 million metric tons higher than the drought-affected harvest a year earlier)[2] and was projected to be even higher in 2011–2012,[3] except that renewed rainfall deficits in 2012 may frustrate that prospect. In any case, the relative recovery of food grain and other agricultural production in recent years is a positive development linked to the growth of investment in the sector (Figure 5.2) and the renewed attention paid to it under the Eleventh Plan. Still, the question about national preparedness to meet the demands of the future remains.

As we write, the Twelfth Plan is being finalized. The 4 percent growth target is retained and agriculture, including agricultural research, is the subject of an ambitious policy and investment framework. The challenge before the national system of agricultural research (public and private) is to position itself for a strong role supporting India's agricultural transformation, discussed in this book. The remaining sections of this chapter offer some guiding thoughts.

Public sector R&D in agriculture

India's NARS is composed of ICAR, central and state agricultural universities (SAU), Krishi Vigyan Kendras (KVK) and other publicly funded entities and programs.

Established in 1929, ICAR was reorganized in the mid-1960s and became one of the world's largest public research establishments (Evenson et al., 1999). Today it is India's apex body guiding and coordinating research and education in agriculture and agricultural extension. ICAR's mandates include crop production, natural resource management, horticulture, plantation crops, fisheries, animal sciences and dairying, and agricultural engineering. The organization has 97 constituent entities, including national and regional institutes, centers, bureaus and project directorates. In addition, it runs 61 networks of 'All India Coordinated Research Projects', oversees the functioning of SAUs, and provides them with selective funding. It also funds and coordinates the activities of the more than 600 KVKs. The General Body of ICAR is presided over by the Union Minister of Agriculture, with a Director General as the chief executive (Figure 5.3). In 2010–2011, ICAR had 25,000 people on its books, including more than 10,000 scientific and technical personnel. The scientists are hired through a competitive selection process by an independent Agricultural Scientists Recruitment Board.[4]

ICAR and other institutions that collectively form the NARS have made critical contributions to Indian agriculture. These led to productivity gains in food and non-food crops during and after the Green Revolution. To mention just a few examples, the publicly bred cotton hybrid NHH-44ICAR

---

2 http://businesstoday.intoday.in/story/foodgrain-production-in-india-2010-11-pm/1/17062.html; accessed July 2012.

3 Advance estimates of Government of India, The Hindu, April 24, 2012 (http://www.thehindu.com/todays-paper/tp-national/article3347238.ece; accessed July 2012).

4 For documentation, see http://www.icar.org.in/en/node/1237 (accessed July 2012) and other resources on the ICAR website.

II/5

Figure
5.3

ICAR Organigram

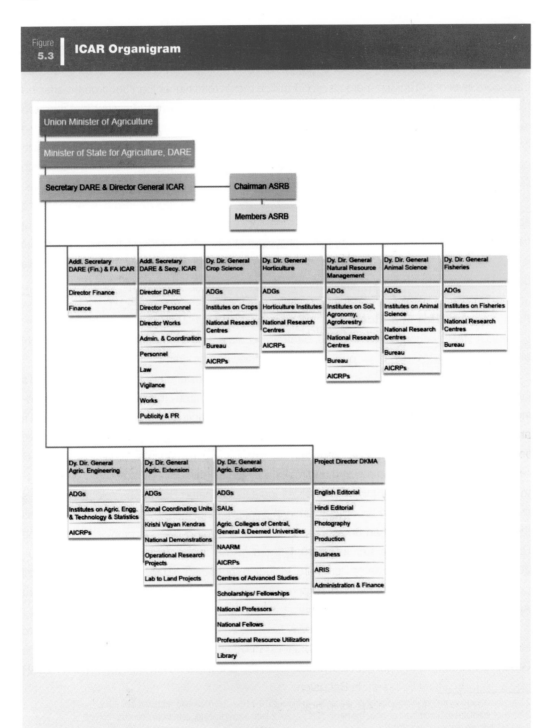

Source: http://www.icar.org.in/en/organization.htm (accessed July 2012).

was farmers' number 1 choice for a decade. NARS is credited with developing and commercializing the world's first hybrid in cotton and castor, while hybrid pigeon pea was developed in collaboration with ICRISAT. The private sector introduced *Bt* cotton in the early 2000s, but the hybrids that carry the transgene are derived from public germplasm under the stewardship of NARS. A recent breakthrough by IARI, the Indian Agricultural Research Institute in New Delhi, has been the rice variety Pusa Basmati 1121 (Box 5.1).

Box 5.1 | Pusa Basmati 1121

Farmers in six northern states of India grew approximately 2.16 million hectares of Basmati rice during the 2011 kharif season. According to satellite data, some 1.34 million hectares were under PB-1121, almost two-thirds of the total. Bred by IARI scientists and released for commercial cultivation in 2003, PB-1121's adoption by farmers proved a runaway success. Its quality attributes include 30 percent elongation on cooking and an aroma that nearly matches that of traditional Basmati. Yields are more than twice those of traditional varieties. In 2009-2010, PB-1121 accounted for nearly half of all Basmati rice exported from India at an estimated value of US$ 2 billion. PB-1121 has reportedly spread to Pakistan, too (Damodaran, 2007; Sud, 2007).

In horticulture, ICAR has improved varieties of mango, citrus, litchi, tuber and root crops, among others, creating plants well suited to different regional and seasonal settings. Over the years it has released hundreds of high-yielding varieties and hybrids of fruit and horticultural crops. Two breakthrough vegetable varieties of IARI, 'Pusa Sawani' (okra) and 'Pusa Ruby' (tomato), played a role in the creation of India's private seed industry: a company that received seeds in the late 1950s, in what was in effect an early public-private partnership, multiplied these and marketed them across the country (Swaminathan, 1986).

Pusa Sawani was the first yellow vein mosaic virus (YVMV)—tolerant okra (bhindi). Pusa Ruby was the first bush variety of tomato (non-staking) that did not require cool growing conditions. (All previously introduced varieties came from Europe or the US, and were only suitable for winter cultivation.) While okra was already popular among Indian vegetables, Pusa Ruby made the tomato affordable and popular. 'Pusa Jwala', meanwhile—also developed by IARI in the 1970s—has a fruit type that remains the benchmark for green chillies to this day. Pusa Ruby was later replaced by 'Selection-20' and Pusa Sawani by 'Parbhani Kranti' and then 'Arka Anamika' as its YVMV-tolerance started to break down. (Selection-20 and Arka Anamika were developed by the Indian Institute of Horticultural Research in Bangalore; Parbhani Kranti by the Marathwada Agricultural University, Parbhani.) In vegetables as in grains, therefore, NARS made contributions that were 'game-changers' at the time, but the system started to lose energy as it matured and as private sector hybrids began to spread after the launch of the New Seed Policy in 1988.

ICAR made important contributions in specialized areas like water technology, seed science, post-harvest technologies, some agricultural machinery, as well as plant disease and pest diagnostics and management, among others. The ICAR website[5] is explicit about much of this work, but fails to provide evidence on the rate and patterns of adoption of the technologies.

The record of SAUs as a group is rather different. The era of SAUs in agricultural education and research began in 1960 with the establishment of Uttar Pradesh Agricultural University at Pantnagar (now G B Pant University of Agriculture & Technology in the state of Uttarakhand). This was followed by Punjab Agricultural University and many others. Currently there are 53 SAUs, including the central and deemed universities. Each runs a number of zonal and crop-specific research stations and also works with KVKs. Initially modeled after the US land grant colleges, the Indian SAUs developed their own modus operandi, with variations across states. Notable differences exist in regard to academic standards, research contributions and impact of the extension work performed by SAUs, whose governing acts and statutes are in some respects now outdated. The financial position of SAUs is a cause of concern. Most are starved of funds, particularly from their own state governments. ICAR contributions are limited to a maximum of 40 percent of their budgets. So the research contribution of SAUs as a group is somewhat modest and has declined over time, although some have worked in collaboration with ICAR and international institutes and produced good results. ANGRAU is a case in point (Box 5.2).

Box 5.2 | High-yielding rice varieties from SAUs

ANGRAU (Acharya N G Ranga Andhra Pradesh Agricultural University in Hyderabad) has made significant contributions to Indian agriculture by releasing modern rice varieties from its research programs. The varieties from Maruteru and Bapatla research stations—including MTU-7029 (released 1979) and BPT-5204 (1986)—occupy nearly a quarter of India's rice area. 'Swarna' (MTU-7029) is the most popular variety, not only in India but in some neighboring countries, too. However, it is now a matter of concern that few new varieties have been released in recent years (http://www.scribd.com/doc/75968285/ANGRAU-Research-Krishna-Godavari-Zone-Maruteru (accessed July 2012) and http://www.fao.org/DOCREP/006/Y4751E/y4751e0b.htm (accessed July 2012)). The variety 'Gontra Bidhan-1', released from Bidhan Chandra Krishi Viswavidyalaya in 2008, has replaced 20 percent of the area planted to Swarna in West Bengal (P Chatterjee, personal communication to PDG, 2012).

Good research requires modern laboratory, office, library and IT infrastructure. In this respect ICAR institutes are in a somewhat better position than the SAUs as a group and perhaps for this

---

5 http://www.icar.org/; accessed June 2012.

reason more productive. But the productivity and relevance of ICAR and NARS as a whole have been the subject of debate for years now. Many observers believe they are declining,[6] giving rise to doubts about India's public agricultural research preparedness. Certainly, the pertinence of NARS innovations has declined. What the system produced 25 and more years ago was significantly better than what was then commonly available; today's releases are rarely breakthroughs. Open markets and the increased availability of products developed by the private sector are part of the story. Before the 1988 New Seed Policy, imports of seed were forbidden. Seeds then became importable under 'open general license'. As a result, domestic research in key crops including vegetables, maize, sunflower and cotton became less relevant, and biotechnology innovations have largely come from abroad. As intimated by NAIP and the Eleventh Plan referred to above, there is a need to re-focus public agricultural research.

Aspects underpinning the perceived decline in relevance (and therefore effectiveness) of NARS include slippages in the average quality of agricultural research in the public sector; sluggishness in the transition from 'old' to 'new' science; problems linked to funding levels and allocation; the problems of size, structure, organization and 'process' detected by prominent commissioned reviews; limited 'co-creativity' with the private sector; and the absence of clear-cut priorities to overcome 'technology fatigue' and address the challenges of rain-fed agriculture. There is also much room for improvement in the delivery of results to farmers' fields.

The quality of research by scientists of Indian public agricultural institutions is lagging behind China and Brazil when measured by the numbers of published papers and citations over identical periods. In numbers of citations, China and Brazil overtook India some ten and five years ago, respectively (Datta, 2010). The Scopus database accessed through the SCImago portal[7] assigns China fifth, India tenth and Brazil eleventh rank in agricultural and biological sciences in 1996–2010 but reports a greater number of citations of documents, and citations per document, arising from Brazilian research. This does not mean that there are not world-class researchers at work in ICAR and the Indian NARS more broadly, but it does point to an apparent broader trend that should be looked at and addressed.

One factor in this context could be ICAR's relatively limited engagement in state-of-the-art crop improvement science and technology, including molecular plant science and biotechnology. 'Molecular breeding'—seen as more effective than 'traditional breeding'—may be defined as the identification and use of associations between genotypic (DNA sequences) and phenotypic (plant trait attributes) variation to select and assemble traits into new crop varieties. 'Traditional breeding', which takes longer to produce results and is less 'precise', is defined as the use of phenotypic data only to develop new and improved varieties (this is discussed in Section 5). The Report

---

6 See 'Making ICAR relevant' (http://www.thehindubusinessline.com/opinion/editorial/article2913431.ece; accessed July 2012) and 'ICAR alert: Wake up now or become irrelevant' (http://www.indianexpress.com/news/icar-alert-wake-up-now-or-become-irrelevant/1189/0; accessed July 2012).

7 http://www.scimagojr.com; accessed July 2012.

of the Agricultural Research and Education Working Group for the 12th Five Year Plan makes the case for renewal with respect to molecular plant science. It argues for the 'judicious integration' of traditional plant breeding with molecular biology and bioinformatics for timely production of high-yielding, nutrient-rich and stress-tolerant/resistant crops (Planning Commission, 2011). The Report also calls for 'interdepartmental platforms' through which NARS research capability could be pooled with that of a range of R&D organizations able to contribute in areas from genomics to health food, precision farming, dryland agriculture and farm mechanization. Recent molecular breeding contributions by IRRI and the Indian public sector are described in Box 5.3.

| Box 5.3 | **Molecular breeding contributions by the international and the Indian public sector in rice** |

The wide-spread adoption of the variety 'Swarna' (MTU-7029) has already been referred to in Box 5.2. However, Swarna is not suitable for India's 12 million rice hectares regularly subject to flooding. So scientists at IRRI developed the flood-resistant variety 'Swarna-Sub1' by inserting the submergence tolerance gene SUB-1 via marker-assisted backcross breeding (MABB). Swarna-Sub1 is becoming increasingly popular in low-lying coastal areas through the efforts of Tamil Nadu Agricultural University and other NARS institutes.

Scientists at IARI and collaborating institutions have successfully used MABB to transfer the genes for resistance to bacterial leaf blight into parental lines of rice hybrids like PRH-10 and others. Work is also at an advanced stage on transferring genes for resistance to blast and for salt tolerance into different varieties of rice. For information on this, see (i) 'Indian farmers adopt flood-tolerant rice at unprecedented rates' (http://www.irri.org; accessed July 2012); (ii) 'Swarna-Sub1: flood resistant variety', The Hindu, September 29, 2011; (iii) 'New rice variety saves the day for rain-hit farmers of Orissa', Financial Express, October 17, 2011; and (iv) 'Marker assisted selection: a paradigm shift in Basmati breeding' (Singh AK et al., Indian Journal of Genetics, 71(2), Special Issue, pages 1-9).

Public expenditure trends in agricultural research and the ratio of spending on salaries and benefits to capital equipment are a cause for concern. The 'intensity' of agricultural research expenditure (i.e., spending as a share of agricultural GDP) has for many years been lower in India than in China and Brazil (Figure 5.4). Increases in public expenditure on agricultural research tend to be spent on salaries, allowances and rising operating costs to the detriment of capital equipment (Vaidyanathan, 2010). This is alarming, since it is unlikely that the research India needs can be pursued without adequate budgetary provisions to upgrade laboratories and capital equipment and outsource technological services where efficient and feasible.

In their detailed state- and crop-wise study of the linkages between public agricultural research expenditure and both TFP and production growth, Chand et al (2011) do not address the institutional reform implications of issues related to the composition of spending, but do conclude

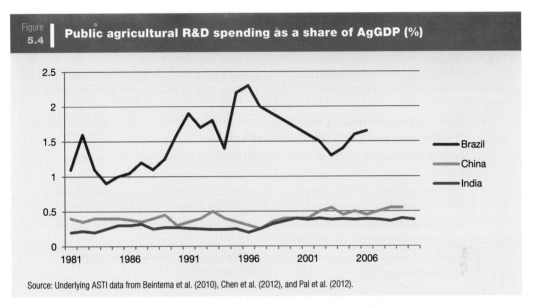

Figure 5.4 | **Public agricultural R&D spending as a share of AgGDP (%)**

Legend: Brazil, China, India

Source: Underlying ASTI data from Beintema et al. (2010), Chen et al. (2012), and Pal et al. (2012).

that more money may be needed for public agricultural research. They find that (i) investments in research generate significant returns,[8] (ii) a one percent increase in TFP growth requires agricultural research investment in India to grow by 25 percent per annum on average, (iii) to attain the target of 4 percent yearly agricultural growth, 'at least one-third of this growth must come through technological innovations and the remaining two-thirds [...] through additional use of agricultural inputs'; and (iv) to attain the 4 percent growth target, 'investments in agricultural research in real terms need to be doubled by 2015 and tripled by 2020 in relation to the investment level of 2002'. For the record, the public agricultural R&D investment level in 2002–2003 was Rs. 36.53 billion (Ramasamy, 2011, based on ICAR's Agricultural Research Data Book, 2009).

We tend to agree with this assessment in general terms—without being able to draw on a comparable modeling exercise of our own—but would suggest that significant increases in public agricultural research expenditure should be predicated on reforms in mission-critical respects, including in particular 'institutional reform', to which we turn next.

There has long been debate about the best structure and organization of ICAR and the NARS. The Swaminathan Committee of 2002 and the Mashelkar Review of 2005—bold and sweeping assessments still frequently cited today—called for far-reaching institutional and organizational change (Box 5.4). But the public research system is difficult to reform. It is rigid and 'set in its ways', governed by processes and rules that have evolved and taken hold over time, and infused by conventions and patterns of decision-making antithetical to entrepreneurship, the taking of risk, and the often necessary pooling of resources across units to get on with given tasks. It is

---

8 According to Chand et al. (2011), the internal rate of return to public research investment in agriculture in India was 46 percent for the period 1985–2006 and 42 percent for 1990–2006. Both of these figures imply significant national economic gains.

II/5

Swaminathan key recommendations (the Steering Group on Agriculture and Allied Sectors for the Tenth Five Year Plan chaired by Professor M S Swaminathan devoted its Chapter 4 to 'Institutional Reforms'):

- Restructure the vast research infrastructure of NARS to meet present and future challenges; stress research quality enhancement and productivity

- Go for breakthroughs in critical areas such as hybrid rice for wide adoption by farmers as in China

- End the proliferation of new research institutes which constrains the resources of existing ones and pushes up establishment costs, leaving little scope for effective R&D work

- Provide for convergence amongst institutes of NARS to make the system more efficient; current results are not commensurate with the size of the infrastructure

- Implement 'national challenge' projects jointly with national and international institutions outside of the NARS

- Engage in cooperative research activities with the private sector

- Address infrastructure and human resource development needs in agricultural education and address the issue of lack of financial support to SAUs

- For a country of India's size, one IARI is not enough; there should be at least one for each major ecological zone.

Mashelkar key recommendations (the ICAR Reorganization Committee chaired by Dr. R A Mashelkar found ICAR a 'poor organization' in terms of research and publication excellence among other aspects and recommended the following):

- Revamp most facets of ICAR—from restructuring the governing body and chain of command to ensuring the participation of stakeholders, including the private sector

- Develop 'scientist-entrepreneurs'; make hiring and shedding procedures for scientists more flexible; provide for deputation of ICAR scientists; allow professionals a greater say in ICAR operations; release ICAR processes to industry

- Name the Prime Minister President of the ICAR Society to send a strong signal about the importance of agriculture and agricultural research to the national economy.

Assessment: The Swaminathan and Mashelkar recommendations (summarized from the respective reports) clearly call for much more than incremental change, but appear not to have been acted on in ways commensurate with the intentions of their authors. Indeed, the Mashelkar report appears to have been rejected, having been severely criticized by ICAR (cf. S. Mishra, 2005. 'What ails ICAR? Ten committees and thirty years on: Nobody knows' at http://www.downtoearth.org.in/author/413; accessed July 2012). The November 2011 Report of the Working Group on Agricultural Research and Education for the 12th Five Year Plan no longer mentions any restructuring of ICAR or the need to reconsider the proliferation of institutes. It does call for more resources for the SAUs and both integration of research efforts across institutes of NARS and cooperation with third-party research organizations.

not that ICAR and the NARS need advice on what to do—they have some of the world's best brains in agricultural research and research management on hand. But there is inertia, as is often the case with mature institutions, and the question is how to overcome it to enable the system to respond as best possible to agriculture's rapidly changing circumstances and needs.

The relatively high rates of return on spending on agricultural research calculated by Chand and co-authors (footnote 8) and the recovery of agricultural production growth in recent times (Section 2) may lead some to ask whether reform of the public agricultural research system is really necessary and worth all the effort. The answer is 'yes' in our view, for at least two reasons. First, farming faces huge unmet challenges (including natural resource stress and climate change, the rapidly spreading problem of diminishing returns on input use, land degradation, and overexploitation of groundwater), and there are numerous opportunities to reduce yield gaps and improve farmer income and consumer choice. Second, agricultural research needs to come up with solutions under circumstances radically different from the staple-grain based sense of purpose that permeated Green Revolution thinking and shaped the intellectual tradition of ICAR and the NARS. Clearly, 'business as usual' no longer applies in a situation in which (i) the role of high-value products (such as vegetables and dairy items) is growing in the food system while that of grains and pulses declines; (ii) rural wage rates are rising and labor can be short at critical times (calling for effort- and labor-saving devices); (iii) agriculture is becoming much more information-driven and commercial; (iv) the private sector is taking over many agricultural R&D activities (although it will not engage in areas where it fails to detect a business case); and (v) advances in modern biology and genetics are offering new opportunities while altering both some of the methods of work and the comparative advantages of the public and the previously non-existent private sector in the vitally important mission of crop and livestock improvement.

In this situation, the point of NARS reform is to establish relevance and enable the system to respond to, and support, the rapid transformation of agriculture now underway. Of the many needs this implies, we will briefly discuss three in the remainder of this section: for creative forms of coexistence with the private sector, for a renewed focus on rain-fed agriculture, and for measures and research preparedness to deal with the problem of leveling-off of productivity growth where it occurs. The vital topic of agricultural extension is the subject of another chapter in this book.

Creative coexistence with the private sector

The private sector is increasingly present at all stages of the value chain, from R&D and the marketing of seed and other inputs to trading, processing and retail. In future, therefore, technology breakthroughs are likely to originate in the private sector (as with *Bt* cotton) or in partnerships between public institutions and crop science or agro-service companies. The latter will undertake a growing part of adaptive and applied research. This will leave public programs to focus more on basic research and the many areas with no real markets such as the stewardship of natural resources or the development of technology for 'pre-commercial' farmers. A mapping of 'needs', the agro-service industry's crop and livestock research offer and, residually, the role of the public sector could help guide research targets, public resource allocation and NARS reform.

In biotechnology, including genomic science, marker-assisted selection and recombinant DNA techniques, cooperation across the 'public-private divide' is essential to stay focused on meeting the market's needs and take advantage of the private sector's skills (see Section 5). But this is still uncharted territory, and the lack of predictability in the regulatory framework for biotech crops does not help. It is essential to have regulations preserving the public's interest. However, the high cost of dealing with uncertain regulatory conditions is a deterrent for private investment in agricultural research and the development of markets for products and inputs that could benefit farmers.

Partnerships are about give-and-take by all involved. The public sector should nonetheless set the agenda, reach out and explore cooperation possibilities. These need to capitalize on the distribution of capabilities: in modern plant breeding, for example, phenotyping is more of a public skill and genotyping heavily a private one. New thinking is needed, as well as better communication to help say the media understand the social gains that can come from public-private cooperation in agricultural R&D. Flexible procedures are needed, too: the public system is somewhat bureaucratic, operating on timescales of its own, and, perhaps more importantly, is unsure how to respond to partnership opportunities in the absence of clear guidelines. Meaningful guidelines can probably only emerge from domestic and international experience. Some good examples in crop and livestock breeding, animal nutrition and health, and perhaps other fields (mechanization, agro-chemistry, fertilizer technology) are therefore needed to help clarify what partnerships can achieve. Some international agricultural research centers now run joint ventures with the private sector in which the parties share intellectual property and agree contributions, obligations and apportionment of benefits. There could be lessons for ICAR to assimilate in this.

Focus on rain-fed agriculture

In terms of area, rain-fed agriculture dominates farming in India. According to *Agricultural Statistics at a Glance 2011*,[9] only about 63 million of the 141 million net sown hectares are irrigated. This distribution deserves special attention in the reform of NARS.

The average annual rainfall in eastern and most parts of central India is in excess of 1000 mm; in the northeast, it exceeds 2000 mm. The country could use this valuable natural resource much better for intensifying agricultural production. The technologies produced by the public system tend so far to be better suited for irrigated agriculture. They do not provide the needed results under rain-fed conditions where, unsurprisingly, the rate of adoption of public technology

---

9 http://eands.dacnet.nic.in/latest_2006.htm; accessed July 2012.

is low. Of course, the right kinds of irrigation solutions need to be propagated, but much of Indian agriculture will remain rain-fed for a long time to come. Strengthening rain-fed agriculture should therefore be a much higher national priority than it currently is. All-out efforts are needed for the development of seeds and the transfer of knowledge, inputs, mechanization and methods of pest control to farmers operating under rain-fed conditions.

It is encouraging that the budget for the 'Bringing Green Revolution to [rain-fed] Eastern India' thrust is due to be increased from Rs. 4 billion to Rs. 10 billion in 2012–13.[10] BGREI is a commendable government initiative that has already led to rice production gains in 2011. In our view, it should be upgraded to 'Mission Mode' on a par with other missions, such as the National Horticulture Mission. BGREI should be resourced for the introduction of hybrid rice, among other aspects, and the aggressive improvement of rain-fed agriculture in general. The work should include hands-on demonstrations and invigorated research-extension-farmer link-ages ('lab-to-land').

Ensuring continued productivity growth

In progressive regions practicing intensive agriculture and achieving high yields, pro-ductivity plateaus are sometimes apparent, prompting one prominent observer, Professor M S Swaminathan, to declare that 'the heartland of the Green Revolution is in grave trouble' (Swaminathan, 2011). Figure 5.5 and Table 5.1 indicate that productivity plateaus in the form of yield and production growth stagnation or reversals have indeed occurred during given periods. But the pattern suggested in Table 5.1 is one of wide disparity of growth during successive phases, including a general deceleration of growth after the initial years of reforms of the 1990s ('PR'), a deceleration in some of the traditional crops and an acceleration in others (onion, potato, cotton, sugarcane) during the years of diversification ('DIV'), and fluctuations, including in some instances strong recoveries, in many key crops overall (Chand and Parappurathu, 2012).

The forces underlying this pattern across time are varied, but in most instances probably economic in nature and (as always in agriculture) influenced by the weather. Where agricultural research and extension are responsible for episodes of yield and production growth stagnation or decline, sub-optimal and imbalanced fertilizer use and the lack of new, superior crop varieties that should be there to replace old ones are often the culprits. A classic example of the latter (according to our sources) is the wheat variety PBW-343 released in 1995. This wheat originally came from CIMMYT where it was known as 'Attila'. In spite of imminent threats from Ug99 and other races of rust, PBW-343 remains the first choice of farmers in north India, whereas Lok-1, released in 1982, remains the leading wheat variety in the central and peninsular region. Dozens of new varieties of wheat, rice and other crops have been and continue to be made available

---

10 Finance Minister's budget speech 2012 (http://indiabudget.nic.in; accessed July 2012).

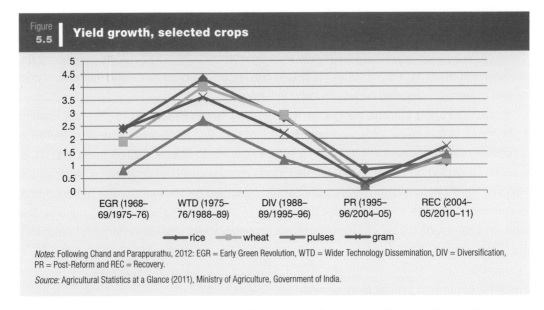

Figure 5.5 | Yield growth, selected crops

Notes: Following Chand and Parappurathu, 2012: EGR = Early Green Revolution, WTD = Wider Technology Dissemination, DIV = Diversification, PR = Post-Reform and REC = Recovery.

Source: Agricultural Statistics at a Glance (2011), Ministry of Agriculture, Government of India.

since the 'workhorses' were released decades ago, but the new creations are in most instances not significantly better than the old material and for this reason do not spread.

There is therefore an urgent need to go back to the basics of breeding in all major crops. Scientists should take a fresh look at the opportunities for step-change and the use of state-of-the-art science and techniques. ICAR, SAUs and other participants in All India Coordinated Research Projects should work together and find ways to align with the private sector to pursue breakthrough varieties and traits. In addition to yield, breeding goals should include improvements in plant architecture, resistance to biotic stresses, and resilience to salinity, drought, heat and other afflictions likely to worsen with climate change. Intensified efforts need to be made to exploit the wealth of genetic resources housed in the National Bureau of Plant Genetic Resources (a constituent of ICAR) and other repositories. So far only a small fraction of the germplasm collections has been characterized for specific breeding purposes. Publicly held germplasm should be shared responsibly with the private sector, against payment of royalties for the varieties developed.

Wheat demand is expected to rise strongly over the coming decades, but wheat yields are feared to be among the worst affected by rising temperatures. The recently launched National Initiative on Climate Resilient Agriculture and, under it, the Centre for Environment Science and Climate Resilient Agriculture at IARI, are commendable steps in the search for solutions to this problem. But given the gravity of the threat, ICAR should also work with the Ministry of Science and Technology, CSIR and private institutions to collectively pursue solutions. A first priority should be the exploration of germplasm and breeding of wheat and other crops for adaptation to warmer conditions. A wheat yield consortium led by CIMMYT is using trait-based hybridization

Table 5.1	**Trend growth rates in production of major crops during various phases of growth (%/Year)**					
crop group	PGR (1960–61/1968–69)	EGR (1968–69/1975–76)	WTD (1975–76/1988–89)	DIV (1988–89/1995–96)	PR (1995–96/2004–05)	REC (2004–05/2010–11)
rice	0.47	1.35	2.75	2.69	0.58	0.60
wheat	4.39	4.75	4.79	3.63	0.85	3.31
maize	4.42	0.18	1.25	2.80	4.30	3.21
gram	-3.19	-2.30	-1.12	3.26	-1.67	5.86
arhar	-1.05	-0.27	2.57	-1.81	-0.20	2.12
groundnut	.33	1.25	.78	-0.34	-2.98	-3.28
rapeseed and mustard	1.69	4.85	5.79	5.34	-1.02	-1.50
soyabean	NA	NA	22.84	20.78	2.66	6.09
onion	NA	NA	2.46	4.57	3.70	13.98
potato	6.88	5.25	5.43	3.19	2.28	9.66
sugar cane	1.14	2.79	2.26	3.70	-0.97	3.26
cotton	1.41	3.19	1.17	5.16	-1.02	9.92

Notes: PGR = Pre-Green Revolution, EGR = Early Green Revolution, WTD = Wider Technology Dissemination, DIV = Diversification, PR = Post-Reform and REC = Recovery.

Source: Chand and Parappurathu (2012), based on Agricultural Statistics at a Glance (2011), Ministry of Agriculture, Government of India.

and biotech applications to improve the performance and regulation of RuBisCo (an enzyme involved in carbon fixation), introduce C4-like characteristics, and improve light interception and photosynthesis. It is hoped that India will seek to be associated with this effort.

In hybrid rice, China has shown that breakthroughs are feasible. Ideotype breeding is the subject of extensive work at IRRI and in China to develop the rice plant's architecture, enhance its physiological efficiency and raise yields. The basic idea is to get higher yields through increased panicle size and weight combined with resistance to lodging. Among several models of superior plant architecture, 'new plant type' (NPT) is one, developed by crossing improved tropical japonica lines with elite indica (Peng et al., 2008). Several second-generation NPT lines at IRRI have out-yielded the indica control varieties. China's 'super' rice breeding project is moving ahead to cross yield barriers, particularly in irrigated rice.[11] India has some NPT lines, but research should be stepped up, possibly in collaboration with IRRI, with a view to developing commercial varieties

---

11 Yuan Longping's rice hybrid DH-2525 has demonstrated yield capability in excess of 13.5 metric tons per hectare (see 'China sets new world record with hybrid rice yield', China Daily, September 19, 2011, retrievable at http://www.chinadaily.com.cn/china/2011–09/19/content_13736806.htm; accessed July 2012).

that could be taken up by private seed companies. Plant breeding is essential to ensure continued productivity growth.

Beyond breeding, what counts are results on the ground—in other words, farmers' yield improvements in addition to considerations such as the resilience of their crops. Public research goods are potential stepping stones on the way to these aspirations. The question is whether superior technology is, in fact, being developed, and if so, if it spreads.

The narrative above suggests the existence of a gap between the varieties developed by the public sector and those used by farmers. What could account for this? It could be the money available, given the link between agricultural productivity and the level of investment in agricultural research. But perhaps organization, incentives and how the money is spent are more important. Or perhaps other factors are at work, inside and outside NARS. Research in the public sector may focus more on knowledge generation than the targeted, demand-driven exploitation of knowledge that farmers need. Or it may operate under funding practices or processes in which priorities match political or institutional considerations rather than farmers' and the market's needs. Perhaps serendipity is insufficiently taken into account in the internal allocation of resources—or it is suffocated by the processes under which candidate innovations are progressed. Luck that favours the prepared is precious in research and must be nurtured as a factor in the creation of innovations that solve concrete and burning problems on the ground.

If there are elements of truth in any of this, it is time for institutional reform. An organization as large as ICAR and the NARS more broadly is constantly evolving and can therefore point to adopted measures of institutional reform at will. But is 'what ails ICAR' being addressed? The Prime Minister seemed to be doubtful when he recently enquired about the spirit, not just the letter, of uptake of reform recommendations that have been made over the years.[12]

Private sector R&D in agriculture

Agribusiness in India is making 'major investments in research and producing innovations that are extremely important to farmers' (Pray and Nagarajan, 2012). Agribusiness on the input side of farming includes seed and biotech companies, the fertilizer and agrochemicals industry, as well as suppliers of farm machinery, irrigation equipment, and implements of different kinds. According to data from the Department of Science and Technology, a total of about 71 companies are currently undertaking agricultural R&D in India: 22 in seeds, 19 in agrochemicals, 10 each in fertilizer and mechanization/irrigation, and 10 in other endeavors, including aspects of production of specific crops (Ramasamy, 2011). Some small enterprises and non-profits are also conducting agricultural research. The seed industry, whose R&D got underway in the late 1960s, is particularly research-intensive. In 1973, the Indo-American Hybrid Seed Company set a milestone

---

12 Prime Minister's address at the Golden Jubilee of IARI, February 20, 2012 (http://pmindia.nic.in/speech-details.php?nodeid=1141; accessed July 2012).

by launching India's first commercial F1 hybrid vegetables (tomato and sweet pepper).[13] Around the same time, the Maharashtra Hybrid Seeds Company (Mahyco) and others started developing proprietary hybrids.[14]

Seed industry

The founding of the National Seeds Corporation Ltd. (NSC) by the government in 1963 and the Seeds Act of 1966 regulating the quality and reliability of seed available to farmers were pivotal events aiding the rise of the private seed industry. NSC's mission was to change the entire picture of seed availability. Its tasks were to produce and multiply seed in large quantities under arrangements with outgrowers, manage seed distribution logistics, transfer know-how to seed operators, and in general guide the creation of a seed industry in the country. NSC did not invest in plant breeding and R&D, relying instead on the publicly bred varieties already available. Private seed operators were still too small to make substantial investments in R&D. They, too, accessed public germplasm, developing their own hybrids as the demand for improved seed (as opposed to farm-saved) grew in the wake of NSC market creation efforts.

In the 1970s, several large seed corporations were set up in different states with financial assistance from NABARD (the National Bank for Agricultural and Rural Development) under the 'National Seeds Project' supported by the World Bank. These corporations ran their operations in much the same way as NSC, focusing on the large-scale processing and distribution of seeds. During the Seventh Five Year Plan (1985–1990), private companies became eligible to draw soft loans under the third phase of the National Seeds Project. This boosted the industry by helping companies strengthen their infrastructure for processing, storage, as well as R&D (Rakesh, 1986). The slow-down in public investment in agricultural research, which started at about that time (Section 2; see also Pray and Basant, 2000), solidified the opportunity for private seed operators to succeed. Policy reforms of the late 1980s, including amendments to the Monopolies and Restrictive Trade Practices Act and the Foreign Exchange Regulations Act, paved the way for large corporates to invest in the seed sector. The 'New Seed Policy' of 1988 liberalized seed imports for the benefit of farmers.

In many kinds of vegetables and some fruits, high-performing private hybrids quickly became farmers' first choice. Superior private hybrids also started to displace publicly bred products in key field crops, including maize, sorghum, pearl-millet, sunflower and cotton. In 1991, the Indian Industrial Policy identified seed production as a 'high priority industry' in regard to foreign investments (Gadwal, 2003). Private research expenditure grew by 70 percent between 1985 and 1995, according to one estimate (Pray and Fuglie, 2001). According to another, it quadrupled between

---

13 www.indamseeds.com/achievements.htm; accessed July 2012.
14 Mahyco's founder, B R Barwale, received the 12th World Food Prize in 1998 and, in the same year, was designated the 'father of India's seed industry' by the Crop Science Society of America.

1986 and 1998, but with greatly varying research intensities across companies, from less than one percent of turnover to more than fifteen in some instances (Gadwal, 2003).

The Indian seed industry today is estimated to be worth around Rs. 80 billion (Rajendran, 2011), with a potential to grow by 60 percent in the next five years.[15] This substantial projected increase is driven by farmers' demand for high-quality seeds and the companies' deepening portfolio of products. The leading companies are increasing their investment in breeding and research by hiring scientists and upgrading their facilities such as greenhouses and laboratories for embryo culture, molecular markers, double haploidy and gene transformation. Next-generation products using these advanced techniques have already arrived and are increasingly popular among farmers. *Bt* cotton illustrates this (Box 5.5).

The incentives of private seed businesses (including large corporates) to invest in research and product development were strengthened with the passage of the Protection of Plant Varieties and Farmers' Rights Act and the Indian Patents Amendment Act in 2001 and of 2005, respectively. The revised Patents Act improved the intellectual property rights regime. By protecting plant varieties and the rights of farmers and plant breeders, the 2001 legislation aimed to encourage the development of new varieties. Factors that cloud the outlook from the private sector's current perspective include uncertainties surrounding the NBRA Bill (see Box 5.5) and the Seeds Bill. The latter would afford protection against the counterfeiting of seed; it is awaiting approval by Parliament after having been cleared by the Union Cabinet in May 2011.

Fertilizers and chemicals

The Green Revolution was primarily the result of the combination of improved varieties and high dosages of fertilizer under irrigated conditions. Pesticides and other agrochemicals played a supporting role. In the early days, fertilizer manufacture was controlled by the state and confined to nitrogenous fertilizers. The Planning and Development Division of the government's first fertilizer factory, at Sindri in Dhanbad District, Jharkhand,[16] engaged in R&D activities that ranged from developing cost-effective manufacturing processes to promoting fertilizer use through trials and demonstrations in farmers' fields. In the mid-1960s the fertilizer sector was opened to private manufacturers. Today it is composed fairly equally of private, public and cooperative operators.

Most nitrogenous fertilizers are manufactured domestically while the bulk of the intermediates for phosphate fertilizers and all potash are imported. India is the third-largest producer and consumer of fertilizers, after China and the US. (China's fertilizer output is more than twice India's.) Although India's average fertilizer use rose from less than 1 kilogram per hectare in 1951 to 143 kilograms in 2007 and is still rising, it is significantly below China's, where at 489 kilograms per

---

15 Nidhi Nath Srinivas, 'Seeds: The one attractive link in the agricultural chain, growing at 20% a year', The Economic Times, June 28, 2011.
16 Established in 1951 and phased out in 2002; later became part of the Fertilizer Corporation of India.

The *Bt* cotton wave and beyond

First approved in 2002, *Bt* cotton signaled the beginning of a new era in Indian agriculture. Seen as the most outstanding breakthrough since the wheat revolution, *Bt* cotton ushered in a new phase of business opportunities for the seed industry. In 2011, over 800 approved *Bt* cotton hybrids marketed by 40-odd companies and licensees were on offer in the country (cf. the site of the Indian GMO Research Information System: http://igmoris.nic.in/; accessed July 2012). Aided by price controls on seed imposed by state governments, the area under cultivation continued to increase, reaching 9.4 million hectares in 2010 and making acreage planted to cotton the world's fourth largest under biotech crops (ISAAA, 2010 http://www.isaaa.org/resources/publications/default.asp; accessed July 2012). In 2011–12, farmers bought an estimated 38 million seed packets of 450 grams each. India's production doubled to about 30 million bales in the first eight years after the introduction of *Bt* cotton. The transgene contributed significantly to yield growth, along with fertilizer, high-quality hybrids and improved crop care (Gruere and Sun, 2012). But the future of agricultural biotechnology is uncertain in India in the face of regulatory and stewardship concerns. The much-discussed *Bt* brinjal, or insect-resistant eggplant, has so far not been approved for commercial production. Biotech events for crops including rice, maize, mustard, sugarcane, groundnut, okra, cauliflower, tomato, papaya and watermelon are at various stages of regulatory review. The traits in question—developed by private companies as well as public institutions (cf. http://igmoris.nic.in/; accessed July 2012)—range from biotic and abiotic stress tolerance to male sterility and other characteristics. The current uncertainties surrounding government policy make it hard to say when some of these products might receive approval. An independent regulatory body is proposed, but the legislation for it—the bill on the NBRA, the National Biotechnology Regulatory Authority—has not so far been passed. The trend toward genetic modification, ushered in by *Bt* cotton, may be waning at present.

hectare it is in many instances excessive.[17] It is worth noting that while India's fertilizer production rose by more than 50 percent between 2001 and 2010, food grain output during the same period increased by only 11 percent. There are two reasons for this imbalance: (i) greater use of fertilizers in commercial non-food crops (including horticultural and plantation crops) and (ii) diminishing returns on the use of a narrow range of fertilizers without proper adaptation to soil fertility status and plant nutrient needs.

The in-house R&D activity of the fertilizer industry pays more attention to innovation in manufacturing processes than quality of product mix and product utility to farmers. There is a need for customized fertilizer solutions for specific crops and locations, slow- and controlled-release fertilizers, water soluble fertilizers suitable for fertigation, micronutrient products, and more. Tata Chemicals have recently introduced customized fertilizers, where nutrient compositions are tailored to produce extra yield in wheat, sugarcane and potato in western Uttar Pradesh. These products, the company claims, add yield while only marginally increasing cost to the farmer—a promising advance that must be accompanied by agricultural extension efforts for full effect. The

---

17 http://data.worldbank.org/indicator/AG.CON.FERT.ZS; accessed July 2012.

fertilizer industry could play an important role in educating farmers on the optimum use of fertilizers as a part of their marketing and promotion activities. Several companies do operate stationary and mobile soil testing laboratories and provide services to farmers (Sharma and Singh, 2006), but needs and opportunities in the field are far from being adequately met. India's new nutrient-based fertilizer subsidy is expected to encourage the development of tailor-made products and help improve the efficiency of nutrient use (see Box 5.6 on nutrient and water use efficiency).

According to one estimate, India loses 18 percent of its crops to pests, diseases and weeds, with damage valued at Rs. 900 billion.[18] Although India is the fourth-largest producer of agrochemicals globally,[19] its domestic deployment of chemical crop protection products is relatively low.[20] The current size of the business is estimated to be more than Rs. 90 billion—larger than seeds, but smaller than fertilizers.

Nearly half the domestic production of 85,000 metric tons—mainly consisting of generics—is exported,[21] and there are also some imports, mainly of active ingredients. There are about 40 manufacturers of agrochemicals in India (among them the public sector company Hindustan Insecticides, Ltd.) and 400 formulators. The domestic companies enjoy a bigger market share as a group than the multinationals, whose strengths lie in specialty products and custom-made farming solutions.

Indian agrochemical companies initially concentrated on developing alternative ways of manufacturing generic and other products not protected under the old patents law. In developing these processes, the manufacturers used both locally available and foreign technical expertise. Although local and foreign firms carry out extensive testing of their products (Pray and Basant, 2000), in reality there has been little investment in R&D by Indian companies. The factors responsible for this include (i) stiff price competition among generics, (ii) market disturbance by spurious products, and (iii) the long lead times needed for product registration. However, under the impact of globalization, Indian companies are finding it necessary to increase investment in research. The discovery of new molecules being both time-consuming and requiring high investment, only large corporates can afford to engage. To bring new and more effective crop solutions within farmers' reach, reform must make the regulatory mechanism more efficient, without compromising on standards.

Farm machinery and irrigation

Water-lifting devices have been the most important element of mechanization in Indian agriculture for many years. Starting with a mere 110,000 electric and fossil-fuel powered pumps in

---

18 http://www.financialexpress.com/news/india-loses-rs-90kcr-crop-yield-to-pest-attacks-every-year/428091/; accessed July 2012.

19 http://www.scribd.com/doc/62516027/Pesticide-Industry-An-Overview; accessed July 2012.

20 http://indiamicrofinance.com/agrochemicals-in-india-2011.html; accessed July 2012. Less than 1 kg/ha compared with 7 in the US and 13 in China.

21 *Bt* cotton has substantially reduced the need for application of certain kinds of insecticides in the production of that crop.

| Box 5.6 | **Nutrient and water use efficiency (NUE and WUE)** |

The efficiency of nutrient and water use in Indian agriculture is poor and a major cause of low crop productivity. In regard to nutrient use, the key concern is the low rate of utilization of nitrogen fertilizer by crops. NUE is a function of nitrogen supply from soil and fertilizer, uptake by the crop, and losses from soil plant systems (Ladha et al., 2005; Shukla et al., 2004). NUE is positively linked to soil organic carbon, which is in decline in the fertile, irrigated north-Indian plains (Dwivedi et al., 2003). In rice-wheat cropping systems, only 30 to 50 percent of nitrogen fertilizer applied is utilized by the first crop and only a small part of the residual is used by the following crop. Denitrification, ammonia volatilization and leaching account for significant losses (Ladha et al., 2005). The quantity and timing of fertilizer use should be in line with attainable productivity. Higher applications are wasted, with negative consequences for the environment and the farmer's purse. Slow-release nitrogen fertilizers offer advantages over the straight use of (subsidized) urea. Simple guides like the leaf color chart developed by IRRI for rice can help farmers apply fertilizer at the right time. The chart has helped the 350,000 farmers thought to be using it to reduce fertilizer use by 25 percent (Ladha et al., 2005). It should be used much more widely. Region-specific research on soil health and nutrient balance is needed to develop the right fertilizer products and applications. Fertilizer formulations must be developed for different agroclimatic conditions and cropping systems. It is not 'one-size-fits-all'.

India also faces major challenges in WUE in both irrigated and rain-fed agriculture. Faulty appliances and other factors waste large quantities of water. Irrigation WUE needs to rise from its current level of 35 percent to 60 percent for surface water and 75 percent for ground water systems (NAAS, 2009). Where used, micro-irrigation systems (MIS) like drips and sprinklers have significantly increased WUE and productivity. Besides saving water, drips improve crop performance by regularly applying water to the root zone. When combined with 'fertigation', they also raise NUE. MIS produce particularly good results in high-value horticultural crops (Soman, 2009). Cost and other limitations currently hinder nationwide MIS use. However, recent development of an efficient, low-cost drip system provides some hope for wider adoption by smallholders (Chakraborty, 2012).

An often overlooked aspect is that low WUE is a much bigger problem in rain-fed agriculture than under irrigated conditions. The eastern states and parts of Madhya Pradesh have generally low crop productivity. A major reason is that the crops suffer from flooding as well as dry spells. Farmers are not yet sufficiently aware of the benefits of shallow tillage and mulching in preventing soil moisture losses. Conservation agriculture would save water and improve NUE, particularly in the Indo-Gangetic Plains (Erenstein, 2009). The practice of harvesting rain water in land profiles and using it for protective irrigation during dry spells is not prevalent, but should be. Smallholders rarely use techniques like planting crops on raised beds to protect them against heavy rains or dry spells (Singh et al., 2010). Great efforts are required to change farming practices in ways that improve WUE.

1951, the country had more than 15 million by 2007, mirroring the prevailing dependence on well irrigation (Ramasamy, 2011). Significant private investment in micro and precision irrigation has started to make an impact in the last ten to fifteen years, as we describe in Box 5.6.

Among other kinds of farm machinery, tractors are the most important. Indigenous manufacture of tractors, public and private, started under technical collaboration arrangements with foreign firms. Import restrictions and strong domestic demand for agricultural and other uses enabled domestic producers to prosper. Over time, Indian manufacturers improved their technologies, some matching international standards. By 2000, India had overtaken the US in overall manufacture of tractors and is now possibly second only to China in total production (Jain, 2002), which by 2010 exceeded 300,000 units, of which one-fifth for export. Following recent mergers and acquisitions, there are currently fourteen tractor manufacturers. Most of these carry out some R&D, while Tractors and Farm Equipment Ltd. (TAFE) reportedly employs 300 engineers and staff at two R&D centers. Mahindra and Mahindra (M&M), the leading manufacturer, provides extension and other services under one roof through its subsidiary Mahindra Shubh Labh.

Agricultural equipment other than tractors has so far mainly been designed, developed and manufactured in the informal sector, in some cases with knowhow and contributions from SAUs and others. By and large, this category of equipment is serviceable but often not of high quality. Much small farm machinery is imported, often from China, and sold and serviced by Indian companies. Many of the tools and types of farm machinery increasingly demanded by small farmers searching for energy-efficient labor-saving devices (including power tillers and tractors in the 15 to 20 hp range) are not readily available. If they were, rental markets for them would likely develop quickly, given the rising cost of labor in many settings. There is a need for robust research on, and design and production of, farm equipment and machinery for this group.[22] Efforts and incentive programs by the public sector and social impact investors to crowd-in private design and production capability of machinery for small-scale producers could make a difference.[23]

Research by small enterprises and non-profits

There are many small and micro-enterprises engaged in the production and marketing of agricultural inputs. They offer a plethora of products ranging from seeds to organic and bio-fertilizers, microbial cultures, bio-pesticides, plant growth regulators, crop performance enhancers, and so forth. Together, they account for a sizable share of the market. Some of their artifacts and inventions comply with professional product standards developed by experts. Others are the result of second-hand processes without any quality assurance. It should be mandatory for all products sold to farmers to have passed the test of proper regulatory control.

Non-profit organizations like the M S Swaminathan Research Foundation, the Barwale Foundation, the BAIF Development Research Foundation, the Nimbkar Agricultural Research Institute and others are also engaged in research and have contributed good publications ranging from environmental services for biodiversity conservation to animal husbandry and genetic engineering. They have their own mandates and work on specific projects, making important

---

22 See http://www.slideshare.net/csisa/escorts-rd-in-farm-mechanisation-icar-cii-meeting-23-may-2011; accessed July 2012 (research and development needs in farm mechanization by Escorts India Ltd, ICAR-CII Meet, May 23, 2011).

23 See statement of Agriculture Minister, Sharad Pawar ('Farm mechanization efforts need to be intensified') in India Current Affairs, December 16, 2011; http://indiacurrentaffairs.org/farm-mechanisation-efforts-need-to-be-intensified-shri-pawar-parliamentary-consultative-committee-of-agriculture-ministry-meets/; accessed July 2012.

contributions that in some instances are the product of associations with government programs or partnerships with commercial firms. The M S Swaminathan Research Foundation has isolated a salt tolerance gene from coastal mangroves that is being transferred into rice. [24]

Summing up

The effects of private sector innovation during the past three decades have recently been re-assessed by Pray and Nagarajan (2012) and can be summarized as follows based on this source: (i) seed/biotech innovations have led to documented increases in yields in key field crops, vegetables and fruit; (ii) proprietary hybrids of pearl millet, sorghum and maize lifted the productivity of these crops in semi-arid settings not well served by the Green Revolution; (iii) proprietary hybrids cover at least 75 percent of the area planted to improved varieties and hybrids overall; (iv) farmers captured substantial economic gains from yield increases in these settings and crops; (v) private research has helped India increase exports of crops, technology, and agricultural inputs such as agrochemicals and machinery; and (vi) private sector R&D has benefited not only the better-off but also poorer farmers, as evidenced, for example, by the spread of improved, privately developed seeds to poor areas, the uptake of vegetable production by some disadvantaged farmers using proprietary seed, and the creation of rural employment that tends to accompany agricultural intensification.

So there is impact from private innovation, but like the public sector, private companies could and should do more; the opportunity is there. Examples include the improvement of rain-fed crops beyond pearl millet, sorghum and rice, or the development of solutions to the mechanization needs of small farmers. If uncertainty regarding the size of the market for innovations deters investment, public-private partnerships and judicious financing solutions involving forms of 'patient money' could be designed to step in.[25] Public-private partnerships have historically been India's method of choice to put seed companies to work. We have highlighted some of this above. An aspect we have not addressed is the special relationship that developed between ICRISAT and the private sector over the years. This cooperation evolved into formal consortia for research in sorghum and pearl millet hybrid parents and for hybrid pigeon pea (FAO, 2005). Three-quarters of the currently available commercial hybrids of sorghum and pearl millet, and all pigeon pea hybrids, were bred using ICRISAT-derived parental lines (Pray and Nagarajan, 2009). A recent public-private cooperation example is IARI's rice hybrid PRH-10 development partnership in which some fifteen private companies and one NGO are involved (Patil and Dadlani, 2010). As we argue in this chapter, more of this is needed, not only in seed systems development but in relation to other categories of inputs as well.

---

[24] See http://www.mssrf.org (accessed July 2012) on this work, which is still in progress.

[25] Government is also offering tax incentives to private R&D investment in agriculture; cf. Finance Minister's Budget Speech, March 16, 2012; http://indiabudget.nic.in/ub2012-13/bh/bh1.pdf; accessed July 2012.

New science: molecular biology and biotechnology

Experts believe that molecular plant science will play a crucial role in the transformation of agriculture in the coming decades. India will be no exception. Molecular plant science aims to accelerate gene discovery, using a wide range of genetic diversity in crop breeding programs, and to stimulate new concepts, strategies and technologies to protect and improve crops. High-throughput automated crop sequencing as part of the genomics revolution has already demonstrated that we can expect step-changes in knowledge and techniques. As with other new technologies with high utility, change in molecular plant science will become faster and applications cheaper.

Twenty food crop genomes have been sequenced and published in peer-reviewed journals so far, including key Indian staples such as rice, wheat, maize, sorghum, soybean and pigeon pea (on the latter, see Varshney et al., 2011). Teams of Indian scientists participated in several of these initiatives, and the Indian government provided financial support. In meeting the food, feed, fiber, fuel and nutritional security challenges ahead in India, it is vital—particularly in view of predicted effects of climate change—to increase investment in frontier technologies such as molecular plant science and biotechnology, as is done in China and Brazil. Both of these countries are embracing the new science opportunities, often partnering with the private sector to achieve early access to innovations and links to markets. The Report of the Agricultural Research and Education Working Group for the 12th Five Year Plan cited in Section 3 has rightly identified genomics and biotechnology as high-priority areas for research. The two currently leading approaches to capitalize on molecular plant science and agricultural biotechnology to improve crops suitable for Indian farmers are molecular breeding and genetic modification of crops.

Molecular breeding

As defined in Section 3, molecular breeding refers to the use of combined genotypic and phenotypic data to develop new and improved varieties of plants. Using the molecular breeding approach, plant traits can be characterized in the form of molecular markers at the DNA level. Joint analyses of molecular markers and trait phenotypes result in the establishment of so-called marker-trait associations (MTAs) or quantitative trait loci (QTLs). Where MTAs explain a large enough percentage of trait variation, molecular breeding is unquestionably more efficient and effective than traditional breeding. Molecular breeding tools are also relevant for the management of plant genetic resources: markers can provide information about genetic distances between varieties or accessions in a collection and can be used to drive and monitor the use of genetic resources in breeding programs. This is especially important for hybrid breeding and the management of heterosis. Markers can also be used to help ensure that enough genetic diversity is present in crops grown in farmers' fields so as to avoid genetically specific epidemics such as that of Ug99 wheat rust.

Numerous examples exist of the successful use of molecular breeding in improving exist-ing varieties with respect to one or a small number of simple traits by means of marker-assisted backcrossing or marker-assisted backcross breeding (MABB). MABB permits favorable traits to be introduced into varieties at twice the speed of traditional approaches. It can be used to intro-duce favorable characteristics from unadapted materials into adapted elite varieties without bring-ing along unfavorable characteristics that in the past prevented breeders from tapping into the corresponding genetic pool.

There are also many examples of molecular breeding applications for the simultaneous selec-tion for larger numbers of often complex traits, based on marker-assisted recurrent selection (MARS), although these examples are mostly found in the private sector. The genotyping technol-ogies that reveal DNA sequence variations have been developed by specialized companies active in analytical technologies that include DNA-based technologies. Most of the developments were originally driven by medical research and later adapted to plant analysis. Molecular breeding for product development is widely used by large private seed companies, in particular to develop the parent lines of their proprietary hybrids. The genotyping that supports these breeding programs often takes place in large, central facilities that can quickly produce huge amounts of high quality marker data at very low cost.

Publicly available crop-specific molecular markers have globally been developed either by private companies, public research institutes, or public-private consortia. The use of molecular breeding has also gained significant momentum in public research institutes. International centers such as ICRISAT, as well as large national agricultural research institutes, have not only developed their own capabilities, but are also sharing their genotyping and phenotyping facilities in providing hands-on training to scientists from other institutes.

Future demographic, environmental and social challenges call for faster development of new crop varieties to meet an increasing number of criteria. Future varieties will have to be more productive, yet require fewer inputs, provide more stable productivity, and meet complex require-ments all the way along the value chain to consumers. Even though plant breeding has been suc-cessful in delivering improved varieties in the past, there is general agreement that more is now required than can be delivered by traditional approaches alone.

Plant genetic resources—the basis of breeding and selection—are sought after by scien-tists because they are the 'raw material' underpinning knowledge of genetic diversity and the function of genes. Developing new and improved varieties from elite genetic diversity will be all the more successful when such materials are well known and characterized. Most importantly, since molecular breeding allows the extraction of favorable attributes from generally unadapted, unfavorable materials, it helps restore the value of unadapted genetic diversity, which traditional breeding has generally failed to recognize, and justifies significant efforts aimed at preserving, collecting, and characterizing biodiversity. Again, considering the magnitude of the task and the fact that private companies possess faster, more reliable and cost efficient molecular technology, there is a need for cooperation between the public and the private sector in the pursuit of socially desired goals.

Genetic engineering and transgenic plants

Genetically modified (GM) plants have by now amply demonstrated their ability to contribute to increased agricultural productivity and enhanced product quality. They are now sown on more than 150 million hectares around the world, up from fewer than 2 million hectares in 2006 (James, 2010)—probably the fastest process of adoption of any new crop technology in the history of agriculture. GM crops are likely to play a significant role in crop improvement in the future, delivering outcomes covering insect, fungal or viral resistance; herbicide and abiotic stress tolerance; and improved nutritional content like pro-vitamin A in 'Golden Rice'. Insect resistance and herbicide tolerance are the prevalent commercial traits today. The 'GM option' is important, offering many possibilities to express traits and alter metabolic pathways, but it should not be taken as a panacea for all challenges that nature and farming present. Still, in the Indian context, transgenic biotechnology could bring about radical improvements in some important commercial crops, including sugarcane, pulses and oilseeds, coarse grains and others. Transgenic biotechnology holds the potential to contribute toward crop productivity gains and crop improvement for smallholder farmers (Anthony and Ferroni, 2012).

After cotton, sugarcane should have been (and may yet become) the next candidate for *Bt* technology in India. Cane is an important cash crop like cotton, occupying nearly five million hectares of land where, on average, 20 percent of production (valued at Rs. 46.5 billion) is lost to insect pests. The crop is susceptible to four major stem borers and one root borer, each of which has been known to cause field-specific losses of up to 35 percent. These insect species are known to be susceptible to Cry and Vip proteins derived from Bacillus thuringiensis, yet there are no serious efforts at present to genetically modify sugarcane to express these proteins. The varieties and planting material are mainly in the public domain, while the genes of interest are the property of the private sector. Developing *Bt* sugarcane to reduce production losses would therefore seem a strong candidate for a public-private partnership initiative.

A similar picture emerges when we look at coarse cereals like sorghum and pearl millet, pulses such as pigeon pea, chickpea and lentil, and oilseeds such as groundnuts, sesame and others that are largely rain-fed crops or grown on post-monsoon residual soil moisture. These crops continue to suffer major losses from lepidopteran insects even though effective solutions could be progressed using Cry genes. Sorghum and pearl millet are affected by the same borers as sugarcane; chickpea and pigeon pea suffer from the same pests as cotton. The technologies exist to control these pests responsibly through transgenic means.

The Working Group Report referred to above recognized this and noted the need to 'strengthen new initiatives in agricultural biotechnology', including the development of GM crops. It suggested a new focus on stress tolerance, gene pyramiding, allele mining and microbial genomics. Given the high potential of transgenic research, it recommended that certification standards and procedures for GM seed should be worked out, and that ICAR should establish a multi-disciplinary team, including social scientists, to create scientific awareness and disseminate science-based information about GM crops. The only shortcoming of the Report in this connection was that it did not spell out the significant synergies that could be achieved if the public sector found ways to join hands with the private sector. We believe that the development of GM crops will intensify, not least through public-private partnerships. The opportunity cost of ignoring particular scientific options has not yet become apparent to everyone involved, but we predict that in due course it will. Increasing awareness of the effects of intensifying challenges such as climate change is likely to accelerate this process.

Conclusion and recommendations

To attain and maintain the targeted agricultural growth rate of 4 percent per year, India needs an effective public and private apparatus of agricultural research. Technology plays a critical role in agricultural growth. Research is required to develop it. The public and the private sector have made important research contributions in the past, raising farm productivity and agricultural growth. But much more is needed to optimize input- and natural resource use efficiency, close yield gaps, protect the environment, and improve income prospects for farmers. Re-invention and renewal must start in the public sector. ICAR and the NARS more broadly must (i) review their priorities and ways of working; (ii) embrace institutional reform; (iii) internalize the tools and techniques of modern biology, biotechnology and molecular plant science; (iv) provide for consolidated approaches and 'critical mass' in genomics science to reap the promises of breakthroughs for agricultural renewal inherent in biotechnology; (v) get organized to address the challenges of rain-fed agriculture and ensure continued productivity growth; (vi) carefully allocate agricultural research expenditure and invest in state-of-the-art facilities without neglecting the opportunities for outsourcing of services; and, finally, (vii) find ways to work with the private sector, recognizing that this is a two-way street. Needless to say, this is an ambitious list, but the challenge of pursuing relevance and therefore effectiveness demands nothing less. NARS institutions have the talent and the potential to enhance their role and upgrade their contributions if these recommendations are suitably addressed.

The private sector, in turn, must do more for less-favored areas and farming types. It particularly needs to deepen its contribution in seeds, soil fertility, precision irrigation and mechanization. Crop- and location-specific fertilizer products and applications supported by soil analysis and digital mapping, for example, could make a dramatic difference to productivity and soil restoration/soil health. The same can be said for the right kinds of mechanization and irrigation technologies and equipment, from better drippers and technology for improved filtration, pressure control, fogging and misting to methods for subsoil irrigation and the uniform application of water, nutrients and other growth substances for plants. Concerns about the value of the market in less-favored areas and farming types could be overcome through public-private partnerships and/or partnerships between non-profit organizations (such as foundations) and commercial firms, where the non-profits would support the for-profits in developing the market. Public policies should encourage and enable such partnerships. There are hurdles to overcome, to be sure, and the stakes are high. From the private sector's perspective, the risks may seem difficult to mitigate and the right financial tools may not be apparent. From the public sector's perspective, there is perceived reputational risk, challenges when it comes to showing impact, and problems dealing with intellectual property rights. Lack of trust is an issue on both sides. But there is no alternative to partnerships as this chapter has argued, and the recommendation is to try harder.

The Quiet Revolution in India's Food Supply Chains

Chapter
6

Thomas Reardon and Bart Minten

Introduction[1]

This chapter focuses on the structural transformation of food supply chains from rural to urban areas in India. It focuses on issues of transformation in the supply stream—'downstream' (in retail), and 'midstream' (in food processing and wholesale)—as the changing market context that will condition and influence the path of agriculture and food security in the coming decades. The emerging evidence and potential impacts on the farm sector of these changes are not discussed here but readers are encouraged to refer to Reardon and Minten (2011b), which focuses on those impacts.

Rural-urban supply chains—and their transformation—are crucial to the food security of urban Indians, estimated at 350 million in 2010 (greater than the population of the US), and about 590 million, or 40 percent of the Indian population, by 2030 (greater than the entire European Union of today).[2] The food supply chain is also crucial to the incomes of hundreds of millions of those employed in rural and urban areas, such as farmers, wholesalers, truckers, processors, and retailers participating in these supply chains.

There have been four principal changes in the rural-urban food supply chains in India over the past several decades.[3]

First, the food supply chain's volume has tripled in three decades: urban food expenditures have tripled (in real terms) over the past 35 years, to USD$45 billion in 2006; these figures are based on a series of representative national household surveys—the National Sample Surveys (NSSs). The trend shows the increased 'urbanization' of the Indian food market: in 1971, urban food expenditures made up about one-quarter of total national food expenditures, by 2006 it had risen to more than one-third. This increase has been nearly entirely supplied by domestic production—as India exports and imports only a very small share of its food economy, and exports more food than it imports (both now and 35 years ago).

---

1 The authors are grateful for funding from Asian Development Bank and USAID/Delhi.
2 Population Census of India, various years. These estimates may be substantial underestimates of urban shares in India, as discussed in Sivaramakrishnan, Kundu, and Singh (2005).
3 The chapter focuses on the chain from farm to retailer/consumer. But there is also substantial transformation in the segment upstream from the farm—the farm input supply segment. For an excellent analysis of that, see Pray and Nagarajan (2011).

Second, the composition of the food supply chains has diversified in three decades. The share of cereal consumption in the urban food basket has declined from 36 percent in 1972 to 23 percent in 2006. In the same period, the share of cereals in the rural areas declined from 56 percent to 32 percent (Indiastat). Just weighing by the urban and rural population (thus abstracting from income differences), roughly 29 percent of India's food economy was in cereals in 2006, compared to about 52 percent in 1972. Yet, the food security debate still tends to focus narrowly on grain, even though non-grain food (e.g., dairy, pulses, fruits, vegetables, meat, and fish) are 71 percent of India's food consumption and important sources of calories, protein, and vitamins. They share center-stage with grains in India's food security situation.

Third, the government's direct role in the marketing of grain output doubled over the three decades—from 12 percent in the 1970s to 24 percent in the early 2000s (our calculations are weighted by rice and wheat shares, and abstracted from grains other than rice and wheat, using parastatal procurement shares of grain output cited in Rashid et al. (2007).

However, the government's role in food marketing as a share of the overall food economy is very small and has stayed nearly constant over the three decades, as evidenced by multiplying the government procurement shares times the share of grains in the food economy—the government's direct role in the Indian food economy was steady, at 6 percent in the early 1970s and growing only to 7 percent by the mid-2000s. The government, as a direct player, is a very minor actor in the Indian food economy. While it has transformed the grain economy by 'parastatalizing' it at the margin—from 12 percent to 24 percent—the grain economy, and the overall food economy, are overwhelmingly operated by the private sector—mainly traditional-private, but increasingly also modern-private.

Fourth, the private portion of India's food economy—that constitutes 93 percent of the food economy—has been transforming, structurally and rapidly, especially over the past decade. The rest of the chapter focuses on this transformation, providing only scant detail here in the introduction. In comparison with traditional market channels (e.g., fragmented, small/traditional processors, shops, wet markets, hawkers, and village brokers), all 'modern'[4] market channels—private-sector-led (e.g., modern retail, food processing, and food service industry) as well as public-sector-led (e.g., parastatal wholesaling, processing, and retail)—show higher annual growth rates than do overall urban food expenditures[5]:

- Modern food retail has been estimated to be growing at 49 percent annually over 2001–2010. (Reardon et al., 2010; 2011a);

---

4 Following Reardon et al. (2009), the term modern refers to "recent" evolutions in the marketplace. If modernization is equated with improved efficiency, then public-sector-led supply chains such as cooperatives and parastatals might fail that test.
5 The processing sector and the food service industry consist of an informal and a formal, modern sector. It is difficult to obtain separate data on their growth rates. However, for both of these sectors, the formal sector is significantly more important than the informal sector in terms of the share of output (for more details, see below).

- The food service industry (e.g., restaurants, fast food, takeaway, cafés/bars, food stalls/ kiosks) have grown overall at 9 percent over 2001–2006, (Euromonitor International, 2007);
- The processing sector (mainly the formal sector, as discussed below) has grown at 7 percent in real terms over 2002–2006 (Ministry of Food Processing, 2008);
- Parastatal grain procurement has grown at 7 percent over 1996–2006 (Reserve Bank of India, Indiastat).

If these rates are compared with growth rates of urban food expenditures of only 3.4 percent annually (in real terms) over the past 10 years (based on the NSSs, NSSO, several years), and of all-India food expenditures of 2.5 percent, the increasing relative and absolute importance of modern channels in food supply chains becomes evident.

The diversification into non-cereals and the structural transformation—away from traditional into increasingly modern market channels—of the food economy are driven by several factors: (i) Urbanization and attendant lifestyle/employment changes, and thus an increase in the opportunity cost of women's time have encouraged the rise of food processing, food-away-from-home, and modern retail. (ii) The effect of urbanization was reinforced with a rapid increase in the ownership of vehicles and 'white goods' (kitchen appliances such as mixers and refrigerators). NCAER (2005) and Ablett et al. (2007) show that from 1995 to the end of the 2000s, the yearly purchase of vehicles increased circa 15-fold, and that of kitchen durables, 4-fold. Most of the increase was in urban areas. The effect was substantial: for example, by 2007, about one-third of the Indian urban population had refrigerators (which allow for less frequent purchases of fresh foods). (iii) While per capita incomes grew only slightly in the 1960s–1980s, per capita grew rapidly since the liberalization of the economy in the1990s. This has increased the food expenditure 'pie' and driven diversification from grains into non-staples, as per Bennett's Law, which states that the share of cereals in total food expenditure will decline with increased income. (iv) A series of reforms of the agricultural and food economy (some of which, like the de-reservation of small enterprises, go beyond the food sector) have encouraged modern private sector players (e.g., retailers, processors, logistics firms, modern wholesalers) to emerge, and in some cases, to procure directly from farmers. These reforms have been progressively enacted over the past decade, in some cases across all of India, while in other cases only partially in some states. These reforms are discussed further in a subsequent Section.

This four-fold transformation of rural-urban food supply chains, driven by the four sets of factors which four sets of factors has ushered in the era of the modernizing food economy in India. This chapter will focus on the fourth change, the structural transformation, proceeding from the 'downstream' segments in the supply chain (retail and food service) to the 'midstream segments' (processing and wholesale). While finding a substantial ferment of change—both in the emergence of the modern sector as well as the transformation of the traditional market channels—it ends with a discussion of persisting constraints to this transformation. As neutral scientists, the authors do not take a position on whether these constraints should be lifted. This is a political decision for the Indian people to make, but they do lay out the factors that they think continue to limit the speed and depth of the transformation of supply chains.

Emerging transformation 'downstream' in India's food supply chains: the take-off of the 'Supermarket Revolution'

The face of food retail is transforming in India's cities, both in the case of stores as well as food service establishments, such as restaurants and fast food chains. This section focuses on the rise of modern food retail in chains of supermarkets, hypermarkets, convenience stores, and neighborhood stores.

An emerging and potentially important factor in the medium-to-long term in the transformation of rural-urban food supply chains in India is the rise of modern food retail. Several points are drawn from work conducted in 2010–2011 (Reardon et al., 2010; Reardon and Minten, 2011a), which analyzes and reviews the recent changes with a unique data set and a substantial set of key informant-interview case studies, and reviews earlier research by Joseph et al. (2008) and others.

Reardon and Minten (2011a) note three surprising observations concerning the rise of Indian modern food retail that make its path somewhat different from other developing countries' recent supermarket revolutions.

First, modern retail in India has developed in three 'waves.' The first wave was of government retail chains, started in the 1960s and 1970s. The second wave revolved around cooperative retail chains, starting in the 1970s and 1980s. The third has been the rise of private retail chains in the 1990s and 2000s. The latter occurred in two phases: the first phase, roughly in the mid-1990s to the early 2000s, was southern India focused, middle class centered, and domestic-foreign joint ventured. It also was very small compared to the second phase. The second phase started in the mid-2000s and has proceeded to the present, with expectations to continue for some years. It has been a pan-Indian development, is mainly in the middle class, but in some formats and places has moved into the lower-middle and upper-working classes, as well as into smaller cities and even rural towns. It has been mainly driven by domestic capital.

The surprise about this first trend is not that there were rises of modern retail starting with state retail, then co-ops, then private retail chains: this has been a common pattern in a number of other developing countries. Rather, the surprise has been that right into the 'take-off' of private modern retail in the past 5–6 years, the state chains (such as Fair Price Shops, FPS) are still a major force (for example, the FPS equivalent to a major chain in itself), as are the cooperative retail chains such as Mother Dairy/Safal (again, equivalent in sales to a major private chain). This means that three different modern retail 'transformative' models vie and compete to diffuse food system changes.

Second, the rise of modern private retail in India in the past 6 years has been among the fastest in the world, growing on average at a nominal growth rate of 49 percent per year over that period, and bouncing back to growth after a dip from the recent recession. The great majority (around 75 percent) of modern private retail occurred in 2006–2010. These numbers closely track a Delhi-specific study by Minten et al. (2010) that shows similar rates.

By 2010, the sales of the leading 20 private chains that sold food were roughly US$5 billion, half of which (US$2.5 billion) came from food sales. This is roughly 5–6 percent of urban food retail (figured roughly with a denominator of US$45–50 billion of total urban food expenditures, as noted above), so it is still a small share of the market. However, that share was much below 1 percent even 6 years earlier, so the change has been rapid. Also, these shares for private retail are corroborated by the findings of Minten et al. (2010) for Delhi, which show that the supermarkets' grain share is 7 percent and fresh produce share is around 4–5 percent.

The Indian numbers are striking given the very recent rise of supermarkets—but of course they are smaller than a number of other Asian cities (such as in Beijing where the share of supermarkets in rice retail has reached 50 percent, which is still much below that of Hong Kong, for example [Reardon et al., 2010]). Moreover, extrapolating at present growth rates of sales, forward to 2020, the share of supermarkets could become 20–30 percent of food retail in the cities. This share might become even higher if there is a liberalization of foreign direct investment (FDI) in the retail sector.

Moreover, this estimate of the private modern retail share is certainly an underestimate of overall modern food retail, which as noted, includes both state and cooperative chains, that may add as much as another 3 percent or 4 percent to the 5–6 percent noted above (with the state and co-op retail sharing the rough estimate, based on a review of the size and number of these chains [Reardon and Minten, 2011a]). Even the share of private retail chains noted above is an underestimate of all private modern retail, as it is only from the leading private chains; beyond those are the small regional and city-specific chains, which may together add another percentage point to the total. The food sales (of all types) of modern chains may be as much as 10 percent of urban food retail—an estimated US$ 4–5 billion. That is still less than one-third of the food exports of India (US$17 billion in 2008 [FAOSTAT]), but at the current rate of growth of modern retail (about twice as fast as food exports), in 5–10 years modern food retail may be a more important 'modern market' than exports for India.

The third surprise is that Indian private retail chain development has unique characteristics: (i) it is driven by domestic capital investment, rather than by FDI-driven retail expansion that is common in many other developing countries; (ii) it is 'early', in terms of usual international patterns, in its diversification into small format stores; (iii) it is 'early' in the penetration of small cities and even rural towns and rural areas (such as in rural business hubs like DCM Shriram Consolidated Ltd.'s (DSCL) Hariyali Kisaan Bazaar and ITC's Choupal Saagar); (iv) it is early in its initial penetration of food markets of the poor and lower-middle class, and of fresh produce retail. These unique factors have helped to propel its rapid diffusion.

Rather than expand on the drivers of the three retail trends, the authors instead lay out what they hypothesize to be ten main determinants of these trends:

- food insecurity and persistent poverty;
- rapid income growth;
- the rise of the middle class in India;
- a rapid increase in expenditures on consumer durables and fast-moving consumer goods (FMCG);
- urbanization—overall, and the growth of tier 2 and 3 cities and towns;
- growth of the rural economy;
- partial liberalization of distribution and FDI;
- the partial liberalization of procurement rights of retailers;
- the transformation of the retail sector via government and cooperative investments from the 1960s–1970s on;
- rapid growth in investable funds from the financial sector, mainly from corporate earnings and from remittances.

The growing importance of modern retail might have important effects on the rural-urban food supply chain: first, modern retail appears to be having emerging impacts midstream and upstream in the food supply chain (Reardon and Minten, 2011b; Minten et al., 2009a).

On the one hand, supermarkets typically offer a large selection of staples, processed, and semi-processed products in their stores, reflecting the relative importance of these products in urban areas. This selection may serve to expand, concentrate, diversify, and modernize the food processing sector.

On the other hand, for their fresh produce needs, modern retailers in India, as elsewhere in developing countries, currently rely on a mix of procurement mechanisms, including sourcing from brokers on wholesale markets in spot transactions, from specialized and dedicated intermediaries on wholesale markets, from their own collection centers, and a bit from their own farms. Given the current small scale of their operations, procurements from wholesale markets through brokers and through wholesale markets are seemingly the most important procurement method. However, some large investors in the modern retail sector (e.g., Reliance) are engaged in peri-urban collection centers where traditional market channels are bypassed, especially for perishables like leafy greens for large city stores. Farmers usually gain in these settings, as transaction costs are reduced due to lower transport costs (collection centers are set up close to producers), faster turnaround (no auctions take place and no waiting for buyers is necessary), reliable weighing, transparent pricing, and immediate payments. However, modern retail establishments often have strict requirements for the produce they purchase, and they typically procure only higher-quality products. This means that when supermarkets procure directly from farms in India, they do so from small or medium rather than marginal farmers, and farmers with more non-land assets such as irrigation.

Second, there is emerging evidence that modern retail charges lower prices than traditional retail which may prove to be a boon for urban food security from the transformation of the retail sector. Relying on their primary survey of traditional and modern retailers in Delhi, Minten et al. (2010) show that prices for a number of products (e.g., rice, wheat flour, vegetables, fruit, edible oil) were lower or at the same level as traditional retailers. They speculate that these modern retailers might deliver these products at cheaper rates possibly because of more efficient procurement systems, better supply chain management, and better in-store and in-distribution-center inventory management and handling. However, the study also shows that there are, in these early stages of modern retail roll-out, important constraints in the quality of delivery, especially in supply chains of fresh fruits and vegetables.

Emerging transformation 'midstream': in the processing sector

The food processing sector has also been transforming in the past two decades. The following trends are salient.

First, there has been increasing consumption of processed food, mirrored by the increasing size of the processing sector. The output of the food processing industry has doubled over 15 years: it has climbed from Rs. 628 billion in 1984–2005, to Rs. 991 billion in 1994–95, to Rs. 1215 billion in 2000–01.

Second, the share of processed food in total food consumption is higher than typically thought—even though the share of moderately- and highly-processed food is still only nearly 50 percent of food in urban areas, and 40 percent in rural areas. Morisset and Kumar (2008) divide the food consumption basket into different levels of processed food and show the following patterns in 2004–05:

- Primary products are products consumed without processing; these products include fruits and vegetables, eggs, and fluid milk at the farm. Non-processed products form only 16.8 percent of food consumption in urban areas, and 15.3 percent in rural areas. In other words, roughly 83–85 percent of Indian food products that are consumed are processed.

- 'First-processing products with low value added' are defined as products that undergo minimal processing, such as de-husking, milling, drying, and grinding. Examples of such products are rice, flour, pulses, spices, and dried fruits. Value addition is estimated at 0–5 percent. These products form 34.8 percent of food expenditure in urban areas and 43.9 percent in rural areas.

- 'First-processing products with high value added' undergo more complicated processing and have a larger value added, of between 5 percent and 15 percent. No ingredients are added, and products are not mixed. Examples are dairy products such as butter and curd, but also meat, fish, and sugar. These products form 38.2 percent of food expenditures in urban areas, and 35.1 percent in rural areas.

- 'Second-processing products' are products that have as an input a first-processed product and to which another product (e.g., a flavor, a preservative, or another ingredient) is added. These include biscuits, bread, ghee, ice cream, and jam. Third-processing food is associated with ready-to-eat food, prepared and packaged meals, and takeout meals. These form 10.2 percent of food expenditures in urban areas and 5.7 percent in rural areas.

Third, processed food consumption rises with income and urbanization. Morisset and Kumar (2008) compare the level of consumption of processed food to income levels in urban areas. Dividing the urban population into 12 income categories, they find that the poorest and richest groups spend about 30 percent and 58 percent, respectively, of their food budgets on a relatively highly-processed food category (i.e., high-value first-processing and second-processing products). This implies that when incomes rise, consumption of processed food increases in importance. This is a typical result globally, including in developing countries (Wilkinson, 2004). Unfortunately, no analysis has been performed on changes in the share of processed food, of various levels of processing, over a longer period in India.

Fourth, there is evidence (Morriset and Kumar, 2008) of changes over sub-sector shares in the composition of the processing sector. Five traditional sectors—oil and fats, grain, sugar, dairy, and tea and coffee—dominate the food processing industry with 80 percent to 85 percent of total processing output, employment, and factories. Of these five large industries, three (oils and fats, grain, and dairy) showed an annual growth rate in output that was larger than the average for the processing sector in recent years, which shows that their relative importance is growing. The processing share is at (or near) 100 percent for grains, oil, and sugar because of the nature of these products, but for other sectors, processing is relatively unimportant. For example, it is estimated that only 2 percent of fruits and vegetables are processed in India (India Brand Equity Foundation, 2006).

Fifth, there is emerging evidence that there is a trend towards consolidation in the Indian food processing industry. This is at least partially driven by the 'de-reservation.' In 1987–88, 18 food processing sub-sectors were 'reserved' for small enterprises (Bhavani et al., 2006, citing DCSSI, 1992). Bhavani et al. note that in the decade 1997–2007, half of these sub-sectors were 'de-reserved,' leaving only pickles/chutneys, some vegetable oils, bread and pastry, sugar confections, and spice grinding as 'reserved,' while tobacco and alcoholic beverages were still 'licensed.' They note that no licensing was required for other sub-sectors. Moreover, FDI in food processing was eased: investments were automatically approved for FDI up to 51 percent of foreign equity or 100 percent of domestic (if by a non-resident Indian, for most subsectors except reserved subsectors, malted food, and alcoholic beverages) . The policy reforms also allowed 24 percent of foreign ownership of small-scale firms. Importantly, food processing machinery could be freely imported and exported, and custom duties on materials and machinery were significantly reduced. The government even provided full tax exemption for the first five years of operations for food processing companies to encourage new investments. Export zones were set up to encourage food processing exports.

De-reservation appears to have been an important factor in driving the change in processing, but there has been no statistical analysis of its impact on the consolidation in processing. There is evidence that the latter was already occurring before de-reservation; one can say that de-reservation and the other recent policies are only a part of the set of factors that are driving structural change in processing.

Based on historical data from the organized and the unorganized segments, Bhavani et al. (2006) argue that there is clear evidence that consolidation is taking place in the organized sector at the factory level, and scaling up (in terms of increases in output per factory) and capacity expansion (of fixed capital per factory) are occurring. The food processing industry is split into two segments—the 'organized' and the 'unorganized'—for statistical and regulatory reasons.[6] The share of the organized segment in total output has been increasing over time, as measured by the value of gross output which has increased from 64 percent in 1984–85 to more than 80 percent in 2000 (Bhavani et al., 2006). Yet 85 percent in 2000–01 (similar to 87 percent in 1985) of processing employment is in the unorganized segment.

Moreover, Reardon and Minten (2011b) argue, using procurement information from Indian supermarkets, that there is a 'symbiosis' between large processors and modern retail. The former have logistics strategies (such as direct delivery to chains, and produce assortment and packaging) that supermarkets desire. Supermarket chains, in turn, tend to select the large processors as suppliers, due to their product assortment, low transaction costs, and scale to supply all stores. The large chains and large processors thus may, in India, as has appeared in Latin America (for example, see Farina, 2002), help each other to develop and win market shares from traditional sector rivals.

The widely differing output and employment shares imply large labor productivity differences between the two segments. They indicate that although the organized segment is increasing its output share, it is expected that the unorganized segment will remain dominant in terms of the

---

6 The organized sector consists of units that employ more than 10 people and use power, or units that employ at least 20 people and use no power. Other units are categorized in the unorganized sector.

number of manufacturing units and employment. But in some industries, there is evidence of an absolute decline in the number of small firms: Das Gupta et al. (2010b) show that for paddy milling (into rice), the share of manual mills (typically very small, village-based, in the informal sector, and dominant in the 1960–1970s) has dropped dramatically over time in central Uttar Pradesh (to a very small number), as have single roller mills (more labor intensive), with the rise of semi-automatic and automatic mills.

Sixth, Bhavani et al. show that from 1984–85 to 2000–01, the total value of fixed assets of the processing sector jumped from Rs. 192 to 469 billion (2.44 times); over the same period, output jumped from Rs. 628 to 1215 billion (1.93 times). The trend is clear that capital-output ratios are increasing, i.e., that there is capital deepening in food processing. By contrast, employment in food processing increased from 10.39 to 12.06 million in the 15 years (1.16 times). Simple calculations show that the capital-labor ratio doubled over the 15 years. The large output growth with little employment growth of the food processing sector mirrors a broader and similar process in Indian manufacturing and services (Dehejia and Panagariya, 2011). It should be emphasized that this technology change was afoot before the liberalization of machinery imports and before de-reservation. The future research should include the statistical analysis of what caused these technology changes, as the policies that are often discussed can only be one part of the picture.

Moreover, capital intensification happened faster in the organized sector, as one would expect: during those 15 years, the share of the organized sector in the fixed assets of the processing sector jumped from 26 percent to 61 percent—even while the share of the organized segment in the number of firms in the processing sector stayed about constant (from 0.53 percent to 0.76 percent, back to 0.52 percent over the 15 years, while the number of firms dropped from 4.66 million in 1985 to 3.85 million in 1995, and rose net to 5.14 million by 2001).

Seventh, there is increasing evidence that private modern processors compete with cooperative processors, apparently displacing them, or at least reducing the cooperatives' market share. The National Dairy Development Board (NDDB) is an example of a cooperative that both processes and also retails. Cooperatives play at least a minor role in milk (with some 10 percent of the market), and play a role in grains, jute, cotton, sugar, areca nuts, fruits, and vegetables. Acharya (1994) estimated that in the early 1990s cooperatives handled about 10 percent of all marketed surplus in the country. However, the importance of cooperatives seems to be declining. For example, in a review of the agribusiness sector in India, Anzec (2005) argues that there are a limited number of successful market-sustainable cooperatives outside the dairy sector.

Not all dairy cooperatives have been successful, and although they might still be growing in absolute numbers, they are increasingly losing market share compared to the private sector in dairy value chains. For example, Sharma and Singh (2007) report that the share of the private sector in (organized) milk processing plants increased from 49 percent in 1996 to 66 percent in 2006. While the private sector and the cooperative sector held about equal shares in milk procurement in the organized dairy value chain in 2006, the private sector is projected to be twice as important in procurement in 2011 (Gupta, 2007).

Eighth, it is not clear what impact the above transformation of the processed food sector has on farmers. There are few survey-based studies on the procurement practices of processing firms (such as Dev and Rao, 2005, for gherkins and palm oil in Andhra Pradesh), and more research is required. It seems that processing firms have specific quality requirements[7] and do not usually buy directly from farmers, but rather seem to rely on dedicated brokers and wholesalers for the procurement of their produce (Singh, 2007). There are, however, exceptions—for example, in grains and oilseed (Singh, 2007). There are also limited requirements for raw produce, especially with regard to fruits and vegetables, which explains why it is often lower-quality products that are procured for processing at lower prices (Fafchamps et al., 2008).

Emerging transformation "midstream": in the wholesale/logistics sector

There have been two sets of trends (one government-driven, one private sector-driven) in the transformation of the wholesale sector in India. These trends are discussed in the two subsections below.

Government-driven transformation of the wholesale sector over the past 50 years

As is the case with retail, Indian government (in particular, at the state level) has had a direct role in the transformation of the wholesale sector, acting as a wholesaler. The government has also played an indirect transformational role through the regulation of and investment in the wholesale sector. Both roles gave rise to trends and policies with echo in many other developing countries at roughly the same time (Reardon and Timmer, 2007).

Direct transformational role: government steps into market as a wholesaler

In the 1960s, the government set up a parastatal—as a wholesale entity—to directly procure grain in place of the traditional wholesaler, who was seen as exploitative of farmers and predisposed to profiteering and speculating. The government set itself up as a substitute to private traders for a part of the market. This direct involvement was aimed at maintaining and controlling reserve stocks, influencing market prices, and subsidizing the poor (by on-selling in its 'forward integrated' retail system, the Fair Price Shops) (Rashid et al., 2007).

Over the past three decades, this role has gradually increased—but only in the grain sector. It is important to keep in mind that the government's direct role in the overall food sector is very

---

7 This is the case of dairy companies such as Reliance and Nestlé, or for suppliers of international companies. For example, Nijjer Agro Foods in Amritsar is supplying processed vegetables to international companies such as Unilever and Nestlé, and buys raw tomatoes and chillies directly from farmers. However, raw fruit for processing into pulp is sourced through independent contractors.

minor—its share in total food expenditure (market and subsistence) has been nearly steady for 30 years at only 6–7 percent of the overall food economy.

But the government's direct role in the grain economy (which constitutes a quarter of the total food economy), as a 'wholesaler' in procurement and a 'retailer' in marketing of grain has doubled over the three decades—from 12 percent of grain output in the 1970s to 24 percent in the early 2000s (calculations are weighted by rice and wheat shares, and abstracted from grains other than rice and wheat, using parastatal (Food Corporation of India) procurement shares of grain output, as cited in Rashid et al., 2007).[8] The authors termed this a 'transformation' role because it is a shift from traditional wholesale, via brokers in villages and wholesalers in mandis, toward an organized system of wholesale procurement.

Rashid et al. (2007) point out that the opposite trend has occurred in other countries in Asia. Using the example of paddy, they point out that from the early 1970s to 2001–03, the share of paddy procured by the government shifted in the following countries: (i) in Bangladesh, from 1.52 percent to 3.11 percent; (ii) in India, from 9.82 percent to 25.26 percent; (iii) in Indonesia, from 3.54 percent to 6 percent; (iv) in the Philippines, from 6.13 percent to 2.68 percent. India is a strong outlier.

Indirect transformational role: government sets up wholesale sector infrastructure and (partially) regulates private wholesale

In order to transform the wholesale sector from the traditional system, the Government of India embarked on a program of investment and regulation to integrate the sector and concentrate it into nodes—away from the fragmented and diffuse structure of the traditional system. The objective was to control, or at least influence, its terms of trade; and to this end, it began developing a public wholesale market system and concomitant marketing regulations, starting in the 1960s.

First, as in many other developing countries in the 1950s–1960s, India built public wholesale markets (mandis). The first several Five Year Plans, starting in 1951, emphasized the construction of wholesale markets, storage structures and warehouses, and transport lines (The Expert Committee on Agricultural Marketing, 2001, known as the 'Guru Report'). It is estimated that in 1947—the year of India's independence—there were only 268 wholesale markets. By 2004, there were 5,964: 2,143 primary "mandis", 2,810 secondary mandis, and 1,011 non-regulated mandis (calculations are based on lists in http://agmarknet.nic.in/).

An example of a primary mandi is the Azadpur Mandi in Delhi. It covers 43 acres, with 438 big stalls of 600 square feet each, and 796 small stalls of 200 square feet each. In recent years,

---

8 The shares noted in the text are government procurement of grain output. Because the marketed surplus rate is less than 100 percent, the shares of government procurement in total marketed surplus of grains are higher than the above shares. Gulati et al. (2011) find that 43 percent of the marketed surplus of rice and wheat was procured in 2008–09.

4 million tons of fruits and vegetables have been exchanged annually. While Azadpur Mandi is the largest in India, it is still only about half the size of the leading wholesale market in Beijing (Xinfadi) in terms of acreage and yearly volumes, and much smaller than large wholesale markets in Latin America (such as those in Mexico City or Sao Paulo).

Second, while this chapter's focus of the wholesale sector's regulation is on the Agricultural Produce Market Committees (APMC) Act, it also emphasizes that it is but one of a series of wholesale sector regulations in India spanning pre- and post-independence, from the 1930s to present. The Report of the Expert Committee on Agricultural Marketing (2001) is important because it initiated a decade of market reforms. It listed 25 regulations on the food sector, enacted from the 1930s to the 1990s. The produce marketing regulation, APMC, which started in the 1960s–1970s, merely fits the general trend. Moreover, a slew of public marketing institutions were set up during the 3rd Five Year Plan in the mid-1960s. Again, the APMCs at the state level merely fit into this general trend of forming government bodies to implement the regulations and to invest in the infrastructure development of the sector.

When the central government enacted the Agricultural Produce Marketing Act (APMA) legislation in the 1960s, each state deliberated and either chose whether to enact a state-specific regulation along the general guidelines and set up an Agricultural Produce Marketing Committee[9] to regulate the wholesale sector and make investments in its infrastructure. Between the 1960s and 1980s, most of the states and union territories adopted APMA and established APMCs. This required that the states build and maintain the 'yards' (mandis) by using market taxes with licensed/registered (but private sector) commission agents (CAs) and licensed traders in the yards. All wholesale trade was then required to pass via the APMC yards, paying a commission to the CAs, a tax to the market (a cess), as well as to off-loaders, loaders, and to 'weighing men' who registered the transaction.

The injunction that all wholesale trade, and thus all purchases from farmers, pass via the APMC mandi, thus implicitly forbade contract farming; collection centers by private retailers, wholesalers, or processors; direct marketing from farmers to consumers; and private wholesale yards. In short, it was aimed at funneling all trade from farmers to consumers through the APMC mandis. This was originally designed to break what was conceived as an exploitative private trader system; there were parallel developments in several other countries such as Turkey, Brazil, and South Africa that had similar regulations (currently still in use in Turkey).

---

9 The committee is empowered to establish markets, control and regulate the admission of traders to the market, charge fees (market, license, and rental fees), issue and renew licenses, and suspend or cancel licenses. Over time, APMCs have emerged as a government-sponsored marketing-services monopoly that prohibits innovations such as contract farming and does not allow traders to buy outside the specified market yards (Acharya, 2004). While the APMC also collects significant revenues from market fees, the infrastructure in most markets is largely deficient, as revenues are often directed toward other ends by the government (Umali-Deininger and Sur, 2007; Fafchamps et al., 2008).

Even though wholesale sector regulation and public investment via the APMC Act represent important aspects of the transformation from the traditional system—and significant controls over the wholesale sector—they are nevertheless more ambiguous and more partial (even before the eventual reforms in the past half-decade) than seems commonly noted in public debate. Several points are relevant:

First, the 2001 'Guru Report' noted that this regulation, as well as institutions and investments, tended to be at the level of the central government and the states down to wholesale markets—but tended to not penetrate to the more local level. To illustrate that point, they noted that as of 2000, nearly all the states had an APMC which regulated to the level of the wholesale mandis, but there were 27,294 rural periodical markets at the more local level—and the committee estimated that only about 15 percent of them functioned "under the ambit of regulation."

Second, of the 28 states, 3 do not have APMCs; in 2 states there has been no adoption of APMC; Bihar had an APMC but then repealed it. Moreover, Tamil Nadu adopted an APMC in 1987 but its original form had already allowed elements that eventually became the reforms (discussed below) such as direct marketing that essentially made the use of the government mandis non-mandatory.

Third, there is ample evidence that not all the mandis were regulated, nor did farmers sell all their output through the regulated mandis in the states where the APMC was enacted. This appears the reason, for example, why only 10 percent of the sampled mandis in Tamil Nadu were APMC-regulated and farmers sold much of their produce to non-APMC licensed buyers (Shilpi and Umali-Deininger, 2008). Surveys in Uttar Pradesh, Orissa, and Maharashtra revealed that 85 percent, 90 percent, and 95 percent of the sampled mandis were APMC regulated; thus even in APMC states in 2005, the regulation and its implementation over mandis was somewhat partial. Farmer surveys also showed that farmers did not sell only via regulated markets (Shilpi and Umali-Deiniger, 2008).

Fourth, even where APMCs are in force, and principal traders from outside the area and private sector actors (like processing firms and supermarket chains) are supposed to buy via the regulated mandi, there is some evidence that the implementation of the regulation is incomplete or skirted. In a survey in Western Uttar Pradesh, Das Gupta et al. (2010a) found that the potato trade had shifted substantially from the Agra mandi to take place at the cold storage. But such examples do not necessarily establish a strong or clear pattern. There are other survey-based studies with opposite findings, such as Fafchamps et al. (2008), who show that farmers sold most of their fruit and vegetables directly to traders in the wholesale markets, especially where regulations require it, but also in cases where the regulations do not require it (such as where the APMC Act was not enacted or had been reformed.)

Fifth, a state's use of the APMC does not necessarily mean that modern retailers cannot become the licensed intermediaries. For example, the Spencer's (chain) collection center in Karnataka, started in 2001, obtained a license as a regular 'yard' of the APMC (as per the authors' field interviews, 2007).

Discussion of the application, implementation, and penetration of the APMC Act emphasizes that there has been a large gap between the image of the Indian wholesale sector—as being highly and completely regulated before reforms—and the reality of the Act's imperfect implementation. The 'state-led transformation' of the wholesale sector from its traditional form has only been partial. Since there is no systematic empirical analysis of this point, however, it remains a general proposition.

Private sector (modern and traditional)-driven transformation of the wholesale sector over the past ten years

APMC reform and entry of the private modern sector into direct wholesale

By the early 2000s, there was increasing criticism of the APMC system—emphasizing both its inadequate performance relative to its purpose, and the narrowness of its purpose relative to the emerging needs for wholesale and logistics services in India's transforming food economy. The first critique of the APMC system's inadequate performance was set out by the 'Guru Committee' (The Expert Committee on Agricultural Marketing, 2001), whose main points are summarized here.

Their report starts by acknowledging that the APMC regulated markets did accomplish some of their original objectives—namely, to redress problems in marketing. The report then goes on to say that the initially promising approach developed into a system that was fraught with problems: (i) APMC has had limited success; (ii) the requirement that trade must pass via the APMC mandi has hurt competition; (iii) licensing gave way to entry barriers; (iv) APMCs were supposed to control unethical practices in the mandis but have often let them persist; and (v) APMCs were supposed to collect taxes/fees and use them for infrastructure development but often did not.[10] The report's findings have been echoed in a number of other reports (Acharya, 2004).

There has been scant empirical research on mandis in India (in terms of representative sample surveys over mandis, traders, retailers, or farmers using mandis), but the few studies done tend to corroborate most of the general critiques noted above, which include:

- The mandi trader system is not efficient (Matoo et al., 2007; Ramaswami and Balakrishnan, 2002; Umali-Deininger and Deininge,r 2001; Thomas, 2003);
- Mandis tend to lack market integration (Palaskas and Harriss-White, 1996);
- The mandi system is plagued by trader collusion (Banerji and Meenakshi, 2004);

---

10 Fafchamps et al. (2008) did an extensive survey of mandis in Maharashtra, Orissa, Tamil Nadu, and Uttar Pradesh, and found the infrastructure and services of the mandis to be generally poor. Moreover, in interviews with the Maharashtra State Agriculture Marketing Board (MSAMB), the Government of Maharashtra noted in March 2009 that there were only 265 APMC mandis in Maharashtra in 2006 (at time of the APMC amendment), and of those, only 5 percent have cold storage, packing areas, and grading facilities.

- There is the hypothesis[11] that the traditional channels to and from mandis are characterized by a high level of physical wastage (Matoo et al., 2007);
- Wholesale market infrastructure for staple and non-staple crops is not very developed.

The majority of wholesale markets are not paved, and there are only a few grading or cold storage facilities. Sanitation facilities are largely deficient, with few public toilets, inadequate drainage, and little or no coordinated pest control (Fafchamps et al., 2008).

Driven by the above critiques and by the segments of political support for reform and market liberalization in general, there have been various reforms in the first decade of the 2000s that affect the wholesale sector.

On the one hand, there have been a series of reforms in the 2000s that indirectly or secondarily affect the wholesale sector, including: (i) the liberalization of FDI in food wholesale (including in 'cash & carry' chains such as Metro), with the clientele required to be registered resellers, not consumers (PlanetRetail, 2008, May 5); (ii) restrictions were removed on domestic and foreign investment (FDI) (up to 100 percent) in bulk handling and storage; (iii) licensing requirements, stocking limits, and movement restrictions were either removed or reduced on wheat, paddy/rice, coarse grains, edible oilseeds, and edible oils; (iv) a warehousing receipts program was initiated; (v) futures and forward markets, and other commodity exchanges were established; and (vi) the Food Safety and Standards Act of 2006 was enacted.

On the other hand, in 2003 major reforms of the APMC system were enacted with the Model Act for State Agriculture Produce Marketing (Development and Regulation). The Amended Act proposes that each state remove the restrictions on farmer direct marketing (under the regulated system, notified products can be sold only at markets to licensed traders), the opening of market infrastructure development to other agencies (especially the private sector), and the establishment of a framework for contract farming.[12] However, the 'Model APMC' Act has been adopted by only about half the states. The 4 states where the reforms are not applicable were previously discussed; of the 28 states, 18 have 'fully' adopted—but not necessarily fully implemented[13] the reforms. There is no systematic information about the extent of implementation among the 'adopters'. In the authors' interviews in Maharashtra, for example, they found that adoption might be complete, but implementation is slow, partial, or complicated. Yet, given the short period in which the Model Act was presented and resisted, the authors rate partial implementation as substantial progress with regard to these reforms. Two of the states, Haryana and Punjab, have adopted parts of the Model Act; there are still 7 states that have not adopted any parts of the Act. Several

---

11 The word "hypothesis" is emphasized because the authors have been unable to find empirical studies of actual wastage in food supply chains in India. Anecdotes, opinions, and assertions abound, and various reports (like Matoo et al., 2007) cite 'key informant' information. But actual surveys on wastage are missing. The reports found show that wastage rates, in this case for potatoes and wheat, are much lower than previously assumed (Das Gupta et al., 2010a; 2010b). The rates might be higher for the 5 percent or so of the Indian food economy that is highly perishable, such as greens. Testing this hypothesis remains a gap in the literature.
12 Accessible at Http://agmarknet.nic.in/amrscheme/apmcstatus08.htm.
13 Accessible at Http://agmarknet.nic.in/amrscheme/apmcstatus08.htm.

of these states are small, but Uttar Pradesh, West Bengal, and Uttarakhand stand out as important agricultural states. There is a correlation between the adoption of the market reforms, the general state of politics of the state and pressure from specific lobbies within the states.

Reardon and Minten (2011b) examine how retail and processing firms enter into direct procurement arrangements with farmers over the past 7–8 years, partly in the aftermath of the APMC reforms in various states, and partly operating within the APMC system, using licenses. Below is a classification of the direct-procurement methods used: Before, or in the absence of APMC reform, companies have set up collection centers by obtaining a license from the APMC; examples are Spencer's in Karnataka for the sourcing of produce, and ITC's Choupal Saagar and e-choupal hub and spoke system for grain procurement in Madhya Pradesh.

After APMC reform in a state, retailers and processors have set up collection centers outside mandis ('off-market') such as retailers/wholesalers Reliance and Metro in Maharashtra and Safal (Mother Dairy) in Uttarakhand, or produce wholesale/ processing companies such as Adani, with controlled atmosphere apple distribution centers in Himachal Pradesh. These collection centers are usually in peri-urban areas (up to a 4 hour drive from tier 1 or tier 2 cities) and source highly perishable products like greens, or semi-perishables, like cauliflower and tomatoes for the city.

The main difference between the state-sanctioned and 'off-market' collection centers is that the latter does not require an APMC license. However, the authors' field research shows that this difference does not necessarily manifest itself as a clear, concrete advantage; for example, in Maharashtra the authors found that off-market collection centers still officially pay a Cess to the APMC, and unofficially, are often forced by the 'network of actors' in the mandi system to pay weighing men, even though they are not officially required; this acts as an 'informal tax' on the supposedly new and liberalized arrangements.

Various 'cluster' platforms such as 'private mandis,' Mega-Food Parks, Integrated Agro-Food Parks, private chains of Rural Business Hubs (such as Hariyali Kisaan Bazaar) and logistic parks. Contract farming is sometimes referred to as "sponsored farming", after APMC reform, such as Godrej in Maharashtra.

The authors did a rapid-reconnaissance version of the inventory (discussed in detail in Reardon and Minten, 2011b) in Maharashtra in 2009 and found that in the short span of three years—from the APMC reform in the state in 2005–06 to early 2009—79 licenses (including 15 retail chains that set up a number of rural collection centers around the state and the others, mainly processors of fruit, grain, and cotton) had been granted for direct marketing through collection centers in rural areas, a number of contract farming schemes, and hundreds of licenses for mandi stalls had been granted to a number of retailers. The authors visited key horticultural areas around Pune and found intensive competition among a number of retail chains with collection centers in the area, but also heard stories of how the traditional mandi actors harass the new actors.

The rise of modern private food wholesale and logistics in India

A modern sector cluster of food logistics/distribution/wholesale companies in India has been emerging rapidly in the past decade. Based on field case studies and a review of the evidence, Reardon and Minten (2011b) trace the rise of this segment, and its 'symbiotic' links to the rise of modern retail in India. The main points are as follows.

First, especially for processed food, refrigerated semi-processed food, and fresh food, today's food retailers in India are increasingly shifting toward the use of modern logistic and wholesale companies (and direct sourcing from manufacturers, as discussed earlier), and away from the use of traditional stockists and general-line wholesalers. These developments follow the general trend found elsewhere in Asia and globally (Reardon et al., 2007; Reardon and Timmer, 2007; Farina et al., 2005 for dairy in Brazil and Argentina, and other illustrations). The underlying reasons are reduced transaction costs, increased consistency and higher quality, safety standards, and regulations. In the long-term in India, this can accelerate the consolidation of processing and logistics/wholesale sectors—a trend that is observed elsewhere, too. Some of the modern logistics companies are establishing backward integration by retailers (such as Future Logistics Solutions Ltd., of Pantaloon/Future Group, and Advanced Logistics Asia of Metro Group). Alternatively, if these companies had been started earlier, then retail would have been a step of forward-integration from them. One example is the Radhakrishna Foodland Company; its retail chain was started, then discontinued, with focus remaining only on logistics.

Moreover, today's retailers now use the services of independent modern logistics companies that were either emerging in the 2000s or have since grown rapidly. Several leading examples include:

- Concor (the Container Corporation of India), for Bharti-Walmart, Pantaloon, and Mother Dairy;
- Agility Logistics (US) with a 2008 turnover of US$7 billion, US$4 billion of which came from outside the US. In 2009, it was investing US$130 million in India and was already a leading modern logistics firm (3plogistics, March 10, 2009, March 10; accessible from: www.agilitylogistics.com). It has several modern retail chains and FMCG (fast-moving consumer goods) companies as clients in India.
- Snowman Frozen Foods (Japan) is said to be the first and largest cold chain-cum-logistics independent firm with a pan-Indian presence. Snowman moves products (mainly dairy, processed foods and pulps, seafood, and meat) in refrigerated trucks from suppliers to cold-storage to retailers (such as Bharti) and processors (www.snowman.in; FnBNews, October 7, 2008).
- Finally, an important development is the emergence of procurement system partnerships between modern Indian retail chains and global chains involved in 'cash & carry' joint ventures with the Indian chains. Examples include Walmart's partnership with Bharti, and

Metro's (Germany) stand-alone cash & carry hypermarkets in India this needs to be made into a full sentence.

Second, the corollary of the above trend is that modern retail's emergence appears to be advantageous for the food logistics/distribution sector in India (FnBNews, October 7, 2008). VcCircle (October 2, 2007), a leading investors site, notes "Logistics see a great potential in the wake of a retail revolution, and so companies are game to capitalise on this high-growth opportunity." The Hindu Business Line (November 22, 2006) also noted that multinational logistics firms are being attracted by the rise of modern retail. It notes: "The port-based container logistics company Gateway Distriparks Ltd (GDL) is foraying into the cold chain logistics segment, expecting a surge in domestic demand for movement of frozen and chilled food in the wake of the boom in the retail sector." This 'symbiosis' is presently a hypothesis since there has been no systematic research, but the trend appears in this direction.

Third, there are other factors that promote third party investments in food logistics. The Economic Times (September 25, 2009; January 29, 2010) noted that while only 20 percent of the logistics sector is 'organized' and the rest is in the informal sector, the share could rise with a reduction in intermediation. This could happen with a shift from multiple warehousing to regional-based logistics under a revised tax regime. Specifically, a rollout of the goods and service tax (GST) is planned that will shift the taxation to a consumer tax. This shift was slated to take place in April 2010, but it appears that it has been delayed for several years. Moreover, as part of the 2009–10 budget, cold chain investments have become tax exempt—a further boost to the sector.

Ferment and change in the traditional wholesale sector

A common view is that India's rural economy is in long-term stasis—by this view, it is an unchanging traditional system where privileged wholesalers dominate trade, village collectors or brokers dominate the interface of the market with the farmer, both extend credit to farmers to 'tie output and credit markets'. Further, there is little storage, minimal competition among different actors, and overall, 'long supply chains' with many 'hands' and many actors, from the farmer to the city. This image is painted accurately and empirically in leading accounts such as described by Lele (1971).

Instead of a stagnant rural market economy, however, there appears to be great ferment and rapid transformation. This appears to not be happening at the same rate or in the same way across rural zones, but instead concentrated in certain areas—primarily in the swaths of agricultural areas within the vast market catchment areas that are within a 6–8 hour drive of tier 1 and tier 2 cities. These areas, which the authors refer to as 'dynamic areas,' may not be typical of traditional and hinterland rural areas, but they include a substantial share of the rural Indian population, and likely account for a large share of the food supply to cities. The authors' recent farm, mill, cold store, and trader surveys in west, central, and eastern Uttar Pradesh and Madhya

Pradesh (Das Gupta, 2010a; 2010b; Reardon et al., 2011c; 2011d) reveal certain key findings that run counter to conventional wisdom about rural markets. General findings from these studies, in addition to findings from comparable surveys in several areas of Bihar and Uttaranchal (Minten et al., 2010b; 2009b) are presented below

First, rural traditional market transformation is much more advanced in certain regions of the study states than in others: for example, the western and central regions of Madhya Pradesh (Malwa Plateau) and western and central Uttar Pradesh are sharply different from the eastern regions of each of those states. The average farm size is larger than in the east (and in the rest of the state); farmland distribution is more unequal (with 30 percent of farmers owning 70 percent of the farmland); land rental share is higher, as are the marketed surplus rates, incomes, livestock holdings, and milk outputs, chemical use, fertilizer use, and credit use, and all other indices of traditional market transformation. In contrast, the eastern parts of these states look more like the conventional image of 'traditional rural India.'

Second, the surveys showed the same types and directions of the differences across the regions; they also were manifest over the farm size strata. The marginal farmer strata (0–1 ha) looked much like 'traditional rural India'—with low marketed surplus, chemical use, credit use, low participation in transformed traditional markets (still selling mainly to village collectors), lower use of cold stores, and so on. In contrast, the small farmers in the dynamic areas and the medium farmers in all zones have assets and display behavior that corresponds to features of the dynamic zones in general. Thus, transformation is differentiated both by zone and farm size—with the smallest and the most hinterland relatively left out from the process—as seen in other developing countries.

Third, it is usually assumed that staple food supply chains are dominated by long chains that lead to inefficiencies. For example, Mattoo et al. (2007) and Landes and Burfisher (2009) argue that in the case of India, most agricultural trade involves a large number of intermediaries. It is argued that this system not only inflates prices but also takes time to move products from farmers to consumers, leading to large transit costs.

However, the authors' surveys show that the role of the village/field broker is quite limited (as a share of market surplus from farmers), as is the role of the local haat. The terms 'dis-intermediation' and 'supply-chain shortening' come to mind; terms that are usually only found in the context of the modern sector, with supermarkets or processors buying directly from farmers. These developments are found in traditional chains as well, with a sharp reduction in the role of the traditional village broker/collector. (Incidentally, this finding is emerging elsewhere in Asia; see Natawidjaja et al., 2007 for Indonesia and Huang et al., 2007 for China.) Most farmers sell the great majority of the marketed surplus directly to the wholesale markets and, to a certain extent, to the mills (grain and soybeans) and the cold stores (potatoes) rather than at the mandis.

Fourth, it is often assumed that farmers in India are typically at the mercy of a field broker, are not very informed about the markets, or get low prices because of tied credit (e.g., Basu, 1986; Bell, 1988; Bell and Srinivasan, 1989; Basu, 2010). But all of the authors' surveys show that fewer than 5 percent of farmers receive any advance payments (such as for inputs at the start of the season) or credit in any form from brokers or wholesalers. The trader surveys present the same picture. In follow-up key informant interviews that help interpret the surveys' results, traders and farmers reported: "Some 15 or 20 years ago it was common for traders to extend credit to farmers; this has nearly disappeared because farmers have many options, roads are better, there is more credit available such as from Kisaan Credit Cards, and they have mobile phones." Das Gupta (2010a) found that a large number of farmers now possess a mobile phone and use it actively to conduct their business. It is estimated that almost 80 percent of the interviewed potato farmers contacted multiple buyers by phone and almost half of the potato farmers in the hinterland of Delhi settled on a price by phone in their last transaction.

Fifth, the surveys (Das Gupta et al., 2010a; Minten et al., 2010b) indicate rapid development (in terms of diffusion and scale increase) of cold stores for potatoes in Uttar Pradesh and Bihar in the 2000s. This has raised farm prices and reduced seasonality for consumers. Direct sales by farmers to traders at cold stores in western Uttar Pradesh have been shown to greatly diminish the role of the mandi in that region—this even applies to a state where the APMC has not been reformed. Ninety-five percent of the farmers in the survey in western Uttar Pradesh were found to be using cold stores in 2009—an increase from 40 percent (estimated by Singh, 2008) in 2000, and from a small minority in the early 1990s (Fuglie et al., 1997). The cold stores were also used by most traders. Moreover, the surveys show that cold storages are increasingly involved in input, output and credit markets.

Conclusion: a quiet revolution in India's food supply chains—emerging but still constrained

The evidence indicates that India's traditional rural areas are rapidly transforming, as the modern sector and traditional supply chains transform. This transformation is mirrored, but in a more concentrated way and at a faster pace, in the urban food economy. This chapter shows that modern and traditional participants from the private sector are the most important actors in this 'quiet revolution'. Further, the government's direct role—as a buyer and a seller—is only 7 percent of the food economy of India. (The common perception is that it is much larger; but that idea is based on the government's role in the grain economy: it has a 25 percent share in grains, which constitute 25 percent of the food economy.) It is the private sector, both modern and traditional, that decides the food security of India.

It is clear that the economic environment and policy changes have been crucial in spurring the 'quiet revolution' in food supply chains. A host of liberalization policies combined with public investments have prompted an avalanche of private investments—by farmers, traders, cold stores, mills, and retailers. Equally clear, however, is that a number of constraints to continued transformation persist. Neutrally presenting these constraints helps to describe the limits of the transformation of supply chains. Whether these constraints are addressed or whether further transformation is encouraged are political decisions that can only be decided by the Indian government and its people.

The first, persistent constraint is asset poverty and policies that do not yet address it sufficiently. These include collective assets—such as poor roads and the lack of electricity in the poorer regions such as eastern Uttar Pradesh. They also include individual assets—such as education, tube-wells, and access to credit. A separate work (Reardon et al., 2011c; 2011d; Rao et al., 2011) shows that the state and cooperative supply of subsidized tube-wells, credit, fertilizer, and seed is heavily biased towards medium and large farmers in the study areas discussed. Broadening infrastructure and the distribution of public goods and services to poorer areas and strata are major challenges.

The second persistent constraint is that unlike most of Asia, perhaps most strikingly China and Southeast Asia, India's continuing constraint on foreign direct investment in food retail means that it still foregoes that source of investment capital and expertise, and the incremental gains to urban food security that retail transformation can bring. This may be a moderate constraint because domestic retail investment is far more vigorous (and fueled by cash-rich conglomerates) than in many other countries.

The third persistent constraint is in the form of numerous, continuing policy-based limitations to direct procurement from farmers by retailers, processors, and modern wholesalers. Among the policy issues are: (i) partial/slow liberalization of wholesale markets (APMC reform); (ii) limits on private sector procurement, storage, and sales to traders: Storage Control Orders under Essential Commodity Act; and (iii) regulatory and fiscal uncertainty and transaction costs, such as double taxation on the inter-state movement of goods.

References

3iNetwork. (2008). India Infrastructure Report 2008 – Business Models of the Future. Infrastructure Development Finance Company, Indian Institute of Management, Ahmedabad, Indian Institute of Technology, Kanpur. New York: Oxford University Press.

3plogistics.com. (2009). Agility Continues its Global Expansion. http://www.3plogistics.com/Agility_2-2009.htm

Ablett, J., Baijal, A., Beinhocker, E., Bose, A., Farrell, D., Gersch, U., Greenberg, E., Gupta, S., and Gupta, S. (2007). The "Bird of Gold": The Rise of India's Consumer Market. San Francisco: McKinsey Global Institute.

Academic Foundation. (2004). State of the Indian Farmer, a millennium study. New Delhi: Ministry of Agriculture, Government of India.

Acharya, S.S. (2004). Agricultural marketing in India. Vol. 17 of Millennium study of Indian farmers. New Delhi: Government of India, Academic Foundation.

Ackermann, Richard. (2011). New directions for water management in Indian agriculture. Background paper for The future of Indian agriculture and rural poverty reduction.

Adhiguru, P., Birthal, P.S., and Ganesh Kumar, B. (2009). Strengthening pluralistic agricultural information delivery systems in India. Agricultural Economics Research Review 22, p. 71–79.

Agarwal, I., Priya, S., and Bhuvaneswai, S. (2005). Contract farming venture in cotton: A case study in Tamil Nadu. Indian Journal of Agricultural Marketing 19(2): 153–161.

Aggarwal, P.K., Hebbar, K.B., Venugopalan, M.V., Rani, S., Bala, A., Biswal, A., and Wani, S.P. (2008). Quantification of Yield Gaps in Rainfed Rice, Wheat, Cotton and Mustard in India. Global Theme on Agroecosystems Report No. 43. International Crops Research Institute for the Semi-Arid Tropics, Patancheru 502 324, Andhra Pradesh, India.

Aggarwal, S. 2008. Digging shallow. Down to Earth.

ASTI (Agricultural Science and Technology Indicators). www. asti.cgiar.org.

Ahluwahlia, Montek. (2011). Prospects and policy challenges in the twelfth plan. Economic and Political Weekly 46(21): 88–105.

AICC (All India Congress Committee). (1949). Report of the Congress Agrarian Reforms Committee (J. C. Kumarappa). New Delhi, India.

Amarasinghe, U.A., Shah, T., Singh, O. (2007). Changing consumption patterns: Implications for food and water demand in India. IWMI Research Report 119. Colombo, Sri Lanka: International Water Management Institute.

Amarasinghe, U.A., Shah, T., Turral, H., Anand, B.K. (2007). India's water future to 2025-2050: Business-as-usual scenario and deviations. Colombo, Sri Lanka: International Water Management Institute. (IWMI Research Report 123).

Amarasinghe, U.A., Sharma, B.R., Aloysius, N., Scott, C., Smakhtin, V., de Fraiture, C. (2004). Spatial variation in water supply and demand across river basins of India. Research Report 83. Colombo, Sri Lanka: International Water Management Institute.

Amarasinghe, U.A., Sharma, B.R., eds. (2008). Strategic analyses of the National River Linking Project (NRLP) of India, series 2. Proceedings of the Workshop on Analyses of Hydrological, Social and Ecological Issues of the NRLP. Colombo, Sri Lanka: International Water Management Institute.

Anderson, J.R. (2007). Agricultural Advisory Services, Background paper for the World Development Report 2008, Agriculture and Rural Development Department. Washington, DC: The World Bank.

Anderson, J.R., and Feder, G. (2004). Agricultural extension: Good intentions and hard realities. World Bank Research Observer, 19(1), 41–60.

Anderson, K., ed. (2009). Distortions to agricultural incentives: a global perspective, 1955–2007. Washington, DC: The World Bank.

Anthony, V.M., and Ferroni, M. (2011). Agricultural biotechnology and smallholder farmers in developing countries. Current Opinion in Biotechnology 23(2).

ANZEC. (2005). Agribusiness Development Support Project. Report prepared for the Department of Agriculture and Cooperation and the Asian Development Bank. Mimeo.

Ashley, J., and Smith, Z.A. (1999). Groundwater Management in the West. Omaha, Nebraska: University of Nebraska Press.

Bagi, F.S. (1981). Relationship between farm size and economic efficiency: An analysis of farm level data from Haryana (India). Canadian Journal of Agricultural Economics 29: 317–326.

Ballabh, V., ed. (2008). Governance of Water. New Delhi, India: Sage Publications.

Banerji, A., and Meenakshi, J.V. (2004). Buyer collusion and efficiency of government intervention in wheat markets in northern India: An asymmetrical structural auction analysis. American Journal of Agricultural Economics 86(1): 236–253.

Barghouti, S., Kane, S., Sorby, K., and Ali, M. (2005). Agricultural Diversification for the Poor: Guidelines for Practioners, Agriculture and Rural Development Discussion Paper 1. Washington DC: The World Bank.

Barrett, J., and Ball, J. (2011). Rethinking Flood Control. The Wall Street Journal.

Barringer, F. (2009). Rising Calls to Regulate California Groundwater. The New York Times.

Basole, A., and Basu, D. (2011). Relations of production and modes of surplus extraction in India: Part I. Agriculture, Economic and Political Weekly 46(42): 41–58.

Basu, J.P. (2010). Efficiency in wholesale, retail and village markets: A study of potato markets in West Bengal, Journal of South Asian Development, 5(1): 85–112.

Basu, K. (1986). One kind of power, Oxford Economic Papers, 38(2): 259-82.

Beintema, N.M., Avila F., and Fachini, C. (2010). Brazil: new developments in the organization and funding of public agricultural research. ASTI Country Note. Washington, DC, and Brasilia: International Food Policy Research Institute and Brazilian Agricultural Research Corporation.

Bell, C. (1988). Credit markets and interlinked transactions. Handbook of Development Economics. Amsterdam: North Holland.

Bell, C., and Srinivasan, T.N.. (1989). Interlinked transactions in rural markets: An empirical study of Andhra Pradesh, Bihar and Punjab. Oxford Bulletin of Economics and Statistics 51(1): 73–83.

Bell, D.E., Sanghavi, N., Fuller, V., and Shelman, M. (2007). Hariyali Kisaan Bazaar: a rural business initiative. Harvard Business School report N2-508-012.

Bhagat, R.B. (2011). Emerging pattern of urbanization. Economic and Political Weekly 46(34): 10–12.

Bhalla, S.S. (1979). Farm size, productivity and technical change in Indian agriculture. Agrarian Structure and Productivity in Developing Countries. Baltimore, MD: Johns Hopkins University Press.

Bhalla, G.S., and Tyagi, D.S. (1989). Patterns in Indian agricultural development: a district level study. Institute for Studies in Industrial Development.

Bharadwaj, K. (1974). Production Relations in Indian Agriculture. Cambridge: Cambridge University Press.

Bhatia, R., Cestti, R., Scatasta, M., Malik, R.P.S., eds. (2008). Indirect economic impacts of dams: case studies from India, Egypt and Brazil. New Delhi, India: Academic Foundation and The World Bank.

Bhavnani, A., Chiu, R.W., Silarszky, P., Subramaniam, J. (2008). The role of mobile phones in sustainable rural poverty reduction. World Bank, ICT Policy Division, Global Information and Communication Department.

Bhavani, T.A., Gulati, A. and Roy, D. (2006). "Structure of the Indian Food Processing Industry: Have reforms made a difference?" in Plate to Plough: Agricultural Diversification and its Implications for the Small holders in India. International Food Policy Research Institute.

Binswanger-Mkhize, H. (2010). Participation and Decentralization for Agricultural and Rural Development: Key to effective implementation of public policies and strategies. Paper prepared for Centennial Group and Syngenta Foundation.

Binswanger-Mkhize, H., and Parikh, K. (2011). "The future of Indian agriculture and rural poverty reduction." Main Report. FAO Stat. Production and yields data. Accessed December 17, 2011: http://faostat.fao.org

Binswanger-Mkhize, H., and d'Souza, A. (2011a). Structural transformation of the Indian economy and of its agriculture.

Binswanger-Mkhize, H., and d'Souza, A. (2011b). India, 1980–2008, structural change at the state level.

Binswanger-Mkhize, H, Pradhan, K., and Singh, S. (2011a). Impact of changing prices and rising wages of Indian agriculture, 1999–2007.

Binswanger-Mkhize, H., de Regt, J., and Spector, S. (2009). "Local and Community- Driven Development: Moving to Scale in Theory and Practice." New Frontiers in Social Policy. Washington, DC: The World Bank.

Binswanger-Mkhize, H., Pradhan, K.C., Nagarajan, H.K., Singh, S.K., and Singh, J.P. (2011b). India 1999–2007: Structural change of Indian agriculture at the village and household level.

Birner, R., and Anderson, J.R. (2007). How to make agricultural extension demand-driven? The case of India's agricultural extension policy. IFPRI Discussion Paper 00729, Development Strategy and Governance Division. Washington, DC: IFPRI.

Birner, R., Davis, K., Pender, J., Nkonya, E., Anandajayasekeram, P., Ekboir, J., Mbabu, A., Spielman, D., Horna, D., and Benin, S. (2006). From best practice to best fit: A framework for analyzing agricultural advisory services worldwide. Development Strategy and Governance Division, Discussion Paper No. 39. Washington, DC: International Food Policy Research Institute (IFPRI).

Birthal, P.S. (2008). Linking smallholder livestock producers to markets: Issues and approaches. Indian Journal of Agricultural Economics 63(1): 19–37.

Birthal P.S., Jha, A.K., and Joseph, A.K. (2006). Livestock production and the poor in India. Memio. Capitalization of Livestock Program Experiences India, New Delhi.

Birthal, P.S., Jha, A.K., and Singh, D.K. (2009). Income diversification among farm households in India: Patterns, Determinants and Consequences. Unpublished manuscript. New Delhi: National Centre for Agricultural Economics and Policy Research (NCAP).

Birthal, P.S., Jha, A.K., Joshi, P.K., and Singh, D.K. (2006). Agricultural diversification in north eastern region of India: Implications for growth and equity. Indian Journal of Agricultural Economics 61(3): 328–340.

Birthal, P.S., Joshi, P.K., and Gulati, A. (2005). Vertical Coordination in High-value Food Commodities: Implications for Smallholders, Markets, Trade and Institutions Discussion Paper No. 85. Washington DC: International Food Policy Research Institute.

Birthal, P.S., Joshi, P.K., Chauhan, S., and Singh, H. (2008). Can horticulture revitalize agricultural growth? Indian Journal of Agricultural Economics 63(3): 310–321.

Birthal, P.S., Joshi, P.K., Roy, D., and Thorat, A. (2007). Diversification in Indian agriculture towards high-value crops. Discussion Paper 00727. Washington, DC: International Food Policy Research Institute.

Birthal, P.S., and Taneja, V.K.. (2006). Livestock sector in India: Opportunities and challenges for smallholders. Smallholder Livestock Production in India: Opportunities and Challenges (P.S.

Birthal, P.S., Taneja, V.K., and Thorpe, W., eds. New Delhi: National Centre for Agricultural economics and Policy Research (NCAP), Nairobi: The International Livestock Research Institute (ILRI).

Blackmore, D. (2010). River Basin Management—Opportunities and Risks. eWater CRC and Asian Development Bank, October 2010.

Bondla, D.J.N., and Rao, N.S. (2010). Resistance against Polavaram. Economic and Political Weekly, August 7, 2010, p. 93.

Briscoe, J., Malik, R.P.S. (2006). India's Water Economy—Bracing for a Turbulent Future. New Delhi, India: Oxford University Press.

Business Line. (2006). Gateway Distriparks acquires 50.1 pc stake in Snowman. November 22. Census of India. Various years. Government of India, Ministry of Home Affairs, Official website, available from www.censusindia.net.

Centennial Group. (2012). India 2039: Transforming agriculture.

CGWB (Central Ground Water Board). (2005). Master Plan for Artificial Recharge to Ground Water in India. New Delhi, India: Ministry of Water Resources, Government of India.

CGWB (Central Ground Water Board). (2006). Dynamic Ground Water Resources of India (as on March 2004). Ministry of Water Resources. Faridabad, India: Government of India.

CGWB (Central Ground Water Board). (2007). Ground Water Regime Monitoring. Ministry of Water Resources. Accessed May 13, 2012: http://wrmin.nic.in/printmain3.asp?sslid=335&subsublinkid=324&langid=1.

CGWB (Central Ground Water Board). (2009a). Report of the Group for Suggesting New and Alternate Methods of Ground Water Resources Assessment. Ministry of Water Resources. Faridabad, India: Government of India.

CGWB (Central Ground Water Board). (2009b). Status Report on Review of Ground Water Resources Estimation Methodology. R&D Advisory Committee on Ground Water Estimation. Faridabad, India: Government of India.

CGWB (Central Ground Water Board). (2011a). Dynamic Ground Water Resources of India (as on March 31, 2009). Ministry of Water Resources. Faridabad, India: Government of India.

CGWB (Central Ground Water Board). (2011b). Water Level Fluctuation [January 2011 Vs Decadal Mean January (2001-2010)]. Ministry of Water Resources. Faridabad, India: Government of India. Retrieved March 31, 2012 from: http://cgwb.gov.in/documents/WL-MAPS/WL-Fluctuation-jan-2011-vs-decadal%20mean-jan%202001-2010.pdf

CWC (Central Water Commission). (2005). Hand book of Water Resources Statistics. New Delhi, India: Government of India, Ministry of Water Resources.

CWC (Central Water Commission). (2009). National Register of Large Dams—2009. New Delhi, India: Government of India, Central Water Commission, Dam Safety Organisation.

CWC (Central Water Commission). (2010). Financial Aspects of Irrigation Projects in India. New Delhi, India: Information Technology Directorate, Information Systems Organisation, Water Planning & Projects Wing.

Chakraborty, S. (2012). A drop to win. Intelligent Entrepreneur, June 2012, p. 32–34.

Chambers, R. (1988). Managing Canal Irrigation. Oxford, England and New Delhi, India: Oxford and IBH Publishing Company.

Chand, R. (2000). Emerging trends and regional variations in agricultural investments and their implications for growth and quity. Policy Paper No. 11, NCAP.

Chand, R. (2001). Emerging trends and issues in public and private investments in Indian agriculture: a state wise analysis. Indian Journal of Agricultural Economics 56(2): 161–184.

Chand, R. (2003). Government Intervention in Foodgrain Markets in the new context. Policy Paper 19. New Delhi, India: National Centre for Agricultural Economics and Policy Research (ICAR).

Chand, R., Garg, S., and Pandey, L. (2009). Regional Variations in Agricultural Productivity: A District-level Study. Discussion Paper NPP 01/2009. New Delhi: National Centre for Agricultural Economics and Policy Research.

Chand, R., and Kumar, P. (2004). Determinants of capital formation and agriculture growth: some new explorations. Economic and Political Weekly 39(52): 5611–6.

Chand, R., Kumar, P., and Kumar, S. (2011). Total factor productivity and contribution of research investment to agricultural growth in India. New Delhi: National Centre for Agricultural Economics and Policy Research.

Chand, R., and Parappurathu, S. (2012). Economic & Political Weekly Supplement xlvil(26, 27), June 30.

Chen, K., Flaherty, K., and Zhang, Y. (2012). China: recent developments in agricultural R&D. ASTI Country Note. Washington, DC: International Food Policy Research Institute.

Chowdury, Subhanil. (2011). Employment in India: What does the latest data show? Economic and Political Weekly 46(32): 23–26.

Collins, J. (1995). Farm size and non-traditional exports: Determinants of participation in world markets. World Development 23(7): 1103–1114.

CAG (Comptroller and Auditor General). (2004). Performance Appraisal of the Accelerated Irrigation Benefits Programme (AIBP). No. 15. New Delhi, India: Union Government.

CAG (Comptroller and Auditor General). (2008). Performance Audit of Implementation of National Rural Employment Guarantee Act (NREGA). New Delhi, India: Government of India.

CAG (Comptroller and Auditor General). (2010a). Performance Audit of the Accelerated Irrigation Benefits Programme (AIBP). No. 4. New Delhi, India: Union Government.

CAG (Comptroller and Auditor General). (2010b). Audit Report (Commercial) for the year ended 31 March 2009. New Delhi, India: Government of India.

Connell, D. (2006). Water politics in the Murray-Darling basin. Leichhardt, NSW, Australia: The Federation Press.

Coppard, D. (2001). The rural non-farm economy of India: A review of the literature. NRI Report No. 2662. London: Natural Resources Institute.

Corbett, S. (2009). A Harvest of Water. National Geographic, November 2009, p. 110.

Damodaran, H. (2007). Pusa-1121 proves a major hit with farmers. The Hindu Business Line, December 28.

Das Gupta, S., Reardon, T., Minten, B., and Singh, S. (2010a). The transforming potato value chain in India: Potato pathways from a commercialized-Agriculture zone (Agra) to Delhi, Chapter 2. ADB-IFPRI. Improved value chains to ensure food security in South and Southeast Asia.

Das Gupta, S., Reardon, T., Minten, B., and Singh, S. (2010b). The transforming rice value chain in India: The rice road from a commercialized-agriculture zone in Western Uttar Pradesh to Delhi. Chapter 3, ADB-IFPRI. Improved value chains to ensure food security in South and Southeast Asia.

Datt, G., Simler, K., Mukherjee, S., and Dava, G. (2000). Determinants of Poverty in Mozambique, 1996–97. FCND Discussion Paper No. 78. Washington, DC: International Food Policy Research Institute.

Datt, G., and Ravallion, M. (2009). Has India's economic growth become more pro-poor in the wake of economic reforms? Washington, DC: World Bank, Policy Research Working Paper 5103.

Datta, S.K. (2010). India, agriculture and ARD. First Global Conference on Agricultural Research for Development (GCARD). Montepellier, March 28–31. Accessed July 2012 www.fao.org/docs/eims/.../India,_Agriculture_and_ARD300310.pdf.

Davis, K. (2006). Farmer field schools: A boon or bust for extension in Africa? Journal of International Agricultural and Extension Education, 13(1): 91–97.

Dev S.M., and Rao, N.C.. (2004). Food Processing in Andhra Pradesh—Opportunities and Challenges', CESS Working Paper No. 57. Hyderabad: Centre for Economic and Social Studies (CESS).

de Silva, H. and Ratnadiwakara, D. (2008). Using ICT to reduce transaction costs in agriculture through better communication: A case-study from Sri Lanka, mimeo, 20.

Dehejia, and Panagariya, A. (2011). "Services Growth in India: Looking Inside the Black Box," Columbia-NCAER Conference on Trade, Poverty, Inequality and Democracy, March 31–April 1, New Delhi.

Deininger, K., and Nagarajan, H.. (2011). Land policy and Land Administration in an Environment of rapid Economic Growth. Background Paper to India 2039: Transforming Agriculture (2012). Washington DC: Centennial Group.

Dev, S.M., and Rao, N.C. (2005). "Food Processing and Contract Farming in Andhra Pradesh: A Small Farmer Perspective," 40(26), June 25: 2705–2713.

Dev, Mahendra. (2012). "Agriculture-nutrition linkages and policies in India." Indira Gandhi Institute of Development Research, Mumbai, India. Working paper WP-2012-006. Accessed on April 17, 2011: http://www.igidr.ac.in/pdf/publication/WP-2012-006.pdf

Deshingkar, P., Kulkarni, U., Rao, L., and Rao, S. (2003). Changing food systems in India: Resource sharing and marketing arrangements for vegetable production in Andhra Pradesh. Development Policy Review 21(5–6): 627–39.

Dharmadhikary, S., Sheshadri, S., and Rehmat. (2005). Unravelling Bhakra: Assessing the Temple of Resurgent India. Report of a Study of the Bhakra Nangal Project. Badwani, Madhya Pradesh, India: Manthan Adhyayan Kendra.

Directorate of Economics and Statistics. (2011). Agricultural Statistics At a Glance 2011. Accessed July 2012 http://eands.dacnet.nic.in/latest_2006.htm

Dolberg, F. (2003). Review of Household Poultry Production as a Tool in Poverty Reduction in Bangladesh and India. PPLPI Working Paper No. 6. Pro-Poor Livestock Policy Initiative. Rome: FAO.

Dubash, N.K. (2007). The Electricity-Groundwater Conundrum: Case for a Political Solution to a Political Problem. Economic and Political Weekly, December 29, 2007.

Dubash, N.K. (2008). Independent Regulatory Agencies: A Theoretical Review With Reference to Electricity and Water in India. Economic and Political Weekly, October 4, 2008.

Duflo, E., and Pande, R. (2007). Dams. The Quarterly Journal of Economics, MIT Press, vol. 122(2), pages 601–646, 05.

Dwivedi, B. S., Shukla, A. K., Singh, V. K., and Yadav, R. L. (2003). Improving nitrogen and phosphorus use efficiencies through inclusion of forage cowpea in rice-wheat systems in the Indo-Gangetic Plains of India. Field Crops Res., p. 399–418.

Erenstien, O. (2009). Leaving the plow behind. In: Millions Fed, David Spielman and Rajul Pandya-Lorch, eds., IFPRI, p. 65–70.

Eswaran, M., Kotwal, Ramaswami, B., and Wadhwa, W.. (2008). How does poverty decline: Suggestive evidence from India, 1983–1999. Bread Policy Paper No 14. Accessed on: http://ipl.econ.duke.edu/bread/papers/policy/p014.pdf

Eswaran, Mukesh, Ashok Kotwal, Bharat Ramaswami and Wilima Wadhwa. (2009). Sectoral labour flows and agricultural wages in India, 1983–2004: Has growth trickled down? Economic and Political Weekly 44(2): 46–55.

Euromonitor International. (2007). Consumer foodservice in India. London: Euromonitor International.

Evenson, R.E., Pray, C.E., and Rosengrant, M.W. (1999). Agricultural research and productivity growth in India. IFPRI paper No. 109.

Evenson, R.E., and Fuglie, K.O. (2010). Technology capital: the price of admission to the growth club, Journal of Productivity Analysis, 33, 173–190.

Fafchamps, M., Vargas-Hill, R.V., and Minten, B. (2008). Quality control in non-staple food markets: Evidence from India. Agricultural Economics 38(3): 251–266.

Fan, S., and Chang-Kang, C. (2005). Is small beautiful? Farm size, productivity, and poverty in Asian agriculture. Agricultural Economics 32(1): 135–46.

Fan. S., Gulati, A., and Thorat, S. (2007). Investment, Subsidies, and Pro-Poor Growth in Rural India, Discussion Paper 00716. Washington, DC: International Food Policy Research Institute.

Farina, E. (2002). Consolidation, Multinationalization, and Competition in Brazil: Impacts on Horticulture and Dairy Product Systems. Development Policy Review 20 (4), September, 441–457.

Farina, E.M.M.Q., Gutman, G.E., Lavarello, P.J., Nunes, R., and Reardon, T. (2005). "Private and public milk standards in Argentina and Brazil," Food Policy, 30 (3), June: 302–315.

Feder, G., and Slade, R. (1986). The impact of agricultural extension: the training and visit system in India. World Bank Research Observer 1(2), 139–161.

Feder, G., Anderson, J.R., Birner R., and Deininger K. (2010). Promises and realities of community-based agricultural extension. IFPRI Discussion Paper 00959, March 2010.

Feder G., Birner, R., Anderson J.R. (2011). The private sector's role in agricultural extension systems: potential and limitations. IFPRI.

Ferroni, M., and Das Gupta, P. (2011). Comments on "The future of Indian agriculture and rural poverty reduction." Email communication, November 18, 2011.

FICCI. (2010). Corporate interventions in Indian agriculture—towards a resilient farming community. New Delhi: Federation of Indian Chambres of Commerce and Industry (FICCI).

FnBNews. (2008). Refrigeration industry in the food sector.

FAO (Food and Agriculture Organization of the United Nations). Corporate Document Repository, Chapter 2, ICRISAT's achievements and impact. Accessed: www.fao.org/WA/RDOCS/TAC/YS272E/y5272e05.htm.

FAO (Food and Agriculture Organization of the United Nations). (2009). Accessed from FAOSTAT: http://faostat.fao.org/site/567/default.aspx#ancor

FAO (Food and Agriculture Organisation of the United Nations). (2010). Accessed from Aquastat Database : http://www.fao.org/nr/water/aquastat/main/index.stm

Foster, A.D., and Rosenzweig, M.R. (2003). Agricultural development, industrialization and rural inequality. Providence, RI: Brown University, Mimeo.

Fourati, K., (2009). Half Full or Half Empty? The Contribution of Information and Communication Technologies to Development. Global Governance, 15, 37–42.

Fuglie, K.O, Khatana, V.S., Ilangatilake, S.G., Singh, J., Kumar, D., and Scott, G.J. (1997). "Economics of Potato Storage in Northern India", Working Paper Series No: 1997–5. Social Science Department. Lima-Peru: CIP.

Fuglie, K.O., forthcoming. Total factor productivity in the global agricultural economy: Evidence from FAO data. In Julian Alston, Bruce Babcock and Philip Pardey, eds., The shifting patterns of agricultural production and productivity worldwide. Midwest Agribusiness Trade and Research Information Center (MATRIC): Iowa State University.

Gadgil, S., and Gadgil, S. (2006). The Indian Monsoon, GDP and Agriculture. Economic and Political Weekly, November 25, 2006, p. 4887.

Gadwal, V.R. (2003). The Indian seed industry: its history, current status and future. Current 84(3), February 2003.

Gahukar, R.T. (2007). Contract Farming for Organic Crop Production in India, Current Science, 93 (12), 1661–63.

Gandhi, V., and Namboodiri, N.V. (2002). Fruit and Vegetable Marketing and Its Efficiency in India: A Study in Wholesale Markets in Ahmedabad Area. Ahmedabad: Memio, Indian Institute of Management.

Gandhi, R., et al. (2009). Digital Green: Participatory Video and Mediated Instruction for Agricultural Extension, Information Technologies and International Development, 5 (1), 1–15.

Glendenning, C.J., Babu, S., and Asenso-Okyere, K. (2010). Review of agricultural extension in India—are farmers' information needs being met? IFPRI Discussion Paper 01048, December 2010.

GOI (Government of India). (2005). Income, Expenditure and Productive Assets of Farm Households. NSS Report No. 497 (59/33/5). National Sample Survey Organization. New Delhi: Ministry of Statistics and Program Implementation, Government of India.

Government of India. (2005). Report of the Task Group on Revamping and Refocusing of National Agricultural Research. Yojana Bhawan, New Delhi: Planning Commission (Agriculture Division), Government of India.

GOI (Government of India). (2006). From Hariyali to Neeranchal. Report of the Technical Committee on Watershed Programmes in India. New Delhi, India: Department of Land Resources, Ministry of Rural Development.

GOI (2006). Some Aspects of Operational Land Holdings in India, 2002–03. Report No. 492(59/18.1/3). National Sample Survey Organization. New Delhi: Ministry of Statistics and Program Implementation.

Government of India. (2008). 11th 5-Year Plan: Inclusive Growth. New Delhi: Planning Commission.

GOI (Government of India). 2008. Common Guidelines for Watershed Development Projects. New Delhi, India: Government of India.

GOI (2010). Climate Change in India: A 4X4 Assessment. The Indian Network for Climate Change Assessment. New Delhi: Ministry of Environment and Forests.

Government of India. (2010). Guidelines for modified 'support to state extension programmes for extension reforms' scheme. Department of Agriculture & Cooperation, Ministry of Agriculture, Government of India.

Government of India, Planning Commission. (2010). Approach to the 12th Plan. New Delhi: Government of India, Planning Commission.

Government of India. (2011). Faster, Sustainable and More Inclusive Growth: An Approach Paper to the 12th 5-Five Year Plan. New Delhi: Planning Commission.

GOI. (Various years). Land and Livestock Holdings in India. National Sample Survey Organization, Central Statistical Organization. New Delhi: Ministry of Statistics and Programme Implementation.

GOP (Government of Punjab). (2009). Net Irrigated Area in Punjab by Source. Punjab: Director of Land Records.

Gruere, G.P., and Sun, Y. (2012). Measuring the contribution of Bt cotton adoption to India's cotton yields leap. IFPRI paper No. 01170, April.

Gujja, B., Joy, K.J., and Paranjape, S. (2010). Babhli Water Conflict: Less Water, More Politics. Economic and Political Weekly, July 31, 2010, p. 12.

Gulati, A., Meinzen-Dick, R.S., and Raju, K.V. (2005). Institutional Reforms in Indian Irrigation. New Delhi, India: Sage (for IFPRI).

Gulati, A., Joshi, P.K., and Landes, M. (2008). Contract Farming in India: An Introduction. Accessed on: http://www.ncap.res.in/contract_%20farming/Index.htm, accessed April 2011.

Gulati, A. (2010). Accelerating agriculture growth—moving from farming to value chains. In Acharya, S. & Mohan R. (eds), India's Economy: Performance and Challenges. New Delhi: Oxford University Press.

Gulati, A., and Ganguly, K. (2010). The changing landscape of Indian agriculture. Agricultural Economics 41(777): 37–45.

Gulati, A., Ganguly, K., and Shreedhar, G. (2011). Food and nutritional security in India: A stock-taking exercise. Report submitted to ICAR. New Delhi: IFPRI.

Gupta, D. (2009). The Caged Phoenix—Can India Fly? New Delhi, India: Penguin Books.

Gupta, K., Roy, D., and Vivek, H. (2006). How do cooperatives select farmers and how do farmers benefit from producer organizations: The case of Milkfed Dairy Cooperative in Indian Punjab", Paper Presented at the Workshop Plate to Plough: Agricultural Diversification and Its Implications for the Smallholders, organized by the International Food Policy Research Institute (Asia Office), New Delhi, and the Institute of Economic Growth, New Delhi at New Delhi, September 20–21.

Gupta, P.R. (2007). Dairy India 2007. New Delhi.

Hazell, P., andHagbladde, S. (1993). Farm-nonfarm growth linkages and the welfare of the poor. In M. Lipton, and J. van der Gaag,eds., Including the poor pp. 190–204. Washington, DC: World Bank.

Hazell, P.B., Headey, D., Pratt, A.N., and Byerlee, D. (2011). Structural imbalances and farm and nonfarm employment prospects in rural South Asia. Washington, DC: World Bank.

Hengsdijk, H., and Langeveld, J.W.A. (2009). Yield trends and yield gap analysis of major crops in the world, Wageningen University, Work document 170.

Hepburn, C., and Ward, J. (2010). "Self interested low-carbon growth in G20 emerging markets." Paper prepared by Vivid Economics and presented at the Emerging Markets Forum 2010 Meeting at Airlie House.

Himanshu, P.L., Mukhopadhyay, A., and Murgai, R. (2011). Non-farm diversification and rural poverty decline: A perspective from Indian sample survey and village study. London, UK: Working Paper 44. London: Asia Research Centre, London School of Economics and Political Science.

Himanshu and Sen Abhijit. (2011). Economic and Political Weekly, vol. xlvi, no. 12. Himanshu, Peter Lanjouw, Abhiroop Mukhopadhyay, and Rinku Murgai. 2011. "Non-farm diversification and rural poverty decline: A perspective from Indian sample survey and village study. Working Paper 44. London: Asia Research Centre, London School of Economics and Political Science.

Hu, R., Yang, Z., Kelly P., and Huang J. (2009). Agricultural extension system reform and agent time allocation in China. China Economic Review, 20, 303–315.

Huang, J., Hu, R., Zhang, L., and Rozelle, S. (2000). China Agriculture Science & Research Investment Economics. pp. 169–203. Beijing: China Agriculture Publisher.

Huang, J., Dong, X., Wu, Y., Zhi, H., Nui, X., Huang, Z., and Rozelle, S. (2007). Regoverning Markets: The China Meso-level Study, Beijing: Center for Chinese Agricultural Policy, Chinese Academy of Sciences.

ICOLD (International Commission on Large Dams). (2003). World Register of Dams. Paris, France.

IIM (Indian Institute of Management). (2008). (1) Studying Gap between Irrigation Potential Created and Utilized in India. IIM Ahmedabad, November 2008. 361 p. (2) Study Related to Gap between Irrigation Potential Created and Utilized. IIM Bangalore, December 2008. 238 p. (3) IPC-IPU Gap Analysis in West Bengal and the North-East. IIM Calcutta, October 2008. 124 p. (4) Study on Issues Related to Gap between Irrigation Potential Created and Utilized. 260 p. IIM Lucknow.

India Brand Equity Foundation. (2006). Food processing. New Delhi: Confederation of Indian Industry.

Indiastat. (Various years). Datanet India Pvt. Ltd. Accessed on: www.indiastat.com.

International Monetary Fund. (2011). World Economic Outlook. Washington, DC: International Monetary Fund.

IWMI (International Water Management Institute). (2002). Innovations in Groundwater Recharge. Water Policy Briefing, Issue 1. IWMI-TATA Water Policy Program, January 2002.

IWMI (International Water Management Institute). (2009). Flexible Water Storage Options: for adaptation to climate change. 5p. (IWMI Water Policy Brief 31). Colombo, Sri Lanka: International Water Management Institute (IWMI).

IWMI (International Water Management Institute). (2010). Influencing irrigation policy in India. 2 p. (Success Stories Issue 6—2010). Colombo, Sri Lanka: International Water Management Institute.

Iyer, R. (2001). Water: Charting a Course for the Future. In two parts. Economic and Political Weekly, March 31 (p. 1115) and April 14, 2001 (p. 1235).

Iyer, R. (2002). The New National Water Policy. Economic and Political Weekly, May 4, 2002, p. 1701.

Iyer, R. (2008). Water: A Critique of Three Concepts. Economic and Political Weekly, January 5, 2008, p. 15.

Iyer, R. (2010). Approach to a new national water policy. The Hindu, October 29, 2010.

Jain, R. C. (2002). Tractor industry in India—present and future. ITMA.

James, C. (2010). Global status of commercialized Biotech/GM crops. ISAAA Brief No. 42. ISAAA, viii + 96.

Joseph, M., Soundararaja, M., Gupta, M., and Sahu, S. (2008). Impact of organized retail on the unorganized sector. ICRIER Working Paper 222. New Delhi: Indian Council for Research on International Economic Relations.

Joshi, P.K., Joshi, L., and Birthal, P.S.. (2006b). Diversification and its impact on smallholders: Evidence from a study on vegetables. Agricultural Economics Research Review, 19 (2): 219–236.

Joshi P.K., Birthal, P.S., and Minot, N. (2006b). Sources of Agricultural Growth in India: Role of Diversification towards High-value Crops, Markets, Trade and Institutions Discussion Paper No. 85. Washington, DC: International Food Policy Research Institute.

Kapoor, R. (2010). Financially sustainable models key to agricultural extension system. BusinessLine (the HINDU), Monday, Jun 21, 2010. Accessed on: http://www.thehindubusinessline.com/2010/06/21/stories/2010062150551300.htm

Kassam, A., Smith, M. (2001). FAO Methodologies on Crop Water Use and Crop Water Productivity. - Expert meeting on crop water productivity, Rome, Italy, 3–5 December 2001, p. 18.

Keller, A., Sakthivadivel, R., and Seckler, D. (2000). Water scarcity and the role of storage in development. VII, 20 p. (Research report 39). Colombo, Sri Lanka: International Water Management Institute (IWMI).

Kenny, J.F., Barber, N.L., Hutson, S.S., Linsey, K.S., Lovelace, J.K., and Maupin. (2009). Estimated use of water in the United States in 2005. Circular 1344, 52p. Reston, Virginia: U.S. Geological Survey.

Kerr, J., in collaboration with Pangare, G., and Pangare, V.L. (2002). Watershed development projects in India: an evaluation. Washington, DC: International Food Policy Research Institute.

Kishore, A. (2004). Understanding Agrarian Impasse in Bihar. Economic and Political Weekly, July 31, 2004, p. 3484.

Kohli, H.S., and Sood, A., eds. (2010). India 2039—An affluent society in one generation. Washington, DC: SAGE Publishing.

Kohli, H.S., Sharma, A., and Sood, A., eds. (2011). Asia 2050—Realizing the Asian Century. Washington, DC: SAGE Publishing.

Krishnan, S., Indu, R., Shah, T., Hittalamani, C., Patwari, B., Sharma, D., Chauhan, L., Kher, V., Raj, H., Mahida, U., Shankar, M., and Sharma, K. (2009). Is It Possible to Revive Dug Wells in Hard-Rock India through Recharge? Discussion from Studies in Ten Districts of the Country. Paper 12 in: Strategic Analyses of the National River Linking Project (NRLP) of India Series 5. Proceedings of the Second National Workshop on Strategic Issues in Indian Irrigation, New Delhi, India, 8–9 April 2009. p. 197–213. Colombo, Sri Lanka: International Water Management Institute.

Kristjanson, P., Krishna, A., Radney, M., and Nindo, W. (2004). Pathways out of Poverty in Western Kenya and the Role of Livestock. PPLPI Working Paper No. 14. Pro-Poor Livestock Policy Initiative. Rome: FAO.

Kulkarni, H., and Shankar, P.S.V. (2009). Groundwater: Towards an Aquifer Management Framework. Economic and Political Weekly, February 7, 2009, p. 13.

Kumar, P. (2006). Contract farming through agribusiness and state corporation. Economic and Political Weekly 41(52).

Kumar, P., Mittal, S., and Hossain, M. (2008). Agricultural growth accounting and total factor productivity in South Asia: A review and policy implications. Agricultural Economics Research Review 21: 145–172.

Kumar, P., Mruthyunjaya, and Birthal, P.S.. (2007). Changing consumption pattern in South Asia., In: Agricultural Diversification and Smallholders in South Asia (P.K. Joshi, Ashok Gulati, and Ralph Cummings Jr. , eds.). New Delhi: Academic Foundation.

Kumar, S. (2008). Raising Wheat Production by Addressing Supply-Side Constraints in India. National Centre for Agricultural Economics and Policy Research. New Delhi, India: ICAR.

Ladha, J.K., Pathak, H., Krupnik, T.J., Six, J., and van Kessel, C. (2005). Efficiency of fertilizer nitrogen in ceareal production: retrospects and prospects. Adv in Agron. 87, p. 85–156.

Ladha, J.K., Pathak, H., Tirol-Padre, A., Dawe, D., and Gupta, R.K. (2003). Productivity Trends in Intensive Rice-Wheat Cropping Systems in Asia, in Improving the Productivity and Sustainability of Rice-Wheat Systems: Issues and Impacts, ASA Special Publication 65.

Landes, M.R. (2008). The Environment for Agribusiness Investment in India. Economic Research Service, U.S. Department of Agriculture, Washington. DC. Available at: http://www.ers.usda.gov/Publications/EIB37/

Landes, M.R., and Burfisher, M.E. (2009). Growth and Equity Effects of Agricultural Marketing Efficiency Gains in India, USDA-ERS Economic Research Report No. 89.

LAO (Legislative Analyst's Office). (2010). Liquid Assets: Improving Management of the State's Groundwater Resources. Sacramento, California: Government of California.

Lei, X. L., Ma, S., Wen, X., Su, J. and Du, F. (2008). Integrated analysis of stress and regional seismicity by surface loading—A case study of Zipingpu Reservoir (in Chinese with English abstract), Geology and Seismology, Vol. 30, No. 4, 1046-1064, Dec. 2008.

Lekshmi, P.S.S., Chandrarandan, K., and Balasubramani, N. (2006). Yield gap analysis among rice growers in North Eastern Zone of Tamil Nadu, Agricultural Situation in India. p. 729–733.

Lele, U. (1971). Grain Marketing in India: Private Performance and Public Policy. New York: Cornell University Press.

Leye, V. (2009). Information and Communication Technologies for Development: A Critical Perspective, Global Governance, 15, 29–35.

Liao, Y., Gao, Z., Bao, Z., Huang, Q., Feng, G., Xu, D., Cai, J., Han, H., Wu, W. (2008). China's water pricing reforms for irrigation: effectiveness and impact. Comprehensive Assessment of Water Management in Agriculture Discussion Paper 6. Colombo, Sri Lanka: International Water Management Institute.

MANAGE (2003). Contract Farming Ventures in India: A Few Successful Cases. Spice, Vol. 1 No. 4, March 2003. Hyderabad, India

Manor, J. (1997). The Political Economy of Decentralization. Washington, DC: World Bank. Ministry of Agriculture, Government of India. 2010. "Agricultural statistics at a glance, 2010." Accessed December 17, 2011: http://eands.dacnet.nic.in/latest_2006.htm

Mattoo, A., Mishra, D., and Narain, A. (2007). From competition at home to competing abroad. Washington, DC: World Bank.

Mellor, J.W. (2003). Agricultural growth and poverty reduction: The rapidly increasing role of smallholder livestock. Keynote address in the International Workshop 'Livestock and livelihoods: Challenges and opportunities for Asia in the emerging market environment. November 10–12. Anand, India: National Dairy Development Board.

Ministry of Agriculture. (2009). Net Area Irrigated From Different Sources And Gross Irrigated Area. New Delhi, India: Directorate of Economics and Statistics, Department of Agriculture and Cooperation, Government of India.

Ministry of Agriculture. (2010). National Seminar on Agriculture Extension Proceedings, New Delhi, February 27–28, 2009.

Ministry of Food Processing. (2008). Annual report 2007–08. New Delhi: Ministry of Food Processing.

Ministry of Statistics and Programme Implementation. (2010). Gross State Domestic Product at Factor Cost by Industry of Origin at 1999–2000 Prices. As on 12-04-2010. New Delhi, India: Government of India.

Ministry of Water Resources. (2001). 3rd Census of Minor Irrigation Schemes (2000–01). Dugwell Scheme. New Delhi, India: Government of India.

Ministry of Water Resources. (2002). National Water Policy. New Delhi, India: Government of India.

Ministry of Water Resources. (2006). Report of the Working Group On Water Resources For the XI Five Year Plan (2007–2012). New Delhi, India: Government of India.

Minot, N., Epprecht, M., Tram Anh, T.T., and Trung, L.Q.. (2006). Income diversification and poverty in Northern Uplands of Vietnam. Research Report 145. Washington, DC: International Food Policy Research Institute.

Minten, B., Reardon, T., Singh, K.M., and Sutradhar, R. (2010). The Potato Value Chain and Benefits of Cold Storages: Evidence from Bihar (India). Report of IFPRI Project for IFAD and the National Agricultural Innovation Project (NAIP) of India.

Minten, B., Reardon, T., and Sutradhar, R. (2010). Food prices and modern retail: The case of Delhi, World Development, 38(12): 1775–1787

Minten, B., Ghorpade, Y., Vandeplas, A., and Gulati, A. (2009). High-value crops and marketing: Strategic options for development in Uttarakhand. New Delhi: Asian Development Bank, International Food Policy Research Institute, and Academic Foundation.

Misra, S. (2005). Delhi Water Supply & Sewerage Services: Coping Costs, Willingness to Pay and Affordability. Mimeo, South Asia Environment and Infrastructure Unit, World Bank.

Mittal, S. (2007). Can Horticulture be a Success Story for India? Indian Council for Research on International Economic Relations, Working Paper No. 197, August 2007.

Mittal, S., Gandhi, S., and Tripathi, G. (2010). Socio-Economic Impact of Mobile Phones on Indian Agriculture. Indian Council for Research on International Economic Relations, Working Paper No. 246.

Mollinga, P.P., Doraiswamy, R., and Engbersen, K. (2004). Capture and Transformation: Participatory Irrigation Management in Andhra Pradesh, India. In: Mollinga, P.P., Bolding, A. 2004. The Politics of Irrigation Reform. Hants, England: Ashgate.

Morisset, M., and Kumar, P. (2008). Structure and performance of the food processing industry in India. Mimeo.

Mosse, D. (2003). The Rule of Water. Statecraft, ecology and collective action in South India. New Delhi, India: Oxford University Press.

Mukherji, A., Facon, T., Burke, J., de Fraiture, C., Faurès, J.-M., Füleki, B., Giordano, M., Molden, D., and Shah, T. (2009). Revitalizing Asia's irrigation: to sustainably meet tomorrow's food needs. Colombo, Sri Lanka: International Water Management Institute. Rome, Italy: Food and Agriculture Organization of the United Nations.

Mukherji, A., Fuleki, B., Shah, T., Suhardiman, D., Giordano, M., and Weligamage, P. (2010). Irrigation reform in Asia: A review of 108 cases of irrigation management transfer. Report submitted to Asian Development Bank, October 2009. 118 p.

Murty, M.V., Singh, R.P., Wani, S.P., Khairwal, I.S., and Srinivas, K. (2007). Yield Gap Analysis of Sorghum And Pearl Millet In India Using Simulation Modeling. Global Theme on Agroecosystems Report No. 37. Patancheru 502 324. Andhra Pradesh, India: International Crops Research Institute for the Semi-Arid Tropics.

Nagaraj, R. (2007). Employment situation in India. GDP and Workforce Distribution. Available at: http://www.un.org/esa/socdev/social/meetings/egm5/papers/Nagaraj_Rayaprolu.ppt#281,28.

Namara, R.E., Upadhyay, B., Nagar, R.K. (2005). Adoption and impacts of microirrigation technologies: Empirical results from selected localities of Maharashtra and Gujarat states of India. Research Report 93. Colombo, Sri Lanka: International Water Management Institute.

Nataraj, P. (2010). City has used up 200 pc of groundwater resources. Bangalore, India: Deccan Herald.

Natawidjaja, R., Reardon, T., and Shetty, S., with Noor, T.I., Perdana, T., Rasmikayati, E., Bachri, S., and Hernandez, R. (2007). Horticultural producers and supermarket development in Indonesia. UNPAD/MSU Report No. 38543. Jakarta: World Bank/Indonesia.

NCAER. (2005). The Great Indian Market: Results from the NCAER's Marketing Information Survey of Households. August 9. Accessed on: www.ncaer.org/downloads/PPT/TheGreatIndianMarket.pdf

NAAS (National Academy of Agricultural Sciences). (2009). Agricultural research preparedness. In: State of India Agriculture, p. 211–238.

NCIWRD (National Commission for Integrated Water Resources Development). (1999). Integrated water resources development. A plan for action. Report of the Commission for Integrated Water Resource Development, Volume I. New Delhi, India: Government of India, Ministry of Water Resources.

NFHS (National Family Health Survey). (2005–06). Government of India.

Neuchâtel Group. (2000). Guide for monitoring, evaluation and joint analyses of pluralistic extension support. Accessed on: http://www.neuchatelinitiative.net/images/guide_for_monitoring.pdf.

Neumann, K., Verburg, P.H, Stehfest, E., and Mueller, C. (2010). The yield gap of global grain production: A spatial analysis, Agricultural Systems, 103, 316–326.

North, D.C. (1990). Institutions, Institutional Change and Economic Performance. Cambridge, UK: Cambridge University Press.

North, D.C. (2005). Understanding the Process of Economic Change. Princeton, NJ: Princeton University Press.

NSSO (National Sample Survey Organisation). (2005a). Access to Modern Technology for Farming, 2003. Situation Assessment Survey of Farmers. NSS 59th Round (January-December 2003). Report No. 499. New Delhi, India: National Sample Survey Organisation, Ministry of Statistics and Programme Implementation, Government of India.

NSSO (National Sample Survey Organisation). (2005b). Some Aspects of Farming. Situation Assessment Survey of Farmers. NSS 59th Round (January–December 2003). Report No. 496. New Delhi, India: National Sample Survey Organisation, Ministry of Statistics and Programme Implementation, Government of India.

NSSO (National Sample Survey Organisation). (2006a). Some Aspects of Operational Land Holdings in India, 2002-03. NSS 59th Round (January–December 2003). Report No. 492. New Delhi, India: National Sample Survey Organisation, Ministry of Statistics and Programme Implementation, Government of India.

NSSO (National Sample Survey Organisation). (2006b). Level and Pattern of Consumer Expenditure, 2004–05. NSS 61st Round (July 2004–June 2005). Report No. 508. New Delhi, India: National Sample Survey Organisation, Ministry of Statistics and Programme Implementation, Government of India.

Ojha, R.K. (2007). Poverty dynamics in rural Uttar Pradesh. Economic and Political Weekly, April 21.

Oxus. (2011). Indian inflation: Populism, politics, and procurement prices. Developing Trends 1(2). New Delhi: Oxus Research Report.

Pal, S., Rahija, M.A., and Beintema, N.M. (2012). India: recent developments in agricultural research. ASTI Country Note. Washington, DC, and New Delhi: International Food Policy Research Institute and Indian Council of Agricultural Research.

Palaskas, T.B., and Harriss-White, B. (1996). The identification of market exogeneity and market dominance by tests instead of assumption: An application to Indian material. Journal of International Development 8(1): 11–23.

Pant, N. (2004). Trends in Groundwater Irrigation in Eastern and Western UP. Economic and Political Weekly, July 31, 2004, p. 3463.

Pant, N. (2008). Some Issues in Participatory Irrigation Management. Economic and Political Weekly, January 5, 2008, p. 30.

Parasuraman, S., Upadhyaya, H., and Balasubramanian, G. (2010). Sardar Sarovar Project: The War of Attrition. Economic and Political Weekly, January 30, 2010, p. 39.

Parikh, K., Binswanger-Mkhize, H.P., Ghosh, P., and d'Souza, A. (2011). Double Digit Inclusive Growth: Not Without Robust Agricultural Growth. Washington DC: India 2039: Transforming Agriculture.

Parthasarathy, R. (2006). Objects and Accomplishments of Participatory Irrigation Management Programme in India—An Open Pair of Scissors. In: Parthasarathy, R., Iyengar, S., eds. New Development Paradigms & Challenges For Western & Central India. Ahmedabad, India: Gujarat Institute of Development Research (Concept Publishing Company).

Patil, S.A., and Dadlani, M. (2010). Successful research—farmers-agribusiness models: IARI experience. Accessed: www.ncap.res.in/AKI%20Workshop/SESS-2/Dadlani.

Pearce, F. (2007). When the Rivers Run Dry: Water, the Defining Crisis of the Twenty-First Century. Beacon Press.

Peng, S., Khush, G.S., Virk, P., Tang, Q, and Zou, Y. (2008). Progress in ideotype breeding to increase rice yield potential. Field Crops Research 108(1): 32–38.

Phansalkar, S., and Verma, S. (2004). Improved water control as strategy for enhancing tribal livelihoods. Economic and Political Weekly, Review of Agriculture, Vol. XXXIX, No. 31, July 31, 2004, p. 3469–3476.

Planning Commission. (2001). Tenth Five Year Plan 2002–2007. New Delhi, India: Government of India.

Planning Commission. (2007). Ground Water Management and Ownership—Report of the Expert Group. New Delhi, India: Government of India.

Planning Commission. (2008). India: Eleventh Five Year Plan 2007–12. Government of India. Oxford University Press (3 volumes).

Planning Commission, Government of India. (2009). Report of the expert group on methodology for estimation of poverty. New Delhi: Tendulkar Report.

Planning Commission. (2010). Mid Term Appraisal of the Eleventh Five Year Plan. New Delhi, India: Government of India.

Planning Commission. (2011). Faster, sustainable and more inclusive growth. Accessed: www.planningcommission.nic.in.

Pray, C.E., and Basant, R. (2000). India—issues in private agricultural research. USDA.

Pray, C.E., and Fuglie, K. (2001). Electronic report from the economic research service private investment in agricultural research and international technology transfer in Asia (with contributions from Josephy G. Nagy, Mumtaz Ahmad, and Rakesh Basant). Economic Research Service, USDA, Report 805, November.

Pray, C.E., and Nagarajan L. (2009). Pearl Millet and Sorghum improvement in India. IFPRI paper No. 00919.

Pray, C., and S. Nagarajan. (2011). Farm Input Industry Transformation in India. Rutgers University, Report to the Gates Foundation.

Pray, C.E., and Nagarajan, L. (2012). Innovation and research by private agribusiness in India. IFPRI paper No. 01181.

Raabe, K. (2008). Reforming the agricultural extension system in India - what do we know about what works where and why? IFPRI Discussion Paper 00775. Washington, DC: Department Strategy and Governance Division, IFPRI.

Rajalahti, R., Janssen, W., and Pehu, E. (2008). Agricultural Innovation Systems: From Diagnostics toward Operational Practices, Agriculture and Rural Development Discussion Paper 38. Washington, DC: The World Bank.

Rajendran, M. (2011). Seeding competition: MNCs are gearing up to grab a bigger share of India's seed market. Business World, January 15.

Raju, K.V., and Gulati, A. (2008). Pricing, Subsidies and Institutional Reforms in Indian Irrigation: Some Emerging Trends. In: Ballabh, V., ed., Governance of Water. New Delhi, India: Sage Publications.

Rakesh, O.S. (1986). Developing an efficient seed programme in India. Proc. First Nat'l Seed Seminar, SAI, New Delhi, December 77–28, p. 13–18.

Ramasamy, C. (2011). "Agricultural R&D in India: Did it go astray?" Paper for Syngenta Foundation India, February 2011. New Delhi: Syngenta Foundation.

Ramaswami, B., and Balakrishnan, P.. (2002). Food prices and the efficiency of public intervention: The case of Public Distribution System in India. Food Policy 27(5–6): 419–436.

Ramaswami, B., Birthal, P.S., and Joshi, P.K. (2006). Efficiency and Distribution in Contract Farming: The Case of Indian Poultry Growers. MTID Discussion paper 91. Washington, DC: International Food Policy Research Institute.

Ranade, R. (2005). Out of Sight, Out of Mind: Absence of Groundwater in Water Allocation of Narmada Basin. Economic and Political Weekly, May 21, 2005, p. 2172.

Ranade, R., and Kumar, M.D. (2004). Narmada Water for Groundwater Recharge in North Gujarat. Conjunctive Management in Large Irrigation Projects. Economic and Political Weekly, July 31, 2004, p. 3510.

Rangachari, R. (2005). Unravelling The Unravelling of Bhakra: A Critique of Shripad Dharmadhikary's "Unravelling Bhakra." New Delhi: Indian Water Resources Society.

Rangi, P. S., and Sidhu, M.S.. (2000). A study on contract farming of tomato in Punjab. Agricultural Marketing 42(4): 15–23.

Rao, C., Srinivasan, J., Das Gupta, S., Reardon, T., Minten B., and Punjabi Mehta, M. (2011). Agri-Services in Andhra Pradesh for Inclusive Rural Growth: Baseline Survey Findings & Policy Implications. Report of IFPRI-PIKA Project on Rural Service Hubs: Business Catalysts for Rural Competitiveness and Inclusiveness. New Delhi: USAID.

Rao, R.J. (2004). Journal of Indian Water Works Association, Jan–Mar., 2004.

Rao, V., and Chotigeat, T. (1981). The inverse relationship between size of land holdings and agricultural productivity. American Journal of Agricultural Economics 63(3): 571–574.

Rashid, S., Cummings, R., and Gulati, A. (2007). "Grain Marketing Parastatals in Asia: Results from Six Case Studies," World Development 35(11): 1872–1888.

Ratnam, B.V., Reddy, P. K., and Reddy, G.S. (2006). eSagu: An IT based personalized agricultural extension system prototype—analysis of 51 Farmers' case studies, International Journal of Education and Development using ICT, 2 (1), 79–94.

Ravallion, M., and Datt, G. (1996). How important to India's poor is the sectoral composition of economic growth? World Bank Economic Review 10 (1): 1–25.

Reardon, T., and Minten, B. (2011a). "Surprised by Supermarkets: Diffusion of Modern Food Retail in India," Journal of Agribusiness in Developing and Emerging Economies 1(2).

Reardon, T., Barrett, C., Berdegué, J., and Swinnen, J. (2009). Agrifood industry transformation and small farmers in developing countries. World Development 37(11): 1717–27.

Reardon, T., and Timmer, C.P.. (2007). Transformation of markets for agricultural output in developing countries since 1950: How has thinking changed? In Handbook of Agricultural Economics, Vol. 3: Agricultural Development: Farmers, Farm Production and Farm Markets, ed. R.E. Evenson and P. Pingali, 2808–2855. Amsterdam: Elsevier Press.

Reardon, T., Timmer, C.P., and Minten, B. (2010). "The Supermarket Revolution in Asia and Emerging Development Strategies to Include Small Farmers," PNAS: Proceedings of the National Academy of Science. December 6 (officially posted online by PNAS before print edition). doi:10.1073/pnas.1003160108

Reardon, T., Minten, B., Punjabi Mehta, M., Das Gupta, S., Rajendran, S., and Singh, S. (2011). Agri-Services in Uttar Pradesh for Inclusive Rural Growth: Baseline Survey Findings & Policy Implications. Report of IFPRI-PIKA Project on Rural Service Hubs: Business Catalysts for Rural Competitiveness and Inclusiveness. New Delhi: USAID.

Reilly, T.E., Dennehy, K.F., Alley, W.M., and Cunningham, W.L. (2008). Ground-Water Availability in the United States. Reston, Virginia: U.S. Geological Survey Circular 1323, 70 p. Also accessible at: http://pubs.usgs.gov/circ/1323/.

Reserve Bank of India. (2009). Handbook of Statistics on the Indian Economy 2008–09. September 15, 2009. Mumbai, India.

Reserve Bank of India. (2011). Annual Report, 2010–2011. Mumbai, India: Reserve Bank of India.

Rivera, W.M. (1996). Agricultural extension in transition worldwide: Structural, financial and managerial reform strategies. Public Administration and Development, 16, 151–161.

Roy, S. (2008). Bihar's curse of embankments. Business Standard. September 3, 2008.

Roy, D., and Thorat, A. (2007). Success in high value horticultural export markets for the small farmers: The case of Mahagrapes in India. World Development 36(10): 1874–1890.

Sainath, P. (1996). Everybody Loves a Good Drought. Calcutta, India: Penguin.

Shah, M. (2008). Rainfed Authority and Watershed Reforms. Economic and Political Weekly, March 22, 2008.

Shah, T. (2007a). Co-management of Electricity and Groundwater: Gujarat's Jyotirgram Yojana. Strategic Analyses of India's NRLP, Regional Workshop, Hyderabad, August 29, 2007.

Shah, T. (2007b). Crop Per Drop of Diesel? Energy Squeeze on India's Smallholder Irrigation. Economic and Political Weekly, September 29, 2007, p. 4002.

Shah, T. (2008a). The New Institutional Economics of India's Water Policy. In: Governance of Water, edited by Vishwa Ballabh. New Delhi, India: Sage Publications.

Shah, T. (2008b). India's Master Plan for Groundwater recharge: An Assessment and Some Suggestions for Revision. Economic and Political Weekly, December 20, 2008, p. 41.

Shah, T. (2008c). Groundwater Management and Ownership: Rejoinder. Economic and Political Weekly, April 26, 2008, p. 116.

Shah, T. (2009). Taming the Anarchy—Groundwater Governance in South Asia. Washington, DC: Resources for the Future.

Shah, T. (2010a). Past, Present and the Future of Canal Irrigation in India. Paper commissioned by the Planning Commission for the Mid Term Appraisal of the Eleventh Plan.

Shah, T. (2010b). Change the Flow of Water Policy. The Economic Times, October 19, 2010.

Shah, T., Giordano, M., and Wang, J. (2004). Irrigation Institutions in a Dynamic Economy—What is China Doing Differently from India? Economic and Political Weekly, July 31, 2004, p. 3452.

Shah, T., Gulati, A., Hemant, P., Shreedhar, G., and Jain, R.C. (2009). Secret of Gujarat's Agrarian Miracle after 2000. Economic and Political Weekly, Vol, XLIV No 52, December 26, 2009, p. 45.

Shah, T., Krishnan, S., Hemant, P., Verma, S., Chandra, A., and Sudhir, C. (2010). A case for pipelining water distribution in the Narmada Irrigation System in Gujarat, India. 25 p. IWMI Working Paper 141.Colombo, Sri Lanka: International Water Management Institute.

Shah, T., and van Koppen, B. (2006). Is India Ripe for Integrated Water Resources Management? Fitting Water Policy to National Development Context. Economic and Political Weekly, August 5, 2006, p. 3413.

Shah, T., and Verma, S. (2008). Co-Management of Electricity and Groundwater: An Assessment of Gujarat's Jyotirgram Scheme. Economic & Political Weekly, February 16, 2008.

Shah, T., Ul Hassan, M., Khattak, M.Z., Banerjee, P.S., Singh, O.P., and Ur Rehman, S. (2009). Is Irrigation Water Free? A Reality Check in the Indo-Gangetic Basin. World Development Vol. 37, Issue 2, February 2009, pp. 422–434. Elsevier.

Sharma, A., Varma, S., and Joshi, D. (2008). Social Equity Impacts of Increased Water for Irrigation. Paper 9 in: Amarasinghe, U.A., Sharma, B.R., eds. 2008. Strategic analyses of the National River Linking Project (NRLP) of India, series 2. Proceedings of the Workshop on Analyses of Hydrological, Social and Ecological Issues of the NRLP. Colombo, Sri Lanka: International Water Management Institute.

Sharma, V.P. (2007). India's agrarian crisis and smallholder producers' participation in new farm supply chain initiatives: A case study of contract farming. Regoverning Markets Innovative Practice series. London: IIED.

Sharma, V.P., and Singh, R.V. (2007). Regoverning markets: The India meso-level study. Ahmedabad, India: Indian Institute of Management.

Sharma, V.P. (2011). India's agricultural development under the new economic regime: policy perspective and strategy for the 12th Five Year Plan. W. P. No. 2011-11-01. Indian Institute of Management: Ahmedabad, India.

Shergill, H.S. (2007). Sustainability of Wheat-Rice Production in Punjab: A Re-examination. Economic and Political Weekly, December 29, 2007, p. 81.

Shukla, A.K., Ladha, J.K., Singh, V.K., Dwivedi, B.S., Balasubramanian, V., Gupta, R.K., Sharma, S.K., Singh, Y., Pathak, H., Pandey, P.S., Padre, A.T., and Yadav, R.L. (2004). Calibrating the leaf colour chart for nitrogen management in different genotypes of rice and wheat in a systems perspective. Agronomy Journal 1621(6): 1606–1621.

Singh. (2002). Contracting out solutions: Political economy of contract farming in the Indian Punjab. World Development 30: 1621–1638.

Singh, A. and S. Pal. (2010). The changing pattern and sources of agricultural growth in India. The Shifting Patterns of Agricultural Production and Productivity Worldwide (Eds. J.M. Alston, B.A. Babcock and P.G. Pardey). Ames, Iowa: The Midwest Agribusiness and Trade Research Center, Iowa State University.

Singh, A.K. (1999). Agriculture extension: impact and assessment. Jodhpur, India: ARGOBIOS.

Singh, J.P., Swanson, B.E., and Singh, K.M. (2006). Developing a decentralized, market-driven extension system in India: The ATMA model, pp. 203–223. In A. W. van den Ban and R.K. Samanta (eds), Changing Roles of Agricultural Extension in Asian Nations. Delhi, India: B.R. Publishing.

Singh, K.K. (2000). Exploration without commitment: the story of Participatory Irrigation Management in India, L.K. Joshi and Rakesh Hooja, eds. Participatory Irrigation Management: paradigm for the 21st century. Jaipur and New Delhi: Rawat Publications.

Singh, K.M., and Swanson, B.E. (2006). Developing a market-driven extension system in India. Annual Conference Proceedings of the Association for International Agricultural and Extension Education, 22, 627–637.

Singh, K.M., Meena, M.S., and Jha, A.K. (2009). Impact assessment of agricultural extension reforms in Bihar, Indian Research Journal of Extension Education, 9 (2), 110–114.

Singh, O.P. (2007). Acceleration of transfer of technology through public-private partnership—a case of Hoshangabad district, Madhya Pradesh, India, National Symposium on plant protection—technology interface.

Singh, S. (2005). Contract Farming for Agricultural Development: Review Theory and Practice with Special Reference to India. CENTAD working paper no. 2, An Oxfam GB Initiative, New Delhi.

Singh, S. (2007). Agribusiness in South Asia: A fact sheet. Make Trade Fair campaign research paper. New Delhi.

Singh, S. (2008). "Supply Chain Management for Fruits and Vegetables in Uttar Pradesh", Study submitted to Uttar Pradesh Council of Agricultural Research, Lucknow.

Singh, V.K., Dwivedi, B.S., Shukla, A.K., and Mishra, R.P. (2010). Permanent raised bed planting of the pigeon pea-wheat system on a typic Ustochrept: effects on soil fertility, yield and water and nutrient use efficiencies. Field Crops Research 116, p. 127–139.

Sivaramakrishnan, K.C., Kundu, A., and Singh, B.N. (2005). Handbook of urbanization in India. New Delhi: Oxford University Press.

Sivaramakrishnan, K.C. (2010). Judicial Setback for Panchayats and Local Bodies. Economic and Political Weekly, August 7, 2010, p. 43.

Smakhtin, V., and Anputhas, M. (2006). An assessment of environmental flow requirements of Indian river basins. IWMI Research Report 107. Colombo, Sri Lanka: International Water Management Institute.

Soman, P. (2009). Improving Water Use Efficiency in enhancing crop productivity. FAI National Seminar, New Delhi.

Sud, S. (2007). Great new aroma. Business Standard, New Delhi, December 18.

Sulaiman V.R. (2003). Innovations in agricultural extension in India. Sustainable Development Department, FAO. Accessed on: http://www.fao.org/sd/2003/KN0603a_en.htm.

Sulaiman, R., Hall, A., and Suresh, N. (2005). Effectiveness of private sector agricultural extension in India: lessons for the new extension policy agenda. ODI Agricultural Research and Extension Network (AgREN) Paper, No. 117.

Sunanda, S. (2005). Farm-firm linkages through contract farming in India, Seminar on Sustainable Contract Farming for Increased Competitiveness, Sri Lanka. Accessed on: http://www.fao.org/uploads/media/Contract_farming_in_India_2.pdf.

Swaminathan, M.S. (1986). Need for an efficient seed programme. Inaugural address, Proc. First Nat'l Seed Seminar, SAI, New Delhi, December 27–28, p. 19–29.

Swaminathan, M.S. (2006). Agriculture cannot wait. The Hindu, May 24.

Swaminathan, M.S. (2011). From Green to Evergreen Revolution. Academic Foundation.

Swanson, B.E., Nie, C., and Feng, Y. (2003). Trends and developments with the Chinese agro-technical extension system. Journal of the Agricultural and Extension Education Association, 10(2), 17–24 (www.aiaee.org/2005/Accepted/066.pdf).

Swanson, B.E. (2009). Changing Extension Paradigms within a Rapidly Changing Global Economy, in Proceeding of the 19th European Seminar on Extension Education. Assisi, Italy. (http://www.agraria.unipg.it/ESEE2009PERUGIA/files/Proceedings.pdf).

Swanson, B.E. and Rajalahti, R. (2010). Strengthening agricultural extension and advisory systems: procedures for assessing, transforming, and evaluating extension systems. Agricultural and Rural Development Discussion Paper 45. Washington, DC: World Bank.

The Economic Times. (2009). Logistics revenues set to cross Rs5 lakh crore in 5 years time.

The Expert Committee on Agricultural Marketing (2001). Marketing Infrastructure and Agricultural Marketing Reforms: Expert Committee Report. Report of the Expert committee to the Ministry of Agriculture. Accessed on: http://agmarknet.nic.in/amrscheme/expcommrepmark.htm.

Thenkabail, P.S., Biradar, C.M., Turral, H., Noojipady, P., Li, Y.J., Vithanage, J., Dheeravath, V., Velpuri, M., Schull M., Cai, X. L., and Dutta, R. (2006). An Irrigated Area Map of the World (1999) derived from Remote Sensing, pp. 74.. Research Report #105. Colombo: International Water Management Institute.

Thenkabail, P.S., Dheeravath, V., Biradar, C.M., Gangalakunta, O.R.P., Noojipady, P., Gurappa, C., Velpuri, M., Gumma, M., and Li, Y. (2009). Irrigated Area Maps and Statistics of India Using Remote Sensing and National Statistics. Remote Sensing 2009, 1, 50–67.

Thomas, S. (2003). Agricultural commodity markets in India: Policy issues for growth. Goregaon, India: Mimeo, Indira Gandhi Institute for Development Research.

Thomas, M.T., and Ballabh, V. (2008). Recovery of Irrigation User Cess and Governance of Canal Systems. Governance of Water, edited by Vishwa Ballabh. New Delhi: Sage Publications.

Timmer, P.C., and Akkus, S. (2008). The structural transformation as a pathway out of poverty: Analytics, empirics, and politics. Working paper 15 (with accompanying technical annexes). Washington DC: Center for Global Development.

Timmer, P.C. (2009). A world without agriculture: The structural transformation in historical perspective. Washington, DC: American Enterprise Institute.

TISS (Tata Institute of Social Sciences). (2008). Performance and Development Effectiveness of the Sardar Sarovar Project.

Tiwary, R. (2010). Social Organisation of Shared Well Irrigation in Punjab. Economic and Political Weekly, June 26, 2010, p. 208.

TRAI. (2011). Information Note to the Press (Press Release No. 13/2011). New Delhi: Telecom Regulatory Authority of India. Accessed on: www.trai.gov.in

Tripathi, R.S. Singh, R., and Singh, S. (2005). Contract farming in potato production: An alternative for managing risk and uncertainty. Agricultural Economics Research Review 18: 47–60.

Tripathi, A., and Prasad, A.R. (2009). Agricultural development in India since Independence: A study on progress, performance and determinants. Knowledge Creation Diffusion Utilization 1(1): 1–31.

Umali-Deininger, D., and Deininger, K. (2001). Towards greater food security for India's poor: Balancing government intervention and private competition. Agricultural Economics 25(2–3): 321–335.

Umali-Deininger, D., and Sur, M. (2007). Food safety in a globalizing world: Opportunities and challenges for India. In Proceedings of the 26th Conference of the International Association of Agricultural Economists, August 12–18, 2006, Queensland, Australia.

USDA (United States Department of Agriculture). (2009a). United States, Summary and State Data. 2007 Census of Agriculture. Washington, DC: National Agricultural Statistics Service.

USDA (United States Department of Agriculture). (2009b). California, State and County Data. 2007 Census of Agriculture. Washington, DC: National Agricultural Statistics Service.

USDA (United States Department of Agriculture). (2010). Farm and Ranch Irrigation Survey. 2007 Census of Agriculture. Washington, DC: National Agricultural Statistics Service.

USGS (United States Geological Survey). (2003). Ground-Water Depletion Across the Nation. Reston, Virginia: U.S. Geological Survey, U.S. Department of the Interior.

Varela, M. (2006). Groundwater Management in Spain: The Way Ahead. European Groundwater Conference.

Varshney, R.K., Chen, W., Li, Y., Bharti, A.K., Saxena, R.K., Schlueter, J.A., Donoghue, M.T.A., Azam, S., Fan, G., Whaley, A.M., Farmer, A.D., Sheridan, J., Iwata, A., Tuteja, R., Penmetsa, R.V., Wu, W., Upadhyaya, H.D., Yang, S.P., Shah, T., Saxena, K.B., Michael, T., McCombie, W.R., Yang, B., Zhang, G., and Yang, H. (2011). Draft genome sequence of pigeonpea (Cajanus cajan), and orphan legume crop of resource-poor farmers. Nature Biotechnology 30(1): 83–92.

Vaidyanathan, A. (2010). Agricultural growth in India—Role of technology, incentives, and institutions. Oxford University Press: New Delhi, p. 170–181.

VcCircle. (2007). Mahindra To Bring In Partners For Logistics. Accessed October 2: www.vccircle. com

Venot, J.P., Turral, H., Samad, M., and Molle, F. (2007). Shifting waterscapes: Explaining basin closure in the Lower Krishna Basin, South India. IWMI Research Report 121. Colombo, Sri Lanka: International Water Management Institute.

Virmani, A. (2004). India's Economic Growth: From Socialist Rate of Growth to Bharatiya Rate of Growth, ICRIER, working paper No. 122.

von Braun, J., Fan, S., Meinzen-Dick, R., Rosegrant, M.W., and Pratt, A.N. (2008). International Agricultural Research for Food Security, Poverty Reduction, and the environment—What to Expect from Scaling Up CGIAR Investments and "Best Bet" Programs. Washington, DC: International Food Policy Research Institute.

WALMI/DSC. (2010). Proceedings of the Multi-State Workshop of Water User's Association for Experience sharing of Best Practices in Participatory Irrigation Management. India: Anand, Gujarat, Water and Land Management Institute/Development Support Centre.

Wang, J., Huang, J., Rozelle, S., Huang, Q., and Blanke, A. (2007). Agriculture and groundwater development in northern China: trends, institutional responses, and policy options. Water Policy 9 Supplement 1, 61–74.

Warr, P. (2003). Poverty and economic growth in India. Economic Reform and the Liberalization of the Indian Economy (K Kalirajan and U Shankar, eds). Edward Elgar, Cheltenham, and Northhampton, MA. p. 185–209.

Weinberger, K., and Genova, C. (2005). Vegetable Production in Bangladesh: Commercialization and Rural Livelihoods. Technical Bulletin No. 33. Shanhua: AVRDC.

Weinberger, K., and Lumpkin, T.A. (2007). Diversification into horticulture and poverty reduction: A research agenda. World Development, 35(8): 1464–80.

Wilkinson, J. (2004). The food processing industry, globalization, and developing countries. Electronic Journal of Agricultural and Development Economics—e-JADE, 1(2), 184–201.

Williamson, O.E. (2000). The New Institutional Economics: Taking Stock, Looking Ahead. Journal of Economic Literature, vol. 38, p. 595–613.

World Bank. (1998). India—Water Resources Management Sector Review. Inter-Sectoral Water Allocation, Planning and Management. Washington, DC: The World Bank.

World Bank. (2000). Health and Environment. Background paper for the World Bank Environment Strategy. Washington, DC: The World Bank.

World Bank. (2001). India—Power Supply to Agriculture. Washington, DC: The World Bank.

World Bank. (2004). Mexico—The 'Cotas': Progress with Stakeholder Participation in Groundwater Management in Guanajuato. GW-MATE.

World Bank. (2006). India—Inclusive Growth and Service Delivery: Building on India's Success. Report No. 34580-IN. Washington, DC: The World Bank.

World Bank. (2007). India—Rural Governments and Service Delivery (In Three Volumes). Report No. 38901-IN. Washington, DC: The World Bank.

World Bank, (2008). World Development Report 2008: Agriculture for Development. Washington DC: The World Bank.

World Bank. (2008a). India—A New Way of Cultivating Rice. Retrieved from the World Bank website: http://web.worldbank.org/WBSITE/EXTERNAL/COUNTRIES/SOUTHASIAEXT/0,,contentMD K:21789689~pagePK:2865106~piPK:2865128~theSitePK:223547,00.html.

World Bank. (2008b). Climate Change Impacts in Drought and Flood Affected Areas: Case Studies in India. Washington, DC: The World Bank.

World Bank. (2008c). Review of Effectiveness of Rural Water Supply Schemes in India. Washington, DC: The World Bank.

World Bank. (2009a). A Hydrogeologic and Socioeconomic Evaluation of Community-based Groundwater Resource Management—The Case of Hivre Bazaar in Maharashtra-India. Washington, DC: GW-MATE, The World Bank.

World Bank. (2009b). Addressing Groundwater Depletion Through Community-based Management Actions in the Weathered Granitic Basement Aquifer of Drought-prone Andhra Pradesh—India. Washington, DC: GW-MATE, The World Bank.

World Bank. (2009c). India—Dam Safety Project. Project Performance Assessment Report. Washington, DC: The World Bank.

World Bank. (2010). Perspectives on poverty in India: Stylized facts from survey data. Washington DC: The World Bank.

World Bank. (2010). Deep Wells and Prudence: Towards Pragmatic Action for Addressing Groundwater Overexploitation in India. Washington, DC: The World Bank.

World Bank. (2011). World Development Indicators (WDI). Washington, DC: World Bank.

Xie, J., Liebenthal, A., Warford, J.J., Dixon, J.A., Wang, M., Gao, S., Wang, S., Jiang, Y., and Ma, Z. (2009). Addressing China's Water Scarcity. Recommendations for Selected Water Resource Management Issues. Washington, DC: The World Bank.

Zomer, R. J., Bossio, D.A., Trabucco, A., Yuanjie, L., Gupta, D.C., and Singh, V.P. (2007). Trees and water: Smallholder agroforestry on irrigated lands in Northern India. IWMI Research Report 122. Colombo, Sri Lanka: International Water Management Institute.

About the Editor and Authors

Editor

Marco Ferroni is an expert in international agriculture and sustainability issues. He joined the Syngenta Foundation for Sustainable Agriculture as its Executive Director in 2008 after a career in multilateral institutions and government. Before joining the Foundation, Dr. Ferroni worked at the Inter-American Development Bank and the World Bank in Washington DC. As Deputy Manager of the Sustainable Development Department of the IDB, he had responsibility for regional sector policy and technical support to the Bank's country departments. As the IDB's Principal Evaluation Officer he assessed the relevance, performance and results of Bank strategies and investments. As a senior advisor at the World Bank he advised on donor relations and directed work on international public goods and their role in development, foreign aid and international affairs. Earlier in his career, he was an economist and division chief for economic affairs and international trade in the government of Switzerland. He holds a doctoral degree in agricultural economics from Cornell University.

Authors

Richard Ackermann served for seven years as the World Bank's South Asia Director for Social Development and Environment. He holds degrees from the California Institute of Technology and the London School of Economics.

Hans P. Binswanger-Mkhize has been a researcher, manager, policy analyst, designer of development programs, and AIDS activist. During his 25 years at the World Bank he has assisted a number of countries in the development of agricultural and rural development strategies and in the design of Community-Driven Development, land reform and HIV/AIDS programs (Mexico, Central America, Brazil, Morocco, Burkina Faso, South Africa, and others). He is a fellow of the American Association for the Advancement of Sciences and of the American Agricultural Economics Association, a recipient of the Elmhirst Medal of the International Association of Agricultural Economists, and is listed in "Who is Who in Economics." He currently lives in South Africa where he is a Visiting Professor in the Department of Agricultural Economics at the University of Pretoria and a free lance consultant. He is also a visiting professor in the College of

Economics and Management at the University of Agricultural Sciences in Beijing. He continues to work on issues of land policy and land reform, community-driven development, agricultural technology, and HIV and AIDS.

Pratap S. Birthal is Principal Scientist at the International Crops Research Institute for the Semi-Arid Tropics (ICRISAT). His research interests include Livestock Development Policy, Agricultural Diversification, Institutional Economics, Agribusiness Management and Impact Assessment. Dr. P. S. Birthal has published more than 60 research articles in peer-reviewed national and international journals, 20 books/policy papers and 7 policy briefs. He has been twice awarded with DK Desai Prize by the Indian Society of Agricultural Economics, and is a recipient of Young Scientist Award from the Indian Council of Agricultural Research (ICAR). He was conferred National Fellowship by the Indian Council of Agricultural Research to work on contemporary issues in livestock research and development in 2005, and was elected Fellow of the National Academy of Agricultural Sciences in 2010. Dr. P. S. Birthal has been a consultant to many national and international research and development organizations, like International Food Policy Research Institute, nternational Livestock Research Institute, and the World Bank.

Partha R. Das Gupta, an agronomist by training with Ph. D. from Saskatchewan, spent early years in the Faculty of Agriculture at University of Kalyani, West Bengal, before moving over to the industry—first joining 'Suttons Seeds', one of the oldest seed companies of India and leaving as its Senior Vice President. After a stint with a public sector corporation in West Bengal, he joined Sandoz India Ltd, to start and head its Seeds R&D in India. Later, he became the first biotech regulatory affairs manager for Asia-Pacific of the company which got transformed first as Novartis, then Syngenta. Since his retirement, he has been consulting—initially for the company and thereafter for Syngenta Foundation for Sustainable Agriculture starting and managing its early operations in India. He has served on several important committees of the government and industry associations, addressed national and international seminars, chaired sessions and contributed articles in books and periodicals.

P.K. Joshi is the director for South Asia at the International Food Policy Research Institute. Previous to this, he held the positions of the director of the National Academy of Agricultural Research Management, Hyderabad, India, and the director of the National Centre for Agricultural Economics and Policy Research, New Delhi. Earlier, Dr. Joshi was South Asia Coordinator at the International Food Policy Research Institute and senior economist at the International Crops Research Institute for the Semi-Arid Tropics in Patancheru. His areas of research include technology policy, market, and institutional economics. Dr. Joshi has also served as the chairman of the SAARC Agricultural Centre's governing board in Dhaka, Bangladesh (2006–08); chairman of the UN-CAPSA governing board in Bogor (2007); and member of the intergovernmental panel on the World Bank's International Assessment of Agricultural Science and Technology for Development (2007–08). He served as a member of the International Steering Committee for the Climate Change, Agriculture, and Food Security Challenge Program, led by the ESSP Science Community and the CGIAR (2009–11). He was also a member of the core group of the Indian government's "Right to Food" National Human Rights Commission and the secretary-general of the Fourth World Congress on Conservation Agriculture. Currently, Dr. Joshi is the trustee of the Trust for Advancement of Agricultural Sciences (since 2008) and secretary of the Agricultural Economics Research Association in India.

Harinder S. Kohli is the Founding Director and Chief Executive of Emerging Markets Forum as well as President and CEO of Centennial Group International, both based in Washington, D.C. He is the Editor of Global Journal of Emerging Markets Economies, and serves as Vice Chairman of the institution-wide Advisory Group of Asian Institute of Technology (Thailand). Prior to starting his current ventures, he served some 25 years in various senior managerial positions at the World Bank. He has written extensively on the emergence of Asia and other emerging market economies, financial development, private capital flows, and infrastructure. He is co-editor of "India 2039: An Affluent Society in One Generation" (2010), "Latin America 2040—Breaking Away from Complacency: An Agenda for Resurgence" (2010), "A Resilient Asia amidst Global Financial Crisis" (2010), "Islamic Finance" (2011), and "Asia 2050—Realizing the Asian Century" (2011).

Bart Minten is a senior research fellow at the International Food Policy Research Institute and the program leader of the Ethiopian Strategy Support Program. He is based in Addis Ababa, Ethiopia. His research focuses on the effects of agri-business development and changes in agricultural marketing on the structure of food systems and value chains. Prior to joining IFPRI, he was a senior research associate for the Cornell Food and Nutrition Policy Program in Madagascar. He also has held the position of assistant professor at the Department of Agricultural and Environmental Economics, KU Leuven, as well as teaching positions at Cornell University and the University of Antananarivo, Madagascar. Bart received his Ph.D. in Agricultural and Resource Economics from Cornell University.

A.V. Narayanan is affiliated with the International Crops Research Institute for the Semi-Arid Tropics (ICRISAT), a non-profit, non-political organization that conducts agricultural research for development in Asia and sub-Saharan Africa with a wide array of partners throughout the world. ICRISAT is headquartered in Hyderabad, Andhra Pradesh, India, with two regional hubs and four country offices in sub-Saharan Africa.

Thomas Reardon is Professor in the Department of Agricultural, Food, and Resource Economics at Michigan State University, USA, where he has researched and taught since January 1992. From 1986 through 1991, he was Senior Research Fellow at the International Food Policy Research Institute (IFPRI), in Washington DC. From 1984 to 1986 he was Rockefeller Foundation Post-Doctoral Fellow attached to IFPRI and also working with ICRISAT in West Africa. He finished his Ph.D. in Agricultural and Resource Economics at University of California at Berkeley in 1984. Dr. Reardon is widely recognized as the leading global expert in links between the "supermarket revolution" and development of agrifood value chains and farms in developing regions, as well as links between agrifood industry transformation and food security. He has also worked extensively on rural nonfarm employment and food security. His work as of May 2010 has 7000 citations in Google Scholar and 1100 in ISI. He is on the World Economic Forum (Davos) Global Alliance Council for Food Security, and has been a food security advisor to the Chicago Council for Global Affairs and an invitee to Davos. He is listed in "Who's Who in Economics," and received a Distinguish Faculty Award from MSU. Dr. Reardon's research work is mainly in India, China, Indonesia, and Nicaragua, including studies of the rise of supermarkets, the transformation of value chains in horticulture, shrimp, dairy, and rice, and the emergence of rural business hubs and other linkages between farmers and the modern food industry.

Anil Sood is Chief Operating Officer of Centennial Group International. In his 30-year career at the World Bank, he occupied many senior positions including Vice President, Strategy and Resource Management, and Special Advisor to the Managing Directors. He has since advised chief executives and senior management of a number of development organizations including the African Development Bank, the Islamic Development Bank, the United Nations Development Program, and the United Nations Economic Commission of Africa, on matters of strategy and development effectiveness. He is the co-editor of "India 2039: An Affluent Society in One Generation" (2010), "Latin America 2040—Breaking Away from Complacency: An Agenda for Resurgence" (2010) and "Asia 2050—Realizing the Asian Century" (2011).

Yuan Zhou is the Head of Research and Policy Analysis at the Syngenta Foundation for Sustainable Agriculture, headquartered in Basel, Switzerland. She advises and supports the Foundation and its partners on policy development on agricultural extension, food security, biodiversity conservation, payment for environmental services, and sustainable land and water management. Before joining the Foundation, Yuan was a researcher at the Swiss Federal Institute of Aquatic Science and Technology (EAWAG), working on water and environmental policies, integrated analysis of water/food/environment relations, and rural development issues. Yuan holds a PhD in Environmental Economics from University of Hamburg in Germany and an MSc in Water and Environmental Resources Management from UNESCO-IHE Institute for Water Education in the Netherlands. She has published in academic journals on a range of topics related to environmental economics, agricultural water management, farmer decision-making process, implications of biofuels for food supply, and economics of desalination and water transport.

Index